D0144115

Mathematical Economics

Mathematical Economics

Kelvin Lancaster

John Bates Clark Professor of Economics
Columbia University

Dover Publications, Inc.
New York

Published in Canada by General Publishing Company, Ltd.,
30 Lesmill Road, Don Mills, Toronto, Ontario.
Published in the United Kingdom by Constable and Company,
Ltd.

This Dover edition, first published in 1987, is an
unabridged republication of the work first published by
The Macmillan Company, New York, in 1968.

Manufactured in the United States of America
Dover Publications, Inc.
31 East 2nd Street
Mineola, N.Y. 11501

Library of Congress Cataloging-in-Publication Data

Lancaster, Kelvin.
Mathematical economics.

Reprint. Originally published: New York : Macmillan, c1968.
Bibliography: p.
Includes index.
1. Economics, Mathematical. I. Title.
HB135.L355 1987 330'.01'51 87-5459
ISBN 0-486-65391-9 (pbk.)

PREFACE

This book is designed to satisfy two great needs in the economics profession. One is for a text suitable for graduate courses in mathematical economics and the other is for a reference book for professional economists who wish to keep in reasonable touch with recent developments in analytical economics. There is no book currently available that fully performs either of these tasks, although there are some excellent books that do part of them.

The incentive to write such a book arose from the author's experience in teaching mathematical economics over several years at the London School of Economics, Johns Hopkins University, and Columbia University. In order to cover the field adequately in graduate mathematical economics courses, students were of necessity referred to a variety of sources that differed widely in degree of difficulty, in approach, and in notation and terminology. The need for a single source with a unified treatment was clearly apparent.

There has been a tremendous expansion in the techniques of economic analysis since about 1950, and the aim of this book is to present those techniques (or as many of them as possible), along with that part of the neoclassical calculus tradition that has not been superseded. The "new mathematical economics" provides many techniques of great power, and in some cases notable simplifications in analysis as compared with earlier methods.

It is presumed that the reader has some acquaintance with elementary calculus. No other prerequisite, except a desire to learn the subject, is called for. Great care has been taken to give an exposition that is complete and rigorous (except in topological methods, where the approach is descriptive and heuristic) and omits no essential proofs or steps in the argument. The completeness and eclectic coverage of the mathematical reviews will make the work useful to those who may already be familiar with much of the field.

To keep the exposition as smooth as possible, the economic analysis has been separated from the purely mathematical material. Clear instructions are given at the head of each chapter as to which sections in the mathematical reviews are necessary for the content of that chapter. The economic analysis covers linear and nonlinear optimizing techniques, input-output, activity analysis, neoclassical and set-theoretic static economic models, modern general equilibrium theory, the Von Neumann and other models of balanced growth, efficient growth and turnpike theorems, and modern stability analysis. The mathematical reviews include discussions of set theory, linear algebra, matrices, linear equations and inequalities, convex sets and functions, continuous functions and mappings (including neoclassical calculus methods), topological ideas, differential and difference equations, calculus of variations, and related topics. The mathematical treatment is thoroughly modern.

The book does not aim to push forward the frontiers of economic analysis, but to transport as many economists as possible up to, or within sighting distance, of those frontiers. There are some new results and several revised, corrected, or expanded versions of existing results, but these are mere by-products of the more important aim of careful and complete exposition.

K. L.

CONTENTS

The book contains twelve chapters, followed by eleven Mathematical Reviews. The reviews are numbered separately, from R1 to R11.

Part I: Optimizing Theory

Mathematical Economics

CHAPTER

1

Introduction

1.1 MATHEMATICAL ECONOMICS

Mathematical economics is not a subject but an area of study within economics, closely affiliated with economic theory. Its scope is changing constantly, since it acts as a port of entry for new analytical techniques imported from mathematics (or engineering, or even other social sciences) on their way into the main body of economic analysis. Yesterday's advanced mathematical economics is today's mathematical economics, and will be tomorrow's economic analysis. We have seen this process happen in the past, as undergraduate courses come to contain what was once regarded as too technical for the ordinary professor of economics, and the process will continue.

The more rapid the rate of import, the larger will be the inventory in transit. In the last twenty years we have had a tremendous flow of new techniques, with a corresponding growth in the scope of mathematical economics. In addition (to change the metaphor), mathematical economics is developing a rate of natural increase. As in other disciplines, economists are developing their own mathematical methods which are something more than simple direct applications of techniques well-known in other fields.

The well-equipped economic theorist must now know much more of mathematics than a few chapters from an advanced calculus text. One of his problems is the variety of different mathematical tools now at his disposal. These come from several areas of mathematics and are nowhere assembled together in a suitable fashion. A course in linear algebra will contain many things of no great interest to the economist and may leave out some things that are important to the economist but peripheral to the mathematician or physical scientist. The same is true of other areas of mathematics. One of the purposes of this book is to assemble and unify as many of these fragments of mathematics as possible.

However, mathematical economics is not just pieces of mathematics, it is the application of these to economic analysis and the development of new techniques to solve new problems. Economists have become genuine innovators in their adaptation of mathematical techniques to their own needs. This book aims to show this process of application, adaptation, and innovation at work.

The history of mathematical economics is yet to be fully written, since few historians of economic thought have been mathematically inclined. Briefly, we can distinguish three chief phases. The first was a period of important individual contributions, almost entirely neglected by the economics profession in general and the Anglo-American branch in particular. This was followed in the thirties by the growth and flowering of neoclassical mathematical economics which continued into the early fifties. The period of fifteen years or so since then has been a period of tremendous absorption of new techniques, leading to what can be regarded as the "new" mathematical economics.

Neoclassical mathematical economics, whose chief tools are the derivative and the equation, can be considered to be digested into the main body of economic theory, even though there remain many technical problems to interest the mathematically inclined. For the new mathematical economics, whose tools are vectors, convexity, sets, and inequalities, digestion has barely commenced. This book is designed to speed the process.

1.2 OUTLINE OF THE BOOK

The book contains, apart from this introduction, eleven chapters on topics in mathematical economics proper (that is, on economic models analyzed primarily from the point of view of their mathematical properties), followed by eleven mathematical reviews[1] designed to cover the required mathe-

[1] The term *reviews* may seem inappropriate, since these are complete expositions. No better term, however, seemed to be available.

matics. The chapters and reviews are designed as a self-contained system, wherein the reviews contain all the mathematics required for the chapters and the chapters illustrate the use of almost all the techniques set out in the reviews. The mathematical background required, with references to the appropriate reviews, is specified at the beginning of each chapter.

The main chapters are grouped into three parts. Part One (Chapters 2 through 5) discusses optimizing theory generally, including linear programming, classical calculus, and Kuhn-Tucker theory. Part Two (Chapters 6 through 9) discusses various static economic models, including input-output (Leontief) models, activity analysis, advanced neoclassical models, set theory formulations, and modern general equilibrium theory. Part Three (Chapters 10 through 12) contains discussions of multisector dynamic models, including Von Neumann and other balanced growth models, optimal growth and turnpike theorems, and stability analysis.

The mathematical reviews (R1 through R11) cover linear algebra, inequalities, convex sets and cones, matrices, functions and mappings, some topological ideas, properties of special matrices, differential and difference equations, calculus of variations, and related topics. In all cases except in the sections involving topological methods and in Review R11 the treatment is complete, with no essential proofs omitted.

Throughout the book, the most complete and rigorous analysis possible is presented. On a few topics the treatment is heuristic only because rigor would require methods beyond the scope of the book or space that is not available.

There are a variety of other topics that might have claimed a place in a book of this kind. Space required selection, but the author believes he has included all topics that are of major importance in the current and recent literature of economic theory.

As a *course text*, the author suggests the following order:

(a) *General Background and Optimizing Theory*. Reviews R1, R2, R3, R4 (omitting Section R4.7), and R8 (omitting Sections R8.7 and R8.8), followed by Chapters 2, 3, 4 (noting only results in Section 4.5), and 5 (Sections 5.1 through 5.5).

(b) *Basic Economic Models*. Reviews R5, R6 (perhaps omitting Section R6.3), and R7 (Sections R7.1 through R7.3), followed by Chapters 6, 7, 8, and 10 (omitting Section 10.5).

(c) *More Advanced Topics* (in any order). (i) Growth Theory: Reviews R8 (Sections R8.7 and R8.8), and R11 (Sections R11.1 and R11.2) followed by Chapters 10 (Section 10.5) and 11; (ii) Stability Theory: Reviews

R10, and R7 (Section R7.4), followed by Chapter 12; (iii) General Equilibrium: Review R9, followed by Chapter 9.

(d) *Tidying Up*. Review R6 (Section R6.3), followed by serious study of Section 4.5 of Chapter 4, Chapter 5 (Sections 5.6 and 5.7), Review 7 (Section R7.5), and Review R11 (Sections R11.3 and R11.4).

From the author's experience, the basic material and some of the advanced topics can be fitted into a two-semester course, given the availability of the present text. The material in the mathematical reviews often exceeds the minimum background required for the relevant chapters and some discretion can be exercised in treating this material in greater or less depth.

1.3 NOTES ON THE LITERATURE

There is no book with coverage similar to this, but the following books in the same general field are valuable additional reading. All are referred to in the text at appropriate places. They are classified as more elementary, at approximately the same level, or more advanced, in comparison with this book. The classification refers only to level of exposition, not to content.

More Elementary

Baumol, *Economic Theory and Operations Research* (Baumol [2])

Baumol gives an excellent conspectus of many of the techniques developed in recent years and of some of their application to economics. It is especially useful for obtaining a general idea of the nature of certain topics not covered in this book. Its chief use is as a survey.

Dorfman, Samuelson, and Solow, *Linear Programming and Economic Analysis* (Dorfman, Samuelson, and Solow)

In the earlier chapters on linear programming and its direct applications, the authors tied their hands by avoiding the use of matrices and vectors. Later chapters contain material of great importance, especially Chapters 12 (growth theory) and 13 (general equilibrium). Chapter 13 is the only attempt at an exposition of general equilibrium theory at a level more elementary than Chapter 9 of this book.

Allen, *Mathematical Economics* (Allen [2])

Allen contains an excellent coverage (Chapters 1 through 9) of single-sector economic models with complex dynamic specifications, as contrasted with the models of this book, which are multisector with simple dynamic specifications. The coverage of linear methods is broad, but the exposition is not in the most desirable form.[2]

Approximately the Same Level

Gale, *The Theory of Linear Economic Models* (Gale [1])

The author regards this as the best exposition of the theory of linear programming among the books listed. Gale also discusses balanced growth models. Gale's notation is not standard, since he uses a row vector where column vectors are more usual, and vice versa. His matrix–vector relationships must be transposed to be comparable with the usage of this book.

More Advanced

Karlin, *Mathematical Methods and Theory in Games, Programming, and Economics* (Karlin [1])

This book covers much ground, in a compressed fashion. Chapters 8 and 9 are filled with miniature gems of mathematical economics, none of which is quite complete as an exposition. Karlin's notes are of interest and his references extensive. This is highly recommended as a next book to read after the present one, although only Chapters 5, 7, 8, and 9 and the three appendices are concerned with topics covered here.

Debreu, *Theory of Value* (Debreu [1])

Debreu gives the most complete account of the modern generalized treatment of production, consumption, and general equilibrium theory. The exposition is rigorous in style but often has important steps in the argument left almost unexplained. It is essential but difficult reading for those who wish to follow up the discussion on general equilibrium given in Chapter 9 of this book.

Morishima, *Equilibrium Stability and Growth* (Morishima [1])

This contains important material on growth theory. It is quite closely argued, and should not be read until after Chapters 10 and 11 of this book.

In addition to the works mentioned, more specialized references are given in footnotes throughout the book. These are not designed to be a bibliography of the subject but merely to lead the reader into further literature in the field. Sometimes the references are to original papers, sometimes to secondary sources. Any issue of *Econometrica* will supply the reader with other applications of the methods set out in this book.

[2] Allen's more recent *Macroeconomics: a mathematical treatment* (Allen [3]) contains the more elementary dynamics of *Mathematical Economics*, with more material on single sector and simple two sector growth models.

PART

I

Optimizing Theory

CHAPTER

2

The General Optimizing Problem

Most of this chapter requires only a general background on sets and functions, such as given in Review R1. Section 2.6 requires additional knowledge of convex and concave functions (Review R8, Section R8.5) and convex sets (Review R4, Sections R4.1 and R4.2).

2.1 INTRODUCTION[1]

Optimizing, a catch-all term for maximizing, minimizing, or finding a saddle point,[2] lies at the heart of economic analysis. In passive economic models (such as general equilibrium studies), we are interested in the optimizing behavior of decision makers in the economy. In active models (such as models of efficient growth), we are interested in finding the optimum ourselves. Much of the history of economic theory has been a process of movement from deterministic to optimizing models. In recent years, this trend has been apparent in the movement from deterministic input-output to optimizing activity analysis models, and from naive deterministic growth

[1] This broad perspective on the general nature of optimizing problems has been suggested mainly by writings in operations research and decision theory. See, for example, Carr and Howe, Chapters 1 and 2.

[2] The discussion of saddle points will not be taken up until Chapter 5.

models to the study of optimal and efficient growth paths. Even in macroeconomic policy-oriented models, where optimizing has been mainly confined to parameter estimation, the trend is toward more sophisticated optimal policy models.

Indeed, optimization subject to constraint has been considered by many as defining the essential nature of economics.[3]

Because of the importance of optimizing, and because the considerable development of optimizing theory in the last twenty years has led to a variety of different techniques of analysis, it seems desirable here to give a general description of the structure of the optimizing problem and of how the different techniques fit into this structure.

2.2 THE GENERAL STRUCTURE

The variables of the problem will be considered to be in the form of a vector in R^n.[4] In addition to this vector, x, we have:

(a) a *feasible set* K.[5] Only $x \in K$ is to be taken into account in the problem.
(b) a single valued continuous *objective function*,[5] $f(x)$, whose value for $x \in K$ is to be optimized.

Thus we can state a typical maximizing problem in formal terms as

Find $x^* \in K$ *such that* $f(x^*) \geqq f(x)$, *all* $x \in K$.

If such an x^* exists, the problem has a *weak global* maximum—weak because it satisfies the weak inequality, global because the inequality is satisfied for all $x \in K$. A global optimum should not be confused with an

[3] For example, the well-known statement in Robbins, page 16:

"Economics is the science which studies human behaviour as a relationship between ends and scarce means which have alternative uses."

This idea, which is virtually a statement of the general optimizing problem, seems to have originally been due to Menger.

[4] We can regard an n-vector x as a function of its indices $i = 1, \ldots, n$, defined by enumeration. If we replace the index by a continuous parameter t, then $x(t)$ can be regarded as the infinite dimensioned analog of a vector in R^n. $x(t)$ is an unknown function of t, not a point to be defined by enumeration of its coordinates. Optimizing problems in which the unknown is a function rather than a point are most likely to fall within the province of the *calculus of variations* or a related approach. Approaches of this kind are discussed in Review R11, but not in this or the next three chapters.

[5] The terminology given is standard in linear programming. We shall use it for all types of optimizing problems.

unconstrained optimum. The latter implies that $K = R^n$. We would have a *strong* maximum if we could find x^* such that $f(x^*) > f(x)$, all $x \in K$.

A weak optimum is equivalent to a nonunique optimum point since any x satisfying $f(x) = f(x^*)$ is also an optimum point. A strong *global* optimum implies a unique optimum.

If we reverse the inequalities we obtain a minimum, weak or strong as the case may be. A minimum for $f(x)$ implies a maximum for $[-f(x)]$. The value x^* is often called simply the *solution* of the optimum problem. To avoid confusion with other claimants for the same name in many economic models, we shall usually call it the *optimal solution* or *optimum point.*

Most calculus techniques cannot solve the problem as set out above, but can only solve a problem of the following kind:

Find $x^* \in K$ *such that* $f(x^*) \geqq f(x)$, *all* $x \in (E \cap K)$, *where E is a neighborhood of* x^*.[6]

Such a point is a weak *local* maximum. We can have a strong local maximum and weak or strong local minima. Some authors use the terms *relative* and *absolute* rather than local and global.

It is obvious that, if $f(x)$ has an optimum at all, it must have a global optimum and that this must also be a local optimum. On the other hand a local optimum is not necessarily global. Our interest is primarily in the global optimum which, after all, is *the* optimum. Thus we are interested in conditions on the structure of the problem that will guarantee that a local optimum is also global. Such conditions (applicable to many economic cases) are given later in the chapter. If they are not satisfied, we may need to adopt *ad hoc* procedures (such as enumerating and comparing all local optima) to locate the global optimum.

The problem itself determines whether the optimum is weak or strong—there is no question of trying to locate the strong optimum if it is essentially weak. It should also be noted that a strong *local* optimum does not imply a unique optimum, since $f(x)$ may take on the optimal value at several distinct points, each being a strong local optimum.

2.3 CONSTRAINTS AND THE FEASIBLE SET

The feasible set, over which the variables are permitted to range, may be defined in any suitable way. In the case of discrete variables, the feasible set

[6] Note that we cannot say "all $x \in E$" in place of "$\in (E \cap K)$." The calculus analysis of the problem depends essentially on the existence of points in E which are not in K.

may even be described by enumeration. Typically, however, the feasible set will be defined by equalities or inequalities involving relationships between the variables. The relationships which define the feasible set are the *constraints* of the problem.

A single constraint defines some set of values of the variables. If there is more than one constraint, the permissible values of the variables must satisfy all constraints, so that the feasible set is the *intersection* of all the sets defined by the individual constraints. In Figure 2–1(a) we have two constraints:

$$x_1^2 + x_2^2 - k^2 \leqq 0, \text{ and } c - x_1x_2 \leqq 0.$$

The first constraints the variables to the disk of radius k, and the second to a rectangular hyperbola with two branches and to the area of the plane cut off from the origin by the hyperbola.

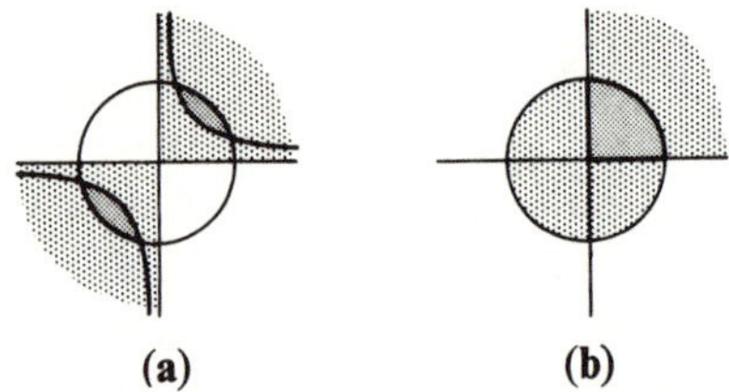

Figure 2-1. Feasible sets.

The feasible region in this case consists of two lens-shaped areas, heavily outlined in the diagram; one is in the positive and one is in the negative quadrant. Typically, we would be interested in only one of these areas, usually because of other constraints of the form $x_1, x_2 \geqq 0$, which would eliminate the area in the negative quadrant.

We can distinguish constraints of the type $x_1, x_2 \geqq 0$ as *direct* constraints on the variables. Usually they appear as *nonnegativity* constraints and can always be put in this form by a simple linear transformation of the variables. We shall refer to the type of constraint represented by the circular and hyperbolic areas as *functional* constraints. A more typical relationship between functional and direct constraints than in Figure 2–1(a) is given by Figure 2–1(b), where the functional constraints give a disk and the direct constraints confine the variables to a quadrant of the disk.

The feasible sets used as examples above are both *bounded*; that is, we could enclose them in a circle of finite radius. If in Figure 2–1(a) the constraints had been only the hyperbola and the direct constraints without the circle, the feasible set would have been *unbounded.*

The existence of constraints gives a feasible set, but the set may be empty. Again referring to Figure 2–1(a), if we shrank the circle sufficiently there would be no points both in or on the circle and on or outside the hyperbola. If the feasible set is empty, we usually refer to the constraints as *inconsistent.*

For some solution techniques, and for a guarantee of a local optimum to be a global optimum, we require the feasible set to be convex. In general, there is no reason why it should be so any more than it should be bounded, another property that leads to useful results. Actually, the feasible sets of both Figure 2–1(a) (with the nonnegativity constraints included) and Figure 2–1(b) are convex. If the direction of the inequality in the circular constraint of Figure 2–1(b) is reversed, the feasible set consists of the positive quadrant with a circular bite taken out of it at the origin, and this is not a convex set.

The boundaries of the feasible set are crucial in optimizing problems, so we need to study them in detail. Consider the simple case of a disk-shaped feasible region defined by the inequality $x_1^2 + x_2^2 - k^2 \leqq 0$. Any point x_1^*, x_2^* in the feasible set must be such that either

$$x_1^{*2} + x_2^{*2} - k^2 < 0,$$

or

$$x_1^{*2} + x_2^{*2} - k^2 = 0.$$

If the first is the case, then we can find a neighborhood of x_1^*, x_2^* such that the inequality is still true for all points in this neighborhood. Then x_1^*, x_2^* is an *interior* point of the feasible set, because it has a neighborhood consisting entirely of feasible points.

If x_1^*, x_2^* gives the second case, satisfying the equality, then there is, in the neighborhood of x_1^*, x_2^*, a point x_1', x_2' such that $x_1'^2 + x_2'^2 - k^2 < 0$, and another point x_1, x_2 such that $x_1^2 + x_2^2 - k^2 > 0$. In this case x_1^*, x_2^* is a *boundary* point, because its neighborhood contains both feasible and infeasible points.

The equation $x_1^2 + x_2^2 - k^2 = 0$ defines the boundary of this particular feasible set. Since the weak inequality is given for the constraint, the boundary is part of the feasible set, so the set is *closed.* If the constraint had been in the form $x_1^2 + x_2^2 < 0$, the boundary itself would not be feasible, and the set would be *open.* In a closed set, every point is either an interior point or a boundary point.

In all our discussions of optimizing, it will be assumed that the constraints are such as to give a *closed feasible set*; otherwise the problem is usually without a solution.[7] This is normally guaranteed by ensuring that there are no strict inequalities in any of the constraints and that the constraints are continuous.

Typically there will be several constraints rather than just one. Consider a feasible set defined by several inequality constraints, and consider a feasible point x^*. We say that a particular constraint is *effective* at x^*, if x^* gives an equality in the constraint, *ineffective* at x^*, if it gives an inequality. The hyperbola constraint $c - x_1x_2 \leqq 0$ is effective at $(c^{1/2}, c^{1/2})$ and ineffective at $(2c^{1/2}, 2c^{1/2})$. If the constraint is defined as an equality it is, of course, always effective. Note that a constraint may be "effective" in a common-sense meaning without being effective in this way. In Figure 2–1(a), for example, the nonnegativity constraints rule out the region in the negative quadrant and so are "effective" in one sense but are never effective in the above technical sense.

2.4 THE GENERAL OPTIMIZING PROBLEM

We shall consider the general optimizing problem to be set out in the following standard format. The problem may be a maximum or a minimum, but we need set out only the maximum case:

$$
\begin{array}{llll}
 & \max f(x) & x = [x_j] & j = 1, \ldots, n \\
\text{S.T.}^{8} & (1)\quad g^i(x) \leqq 0 & & i = 1, \ldots, m \\
 & (2)\quad x_k \geqq 0 & & k \in S, \text{ where } S \text{ is some subset of the indices } 1, \ldots, n.
\end{array}
$$

$f(x)$ is the objective function, x the vector of the n variables of the problem. The constraints (1) are the functional constraints, the constraints (2) the direct constraints. The functions f, g^i are assumed, unless otherwise stated, to be continuous of class C^1 or C^2, as required.

It is convenient to have all the inequalities in the same direction in a standard form. If we have a constraint of the form $\phi^i \geqq 0$, we can always write $g^i = -\phi^i$ to obtain the standard form. The direction of the inequality which is chosen as standard is clearly arbitrary. The choice above for a maximum [with the inequalities in (1) reversed for a minimum] fits the context of most economic problems in a natural way, and also fits the linear programming convention. For historical reasons, the inequalities are more usually

[7] Consider the problem of maximizing $f(x) = x$ over the open feasible set $K = \{x \mid 0 < x < 1\}$. For every $x' \in K$, however close to 1, there is always another $x \in K$ which is greater. Thus $f(x)$ is bounded (<1), but has no maximum because it cannot attain its upper bound.

[8] S.T. will be used throughout as an abbreviation for "subject to."

written the other way for problems using Kuhn-Tucker methods,[9] but we shall break with this practice and preserve the above convention for all problems.

The functional constraints can always be put in *inequality* form by noting that a constraint of the form $\phi(x) = 0$ is equivalent to the two inequality constraints $\phi(x) \leqq 0$ and $-\phi(x) \leqq 0$. For a very important case of the general problem, *all* the constraints are equalities and it is then easier to drop the inequality form.

The direct constraints (2), which may or may not be present (S may be empty), will always be written in the nonnegative form, even in a minimum problem where the inequalities on the functional constraints are reversed. All direct constraints can be put in the nonnegative form by writing $x'_k = -x_k$, if the natural formulation gives $x_k \leqq 0$, or by writing $x'_k = x_k - b$, if the natural formulation gives $x_k \geqq b$.

The inequalities in both the functional and direct constraints are assumed always to be weak inequalities to ensure that the feasible set is closed.

An optimum problem need not, in general, have a solution. The problem, $\max x_1 + x_2$, S.T. $x_1 - x_2 \leqq 0$ with an unbounded feasible set, has no solution since, for every x, we can find another feasible x giving a greater value for the objective function. However, we do have a guarantee that the problem is worth pursuing in a large class of cases:

Weierstrass's Theorem

A continuous function defined over a nonempty closed bounded set attains a maximum and a minimum at least once over the set.[10]

Since we can usually take the objective function to be continuous and the feasible set to be closed, the boundedness of the feasible set is the only condition which is not ensured. In many cases, the feasible set will be bounded, but the property may not be apparent without examination in detail. It should be emphasized that the Weierstrass theorem gives *sufficient* conditions. Problems which do not satisfy the conditions may still give an optimum.[11] Even feasible sets that are not closed may give an optimum, provided the set contains at least a suitable part of the boundary, if not the entire boundary.

The optimum may, of course, be trivial. $f(x) = c$ for $a \leqq x \leqq b$ (x a scalar) has a maximum and a minimum, both c.

[9] See Chapter 5.

[10] If $f(x)$ is a continuous function and the feasible set K is compact (closed and bounded), the set $\{f(x) \mid x \in K\}$ is also compact. Since $f(x)$ is a scalar, this set must be a closed interval (or several closed intervals). Thus $f(x)$ has a maximum at the upper bound of the interval (or one of the intervals) and a minimum at the lower bound.

[11] If $f(x)$ is an *increasing* function of x, it is obvious that $f(x)$ will have a maximum if the feasible set is bounded *above*, even if not below.

2.5 THE GENERAL SOLUTION PRINCIPLE

An optimal point must be in the feasible set but may be either an interior point or a boundary point. If it is an interior point, it has a neighborhood of feasible points and must be an optimum relative to those neighborhood points. Such a point must satisfy the ordinary calculus conditions for an unconstrained optimum, that is, it must be a critical point ($f_j = 0$ for all j) of f. A boundary point has a neighborhood which includes infeasible points as well as feasible ones, and it is not possible to say that such a point must be optimal relative to its neighborhood. A boundary optimal point need not be a critical point of f.

Thus we have a general solution principle.

The solution to the general optimum problem, max [*or* min] $f(x)$ *for* x *over some closed feasible set* K *will, if it exists, be some point* x^* *which is*: (*a*) *a critical point of* $f(x)$: *or* (*b*) *a boundary point of* K (*or both*).

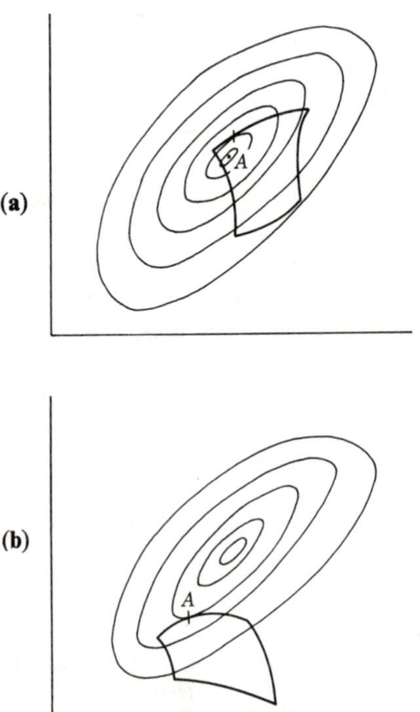

Figure 2-2. Interior and boundary optima.

We shall distinguish the two types of optimal point as *interior* and *boundary* optima. These are illustrated in Figures 2–2(a) and 2–2(b). In both figures the elliptical curves are level lines or contours of the objective function, the heavily outlined region is the feasible set. The optimal point A is an interior optimum in Figure 2–2(a), a boundary optimum in Figure 2–2(b).

In principle, an optimum problem can be solved by finding the critical points of $f(x)$, then computing the values of $f(x)$ along the boundary, finally choosing the point giving the maximum or minimum of $f(x)$. As a practical matter, this task is usually too great, but some *ad hoc* procedure based on this principle may have to be used if the problem does not fit one of the various simplifying cases.

If $f(x)$ is not differentiable everywhere, as has been assumed, the points at which it is not differentiable will need to be examined in addition to the critical and boundary points.

2.6 CONDITIONS FOR A GLOBAL OPTIMUM[12]

Since we are usually interested in a global optimum and since many techniques discover only local optima, conditions which guarantee that a local optimum is also a global optimum are of great value. In general, there are no properties of a local optimum itself that will tell whether it is global or not, but there are properties of the general structure of the problem itself that can guarantee that *every* local optimum is also global. These are *sufficient* conditions based on the structure of the problem, but not necessary conditions.

The particular conditions set out below are of special importance because they are satisfied by most typical optimum problems in economics.

For a problem concerned with optimizing a continuous function $f(x)$ over a closed feasible set K, every local optimum is also a global optimum if: (*a*) *f is a concave function for a maximum, or a convex function for a minimum:* and (*b*) *K is a convex set.*

Typical maximands in economics, such as regular production functions, are concave, and the feasible set is very often convex, so that these conditions have wide applicability. It is the fulfillment of the above conditions in such classic examples of optimization in economics as the theory of consumer choice and the theory of production that have eliminated problems of

[12] This section requires some knowledge of the simplest properties of convex and concave functions (See Review R8, Section R8.5) and convex sets (See Review R4, Sections R4.1 and R4.2).

local and global optima. In a complex model, it may be necessary to examine the optimizing problem carefully to determine whether the conditions are fulfilled.

To show that the conditions suffice to ensure that a local optimum is global, suppose, to the contrary, that $f(x)$ is concave over a convex feasible set K and has a global maximum at x^* and another local maximum at x' that is not a global maximum, both x^* and x' being feasible. Then, from the definition of a concave function, we have

$$f[kx^* + (1 - k)\,x'] \geqq kf(x^*) + (1 - \text{k})\,f(x') \qquad 0 < k < 1.$$

Since x^* gives a global maximum, but x' does not, $f(x^*) > f(x')$. Hence

$$kf(x^*) + (1 - k)\,f(x') > f(x') \qquad 0 < k < 1,$$

so that

$$f[kx^* + (1 - k)\,x'] > f(x') \qquad 0 < k < 1.$$

Now consider a neighborhood of x' small enough to exclude x^*. $x = kx^* + (1 - k)\,x'$ is in this neighborhood for some $k > 0$ sufficiently small, and x is feasible because K is convex. But we have just shown that $f(x) > f(x')$, contradicting the statement that x' gives a local maximum. Thus x' cannot be a local maximum unless $f(x') = f(x^*)$, that is, it is also a global maximum.

It is implicit in the above argument that x, the convex linear combination of x^* and x', also gives a global maximum if x^* and x' do, since $f(x) \geqq kf(x^*) + (1 - k)\,f(x') = f(x^*) = f(x')$. Thus a problem satisfying the global optimum conditions either has a *unique* global optimum, or an infinite set of such optima.

If $f(x)$ is *strictly* concave, x^*, x' are different global optima, and x is the convex combination of x^*, x'; as before we have

$$f(x) = f[kx^* + (1 - k)\,x'] > kf(x^*) + (1 - k)\,x' = f(x^*) = f(x').$$

But $f(x^*)$ is a global optimum, so that $f(x^*) \geqq f(x)$, contradicting the above. Therefore x^*, x' must coincide, and the optimum is unique. Thus we can state the following:

If f is strictly concave over a convex feasible set the global optimum is unique.

The above conditions can be strengthened to cover important cases in economics in which the objective function is not *concave* but is *concave-contoured* (see Review R8, Section R8.6) and increasing. Utility and welfare functions, and increasing-returns production functions, fall into this category. *It is sufficient for satisfaction of the global optimum conditions for a maximum that $f(x)$ be a positive monotonic transformation of a concave function, and K a convex set.*

The extended condition follows directly, since if $f(x)$ is a positive monotonic transformation of some function $F(x)$, $F(x^*) \geqq F(x)$ implies $f(x^*) \geqq f(x)$ and conversely, so that the maximum of $f(x)$ occurs at the same point as the maximum of $F(x)$. Any concave-contoured function $f(x)$ which is an increasing function of every component of x can be expressed as a positive monotonic transformation of some concave function.

The arguments given above are for a maximum. By putting $\phi(x) = -f(x)$, the concavity condition on f becomes a convexity condition on ϕ, and the maximum becomes a minimum.

2.7 IMPORTANT SPECIAL CASES

Two special cases of the optimizing problem exist for which there is a well-developed theory, and most other cases are handled by methods that depend heavily on either or both of these. They are the *linear programming problem* and the *classical calculus problem.*

The Linear Programming Problem

In this problem, the objective and constraint functions are linear, and nonnegativity constraints are a prominent feature.

Since the objective function is linear, it has no critical points, so all optima are boundary optima. Since the constraints are linear, the feasible set is convex. The linear objective function is both concave and convex, so that maxima and minima are global. As we shall see in the following chapter, the special solution property of the linear programming problem is that only a finite number of the boundary points need be examined.

If a solution to a linear programming problem exists, it can, in principle, be solved explicitly. For this reason, linear programming has opened the way to direct numerical solution of practical optimum problems if these are in the form of, or can be approximated by, a linear programming problem.

Linear programming has had an enormous impact on operations research and on other fields in which actual numerical solutions of optimizing problems are required. The applied economist is also interested in the solvability of linear programming problems, but the economic theorist is less interested in this feature. The impact of linear programming on analytical economics has been primarily through the new light that linear programming theory has thrown upon the nature of optimal problems generally, and especially on the nature of prices in typical economic optimizing problems.

The theory of linear programming has given many insights into the properties of more general optimizing problems with inequality and nonnegativity constraints. The repercussions of the theory go well beyond the limits of the relatively restricted class of problems fitting the strict linear programming format.

The Classical Calculus Problem

Here the objective and constraint functions must be continuous and differentiable, but otherwise are unrestricted in form. There must, however, be equalities in all the functional constraints and no nonnegativity constraints.

The study of this problem preceded that of the linear programming problem by well over a century, and the techniques evolved that depend on calculus concepts (which the linear programming techniques do not) have had wide application in many fields and have been the foundation of neoclassical mathematical economics.

Since all the constraints are equalities, all are effective at all points in the feasible set so that the interior of this set is empty. As in the linear programming problem, only boundary optima exist. However, the special linear programming property, which states that only a finite number of boundary points need be examined, does not hold. Nor are the conditions for a global optimum satisfied, so that local optima may have to be compared with each other.

Strictly speaking there is no direct solution technique. What the standard methods give us are properties that must be satisfied by optimal points. In general, we may not be able to discover exactly which points satisfy these conditions. To the economist, the descriptive properties of the optimal points, rather than their exact location, have proved to be of the greatest value.

Extensions of the Classical Calculus Problem

A combination of insights derived from linear programming theory (and mathematically related game theory) with the classical calculus approach have enabled us to extend the latter to cover cases in which there are nonnegativity constraints, inequalities in the functional constraints, or both. As in the strict classical calculus method, we obtain properties of optimal points, rather than the optimal points themselves.

2.8 DIRECT SOLUTIONS OR OPTIMAL CONDITIONS?

One's approach to an optimizing problem is conditioned from the beginning by what one is looking for and what one regards as a solution to the problem. Consider a simple optimizing problem—that of maximizing output for a given outlay on inputs, the production function and other relevant data being specified.

A production executive would presumably want a *direct solution*, i.e., a quantitative specification of exactly what inputs would be used, in what proportions, and what output would result. If not a numerical solution, he would at least want a direct algebraic solution that would provide numerical answers if numerical values of parameters were inserted. He would find a statement about certain of the dual variables being zero or about the ratio of partial derivatives of the production function distinctly unhelpful.

An economic theory, on the other hand, would be chiefly concerned with more universal characteristics of the solution, those that are specifically independent of the numerical values that are the subject of direct solution. In other words, given that several firms have actually reached a direct solution of the above problem, the economist is interested in the properties of these solutions common to all the firms. These properties of the solution are the *optimum conditions*.

The optimum conditions provide a method for *recognizing* an optimal point and giving its properties, but do not necessarily give a *search procedure* for finding it. For many years economists were sadly disappointed to find that firms did not think in marginal terms, a disappointment due to confusion between optimal conditions and efficient direct solution techniques. Actually, it is only since World War II that efficient direct solution techniques have existed for the kind of optimizing problem faced by a typical firm.

The difference in approach to what is required of a solution technique results in a divergence between the operations research, quantitative decision procedure approach to optimizing problems, and the mathematical economics approach. The latter is related primarily to the interests of economic theory and analysis and so is chiefly concerned with optimum conditions.

As a result, we do not discuss in this book a variety of optimizing techniques which are primarily designed for problem solving or for numerical solutions and whose theory does not provide special insights in economic problems.[13]

[13] These methods are discussed in the operations research and decision theory literature. See, for example, Baumol [2] for an introduction, Hadley [3] for more advanced treatment. Dynamic programming has affiliations with the calculus of variations. See, for example, Bellman and Dreyfus, Chapter 5.

Among these techniques are *quadratic programming*, a relatively simple solution technique for problems with linear constraints and nonlinear objective functions that can be well-represented by quadratic functions; *integer programming* which is concerned with problems in which some or all of the variables can take on discrete values only; *concave* and *gradient programming*, solution techniques for problems of the extended classical type; and *dynamic programming*, a solution technique for optimizing problems in which the constraints are of a sequential kind. Both integer programming and dynamic programming have potential uses in theoretical models.

EXERCISES

1. Five functions of x_1, x_2 are defined as follows:

$$\begin{aligned} \phi^1 &= 2x_1 - 3x_2 & \phi^4 &= 3x_1 + 2x_2 \\ \phi^2 &= x_1 - 3x_2 & \phi^5 &= x_1 + 2x_2 \\ \phi^3 &= x_1 + x_2 \end{aligned}$$

Three feasible sets are defined by the following inequalities:

$$\begin{aligned} &K_1\colon\ \phi^1 \geqq -3, \quad \phi^2 \geqq 0, \quad \phi^3 \leqq 3, \quad \phi^4 \leqq 6, \quad \phi^5 \geqq 2 \\ &K_2\colon\ \phi^1 \geqq -3, \quad \phi^2 \leqq 0, \quad \phi^3 \leqq 3, \quad \phi^4 \leqq 6, \quad \phi^5 \geqq 2 \\ &K_3\colon\ \phi^1 \leqq -3, \quad \phi^2 \leqq 0, \quad \phi^3 \leqq 3, \quad \phi^4 \geqq 6, \quad \phi^5 \leqq 2. \end{aligned}$$

One of these sets is empty, one is a triangular region, the other a quadrilateral region. Use algebraic or graphical methods to discover which is which.

2. A local optimum with all variables positive is known to exist for the problem: max $x_1^a x_2^{(1-a)}$; S.T. $x_1 \leqq l_1^b$, $x_2 \leqq l_2^c$, $l_1 + l_2 \leqq L$, where L is a constant. For what values of a, b, c are the global optimum conditions satisfied?

CHAPTER

3

The Theory of Linear Programming

This chapter requires familiarity with linear algebra (Review R2), linear equations and inequalities (Review R3), and convex sets (Review R4, Sections R4.1 through R4.4). The exposition is, however, designed to draw on the more advanced aspects of these topics as little as possible.

3.1 INTRODUCTION[1]

A linear programming problem is a special case of the general optimum problem in which the objective and all constraint functions are linear. Inequalities in the functional constraints and nonnegativity constraints on the variables are not only permissible but typical. The very importance of linear programming has been based on its ability to handle these problems.

[1] An exposition of linear programming theory along similar lines is given in Gale [1], Chapter 3. The author's exposition follows that of Gale to a considerable extent. Note Gale's notation: He transposes matrix-vector relationships relative to conventional usage. An advanced statement of the theory, with emphasis on the relation to game theory (not discussed here) is given in Goldman and Tucker.

Most texts on linear programming are directed toward computation, and the exposition of the theory is not ideal for our purposes.

We could write the standard maximum problem in the linear programming case as

$$\max f(x) = \sum_{j=1}^{n} c_j x_j$$

$$\text{S.T.} \quad \sum_{j=1}^{n} a_{ij} x_j \leqq b_i \qquad i = 1, \ldots, m$$

$$x_j \geqq 0 \qquad j = 1, \ldots, n.$$

It is a great convenience, however, to use matrix-vector notation throughout the analysis of linear programming, so we will write the *standard maximum problem* in the form

$$\max cx$$

$$\text{S.T.} \quad Ax \leqq b, \qquad x \geqq 0,$$

where c is a row vector of order n, b a column vector of order m, and A an $m \times n$ matrix.

In the standard program, the number, m, of constraints may bear any relation to the number, n, of variables. Typically we shall have $m > n$.

Sometimes we may wish to write a linear program in the alternative *canonical form*:

$$\max cx$$

$$\text{S.T.} \quad Ax = b, \qquad x \geqq 0.$$

(The matrix A is not, of course, the same matrix as in the standard form.)

A standard program can always be converted to canonical form by using the device of *slack variables.* If we write $z = b - Ax$, then $z \geqq 0$, if the constraints are satisfied. The constraint systems

$$Ax \leqq b, \qquad \text{and} \qquad Ax + z = b, \qquad z \geqq 0$$

are equivalent. The canonical version then has $n + m$ variables x, z.

We can also convert a problem in canonical form into one in standard form by using the equivalence of the two systems of constraints $Ax \leqq b$, $-Ax \leqq -b$ with the constraints $Ax = b$.

Before analyzing the linear programming problem further, we note that, since the objective function is linear, it has no critical points. We can state immediately from the general optimum principles that

The optimum of a linear programming problem will occur at a boundary point of the feasible set.

3.2 THE FEASIBLE SET

Since the constraints are written as equalities or as weak inequalities, we can take the feasible set to be *closed.* It may, however, be empty, since an arbitrary set of constraints may leave no feasible point, in which case we refer to the program as *infeasible*. The feasible set may not be *bounded.* It is, however, *convex*. The convexity follows from the convexity of individual linear constraints and the convexity of the intersection of convex sets.[2]

We can demonstrate the convexity of the feasible set more directly. Let x'', x' be feasible points. (These are often called feasible solutions, but this can lead to confusion between a solution of the constraints and the solution of the problem.) We have

$$Ax' \leqq b, \qquad Ax'' \leqq b.$$

Now consider x, a convex combination of x', x'':

$$x = kx' + (1 - k)\,x'' \qquad 0 \leqq k \leqq 1.$$

Substituting for x in the constraint system,

$$\begin{aligned} Ax &= A[kx' + (1 - k)\,x''] \\ &= kAx' + (1 - k)\,Ax'' \\ &\leqq b. \end{aligned}$$

Thus if x', x'' are feasible their convex combination is feasible and the feasible set is convex.

The objective function is linear and so may be taken as both concave and convex. Since the feasible set is convex, the conditions set out in Chapter 2 for a global optimum are satisfied for both a maximum and a minimum.[3] Thus we can state:

An optimum of a linear programming problem is a global optimum.

We know that the optimum of a linear programming problem is global, and that it occurs at a boundary point of a closed convex feasible set. From convex set theory we know that there is a special class of boundary points, *extreme points*, which have the property that they cannot be expressed as convex combinations (a convex combination is a linear combination with nonnegative weights, in which the sum of the weights is 1) of other points in

[2] Or we can use Result (6) of Review R4, Section R4.2, which gives the result directly.

[3] See Section 2.7.

the set, but that all other points can be expressed as convex combinations of extreme points.[4]

Suppose we have a maximum problem and consider any boundary point, not an extreme point. This point x can be expressed as a convex combination of some number of extreme points x^k. We choose the minimum number of extreme points necessary to define x, so that none have zero weight. There must be at least two.

$$x = \sum a_k x^k \qquad a_k > 0, \ \sum a_k = 1.$$

Now consider cx. We have

$$cx = c(\sum a_k x^k) = \sum a_k(cx^k).$$

Choose that x^k for which cx^k is the maximum. If several give the same maximum value, choose one. Denote this maximum value by v. Now the right-hand side of the above equation cannot be reduced, and might be increased, if we substitute v for every cx^k. Thus

$$cx \leqq v.$$

There are two possibilities:

(a) The maximum, v, is not reached at *all* the extreme points used to define x, $cx^k < v$ for at least some k, $cx < v$ and x certainly is not a maximum.
(b) The maximum is the same at all the extreme points so that $cx = v$. If x is optimal, so are all the extreme points and, by extension of the argument, so are all other convex combinations of these extreme points.

We have shown that no point in the feasible set, other than an extreme point, can be optimal unless it is defined by a convex combination of extreme points all of which are optimal.

We can now state the following:

The optimum of a linear programming problem is reached either at an extreme point or at a set of extreme points. In the latter case, all convex combinations of the extreme points are also optimal.[5]

This is an important result for solution of a linear program. Since the feasible set is a convex set bounded by linear constraints, all extreme points are at intersections of the linear boundaries. There are, therefore, only a *finite* number of extreme points. Since all optimal points are extreme points

[4] See Review R4, Section R4.4.

[5] An alternative proof can be derived from noting that, if x^* is optimal, the hyperplane $cx = cx^*$ must be a supporting hyperplane of the feasible set and so must include an extreme point. See Theorem (1) in Review R4, Section R4.4.

or convex combinations of them, we can search for a solution by examining the value of the objective function only at these points. Since the number of extreme points is finite, the problem can always be solved in a finite number of steps.

These characteristics of a linear programming problem are illustrated in Figure 3–1. This is a simple example in two variables. There are three linear constraints, C_1, C_2, C_3 (the direction of the inequalities is assumed to be such that the variables must lie on the origin side of the constraint boundaries), giving the heavily outlined feasible set $OE_1E_2E_3E_4$ when combined

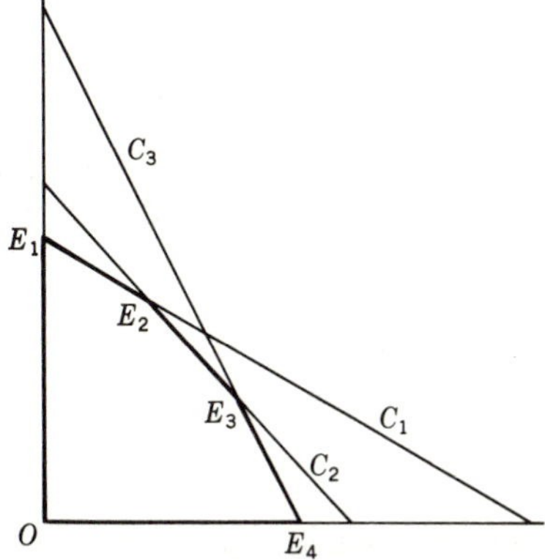

Figure 3-1. Feasible set for a linear program.

with nonnegativity constraints. The points E_1, E_2, E_3, E_4 are the extreme points. It is easily seen that none of these extreme points lies on any line joining two points in the feasible set, that every other boundary point is on a line joining extreme points, and that every interior point is on a line joining boundary points.

In the problem illustrated, the problem would be solved simply by evaluating the objective function at each of the five extreme points and choosing that which was optimal.

All solution techniques for linear programming are search techniques, and all these techniques proceed by taking an extreme point, evaluating the objective function at it, then taking another extreme point, and so on. *Efficient* solution techniques, of which the *simplex method*[6] is the best known,

[6] The simplex method made linear programming a computational tool. It was originated by Dantzig. See Dantzig. The method is described in Gale [1], Chapter 4, and all books on linear programming. The *revised simplex method*, using the inverse of the feasible basis, is most often used for computer work.

provide some rule by which one avoids looking at extreme points that give lower values for the objective function than the particular extreme point one is currently evaluating. In this way, the steps in the search procedure can avoid going backward, and the optimum can be recognized when reached. The more subtle aspects of solution technique are chiefly concerned with choosing a good extreme point from which to start, since the number of steps to solution depends on the point from which the process starts.

To *recognize* the optimum when it is reached, direct computation is not necessary, as we see in the next section.

3.3 DUALITY

The idea of the *dual* of a linear program and the linear programming theory associated with the notion of duality has been of the highest importance in economic analysis for the light it has thrown on the nature of prices. No idea since marginal analysis has been as important for the fundamental theory of price.

In the standard maximum problem,

$$\begin{aligned} &\max cx \\ \text{S.T.}\quad &Ax \leqq b, \qquad x \geqq 0, \end{aligned}$$

the vectors c, b and the matrix A are the *data* of the problem. Using the same data, but a different set of variables, we can form another problem

$$\begin{aligned} &\min yb \\ \text{S.T.}\quad &yA \geqq c, \qquad y \geqq 0. \end{aligned}$$

This second problem is the *dual* of the first, which is called the *primal* when the two programs are being considered together.

In forming the dual from the primal, the following changes are made:

(a) If the primal variables are written as a column vector of order n, the dual variables are written as a row vector of order m, and vice versa.
(b) The elements of the constraint vector of the primal program become the weights of the dual objective function, and the weights of the primal become the constraints of the dual.
(c) A maximum problem becomes a minimum problem, and vice versa.
(d) The direction of the constraint inequalities is reversed.
(e) The nonnegativity constraints on the variables are *not* reversed.[7]

[7] This list refers to the dual of a standard program. The dual of a canonical program differs in (d) and (e).

The duality relationship is symmetric. If we take the dual of a standard program, then the dual of the dual, we are back with the original problem. Thus there is just one pair of problems which stand in dual relationship to each other.

It is important to realize, from the outset, that duality is first of all *a formal mathematical relationship*. Given any program, we form the dual by following the rules above. If the primal program represents a substantive problem, we naturally expect the dual to possess some appropriate interpretation. Whether this interpretation is easy or not, and whether it represents an interesting problem or not, makes no difference to the existence of the dual problem as a formal property of the primal problem itself.

Why introduce the dual at all? We can start to answer this question by stating, then proving some part of, two theorems of great importance in linear programming theory.

Duality Theorem

A feasible vector x^* of the primal problem is optimal, if and only if the dual has a feasible vector y^* such that $cx^* = y^*b$. The vector y^* is then optimal for the dual.

Existence Theorem

The primal and dual have optimal solutions, if and only if both have feasible vectors.

If we accept the duality theorem for the moment, then it follows that, if either program has an optimum, so does the other. Obviously both can have optimal solutions only if both have feasible vectors, proving the necessity part of the existence theorem.

A not-quite-complete proof of the sufficiency part of the existence theorem can be derived as a consequence of the following:

Fundamental Lemma

If x, y are feasible for the primal and dual respectively, then the following relationship holds; $cx \leqq yAx \leqq yb$.

To prove the lemma, we note that the primal constraints give $Ax - b \leqq 0$. Since $y \geqq 0$, if it is feasible, $y(Ax - b) \leqq 0$, giving

$$yAx \leqq yb.$$

Using the dual constraints $yA - c \geqq 0$, and the nonnegativity constraints on x, we have for feasible y and x,

$$cx \leqq yAx.$$

The lemma follows from putting the two inequalities together.

Now return to the existence theorem itself. From the lemma, if y' is *any* feasible vector in the dual,

$$cx \leqq y'b$$

for all x feasible in the primal. Thus the set of $v = cx$ such that x is feasible in the primal is a continuous closed set, bounded (at least from above), and has a maximum. Thus the primal has an optimum if both programs are feasible. The same type of argument can be applied to the set $v' = yb$, so that the dual has an optimum if both programs are feasible.

In the duality theorem, the sufficiency part is easy to prove, using the fundamental lemma. Let x^*, y^* be feasible solutions such that $cx^* = y^*b$, and let x be *any* other vector which is feasible for the primal. Then

$$\begin{aligned} cx &\leqq y^*b \qquad \text{(fundamental lemma)} \\ &\leqq cx^*, \end{aligned}$$

so that x^* is optimal for the primal. The optimality of y^* in the dual follows from similar reasoning.

The proof of the necessity part of the duality theorem, that if x^* is optimal for the primal there is a feasible dual vector y^* such that $cx^* = y^*b$, is quite a difficult matter at this stage of the analysis. It is much easier to prove by the use of the properties of the Lagrangean function, so we shall give the proof in Chapter 5.[8]

The duality theorem enables us to check whether a given pair of feasible vectors is optimal or not, but we must examine the primal and dual simultaneously. In one sense, therefore, the duality theorem gives an optimum condition. Normally, however, we regard the optimum conditions of linear programming to be given by the equilibrium theorem, which is discussed in the next section. The examples below illustrate the use of the duality and existence theorems in simple problems, but their importance is theoretical rather than in direct application to solution methods.

Example 1. Show that the points $x^* = (8, 0)$, $y^* = (0, 3)$ are optimal for the dual pair of problems,[9]

$$\begin{array}{llll} & \max 3x_1 + 2x_2 & & \min 2y_1 + 8y_2 \\ \text{S.T.} & -2x_1 + x_2 \leqq 2 & \text{S.T.} & -2y_1 + y_2 \geqq 3 \\ & x_1 + 2x_2 \leqq 8 & & y_1 + 2y_2 \geqq 2 \\ & x_1, x_2 \geqq 0 & & y_1, y_2 \geqq 0. \end{array}$$

[8] See Section 5.7. More direct proofs are available. See Gale [1], Theorem 3.1.

[9] The reader should satisfy himself that these problems are, in fact, dual.

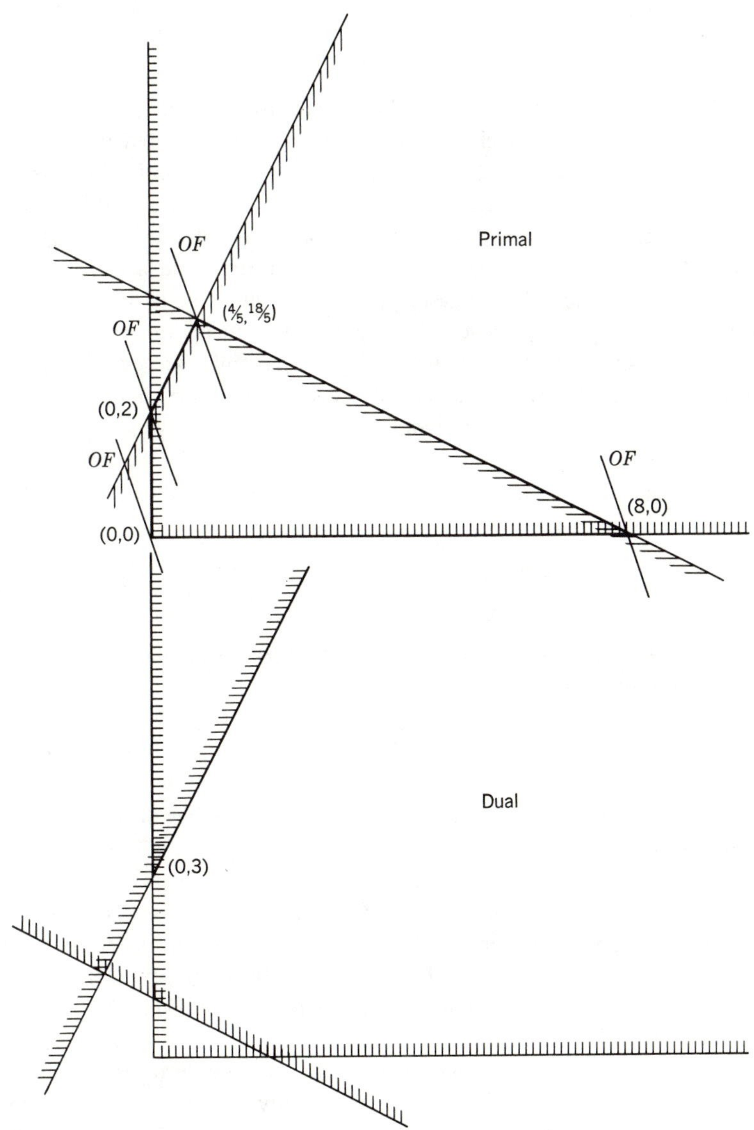

Figure 3-2. Graphical solution of Example 1.

First we must check for feasibility. We certainly have x^*, y^* nonnegative. Substituting for x^* in the primal, the left-hand sides of the constraints are $-16, 8$, respectively, so x^* is feasible. Substituting for y^* in the dual, the left-hand sides are 3, 6, so y^* is feasible. $cx^* = 3 \cdot 8 + 2 \cdot 0 = 24$, $y^*b = 2 \cdot 0 + 8 \cdot 3 = 24$, so x^*, y^* are optimal.

Since both primal and dual problems have two variables, this is one of the rare examples in which both problems can be illustrated graphically. This is done in Figure 3–2. The shading along the constraint boundaries shows which side of the boundary is feasible, the feasible set itself is heavily outlined. The feasible set for the primal has four extreme points, (0, 0), (0, 2), (4/5, 18/5), (8, 0). The objective function, drawn in as OF at each extreme point, clearly lies furthest from the origin at the point (0, 8). For two variables we can always solve a linear program by this graphical technique.

The feasible set of the dual in this case has only one extreme point, (0, 3). Note that the set is unbounded above, but this does not give rise to problems because we seek only a minimum.

Example 2. Does the following problem have an optimal solution?

$$\begin{aligned} &\max x_1 + x_2 \\ \text{S.T.}\quad &-2x_1 - x_2 \leqq 2 \qquad x_1, x_2 \geqq 0. \\ &-x_1 - x_2 \leqq 1 \end{aligned}$$

We use the existence theorem, and merely check for feasibility in the dual. The dual constraints are

$$\begin{aligned} -2y_1 - 2y_2 &\geqq 1 \qquad y_1, y_2 \leqq 0. \\ -y_1 - y_2 &\geqq 1 \end{aligned}$$

It is immediately obvious that no y can be found satisfying these constraints. The dual is infeasible so the primal has no optimum.

The reason the primal has no optimal solution is, of course, that the feasible set is unbounded above. For every x satisfying the constraints, there is some other feasible x giving a higher value for the objective function. This was obvious from the first, but we were interested in showing the existence theorem at work!

In complex examples, boundedness may not be easy to assess, so the existence theorem can be useful. The infeasibility of the dual means either infeasibility of the primal or a primal feasible set unbounded in the direction of the optimum.

3.4 THE OPTIMUM CONDITIONS

As has been pointed out, the duality theorem gives an optimum condition, but this is in terms of the value of the objective function. We think of optimum conditions in economic analysis as more typically relating constraints and variables (for example, price equals marginal cost). The closest linear programming equivalent is the following theorem:

Equilibrium Theorem for Linear Programming

(a) If x^*, y^* are feasible for the primal and dual, they are optimal, if and only if

$$(1)\ y_i^* = 0 \qquad \text{whenever} \qquad \sum_j a_{ij}x_j^* < b_i,$$

$$(2)\ x_j^* = 0 \qquad \text{whenever} \qquad \sum_i a_{ij}y_i^* > c_j,$$

that is, if the kth variable in one program is zero whenever the kth constraint in the other program is ineffective.

(b) The optimal point (or one of the optimal points in the case of a weak optimum) will always be such that the number of nonzero variables in each program will be no greater than the number of constraints in that program.

Proof of Part (*a*) follows from the duality theorem and the fundamental lemma. To show necessity, we note that, if x^*, y^* are optimal

$$cx^* = y^*Ax^* = y^*b.$$

From the first equality it follows that $y^*Ax^* - cx^* = 0$, so that $(y^*A - c)\,x^* = 0$. Since y^* is feasible, $y^*A - c \geqq 0$, and since x^* is feasible, $x^* \geqq 0$. Thus every term in

$$(y^*A - c)\,x^* = \sum_j (\sum_i a_{ij}y_j^* - c_j)\,x_j^*$$

must be zero, and either

$$(\sum a_{ij}y_j^* - c_j) = 0 \qquad \text{or} \qquad x_j^* = 0.$$

By using the second equality and arguing in the same way we prove the necessity of condition (2) of the theorem.

To prove sufficiency, suppose x^*, y^* are feasible and satisfy conditions (1) and (2). Then every term in the expansion

$$\sum_j (\sum_i a_{ij}y_j^* - c_j)\,x_j^*$$

is zero, and every term in the expansion

$$\sum_i \left(\sum_j a_{ij}x_j^* - b_i\right) y_i^*$$

is also zero. Thus

$$cx^* = y^*Ax^* = y^*b,$$

so that x^*, y^* are optimal.

The equilibrium theorem is important for two reasons. First, it enables us to verify the optimality of a primal solution, even when we are not given the optimal dual solution. Second, and much more important for our purposes, it leads to extremely important interpretations of the optimum conditions in economic models that are in linear programming form. Both these uses are best illustrated by example.

Example 1 (Solution check). Show that $x^* = (8, 0)$ is optimal in the program

$$\begin{aligned} &\max 3x_1 + 2x_2 \\ \text{S.T.} \quad &-2x_1 + x_2 \leqq 2 \qquad x_1, x_2 \geqq 0. \\ &x_1 + 2x_2 \leqq 8 \end{aligned}$$

This is the same program as used in Example 1 of the preceding section, but this time we are given only the presumptive optimal of the primal.

We proceed as follows: Putting the values (8, 0) into the primal constraints, we see that the first constraint is ineffective. From the equilibrium theorem, it follows that $y_1 = 0$, if (8, 0) is optimal. The presumptive optimal has $x_2 = 0$, so we will expect the second dual constraint to be ineffective and the first to be effective. Given $y_1 = 0$, this immediately gives $y_2 = 3$, so (0, 3) is the presumptive optimal point for the dual. Using the duality theorem, we can then show, as in the previous section, that the pair of solutions (8, 0), (0, 3) is, in fact, optimal, thus solving the problem.

Example 2 (Economic Interpretation). We have a linear model of production in which there are n outputs x_j and m inputs b_i. These are related by constant production coefficients a_{ij} which give the amount of the ith input required to produce a unit amount of the jth output. Then $\sum_j a_{ij}x_j$ gives the total amount of the ith input necessary to produce the output mix x, and Ax gives the vector b of inputs necessary to produce this output mix.

We are given a vector of product prices p and a vector of total available resources $\bar{b}$. We wish to examine the properties of optimal production, defined as max px. This is a linear programming problem

$$\begin{aligned} &\max px \\ \text{S.T.} \quad &Ax \leqq \bar{b}, \qquad x \geqq 0. \end{aligned}$$

The dual must be

$$\min y\bar{b}$$
$$\text{S.T.} \quad yA \geqq p, \qquad y \geqq 0.$$

First of all, let us interpret the dual. From the duality theorem we have $px^* = y^*\bar{b}$. Now px^* has the dimension of value (price multiplied by quantity), so we expect $y^*\bar{b}$ to have the same dimension. Since $\bar{b}$ is a quantity vector, y is some kind of price vector, the prices being prices of inputs. Because of this price dimension of the dual variables in economic problems of this kind, the dual variables are frequently referred to as *shadow prices.*

Now consider the dual constraints, each of which is in the form

$$\sum_i a_{ij} y_i \geqq p_j.$$

Since a_{ij} is the amount of the ith input required to produce a unit of the jth output, $a_{ij}y_i$ is the value of the ith input required to produce a unit of the jth output, and $\sum a_{ij}y_i$ is the total value of inputs required to produce a unit of the jth output, all inputs being valued at the shadow prices y.

Thus we can interpret the dual in the following way (it is important to realize *we do not have to follow this particular interpretation*, it is merely consistent with the model). It is the answer to the question, what is the least value that should be attached to the available resources $\bar{b}$, given that the option of converting resources into products and selling the products is also available? The constraints of the dual express the fact that, if the value of inputs incorporated into a product were less than the price of the product, it would be more profitable to produce and sell the output than to sell the resources. At optimal values x^*, y^* the economy or firm would be indifferent between using the resources and selling the output for px^*, or selling the resources at prices y^* for total receipts $y^*\bar{b} = px^*$.

It is now easy to see the implications of the equilibrium theorem which states, in the context of this model, that:

(a) Any resource which cannot be entirely used up in producing the optimal output mix will receive a zero shadow price or optimal valuation.
(b) No product whose unit cost of production exceeds its price (when inputs are valued at optimal shadow prices) will be produced at the optimum.[10]

In other words, resources in excess supply will be free goods, and processes that make losses will not be used, if the shadow prices are actual

[10] The last statement holds for at least one optimal point if the optimum is not unique.

prices. Since these relationships correspond to those of *equilibrium* in a competitive economy, we have the name for the theorem.

We shall give more substance to the term shadow prices when we give a more complete interpretation of the dual variables in a later section.[11]

3.5 BASIC SOLUTIONS

We have already shown that a standard program can be put into canonical form by inserting slack variables, and that a canonical program can be put into standard form. For some purposes it is more revealing to discuss a program in canonical form, and in some economic problems the problem is originally formulated this way.

Consider a canonical program

$$\max cx$$
$$\text{S.T.} \quad Ax = b, \qquad x \geqq 0.$$

The constraints form a system of m equations in n variables, with m assumed to be less than n. We know from ordinary linear equation theory that, if A has rank m (that is there is linear dependence among any set of more than m columns), we can put any $n - m$ of the variables equal to zero and then solve uniquely for the remaining m. Any such solution to the equation system will be referred to as a *basic solution.* Clearly, since the equation system is linear, any linear combination of solutions is also a solution, although not generally a basic solution.[12]

Now a basic solution of $Ax = b$ may not be feasible, since we may not have $x \geqq 0$. If we do have $x \geqq 0$, such a basic solution is a *basic feasible solution.* They will, in general, be other feasible solutions which are not basic, and there may be several basic feasible solutions. The following is the fundamental theorem on basic feasible solutions:

If $Ax = b$, $x \geqq 0$ has a feasible solution, it has a basic feasible solution.

The proof is inductive. Suppose we have a nonbasic feasible solution x, that is, x is nonnegative and has k $(> m)$ nonzero variables. Write the jth column of A as A^j, so that

$$Ax = \sum_{j=1}^{n} A^j x_j.$$

[11] See Section 3.7.

[12] See Review R3, Section R3.4.

Without loss of generality we can take the k nonzero variables to be the first k. Since the remaining variables are zero,

$$\sum_{j=1}^{k} A^j x_j = Ax = b.$$

But the columns A^j are vectors of order m. For $k > m$, the set of vectors $A^j, j = 1, \ldots, k$ must be linearly dependent, so that we can find weights, not all zero, such that

$$\sum_{j=1}^{k} v_j A^j = 0,$$

where at least one v_j can be taken as positive.

There is a one to one correspondence between the elements v_j and x_j. Choose the maximum of the ratios v_j/x_j and denote this by θ. Using the equations above for x, v we can form the linear combination with weights $1, 1/\theta$

$$\sum_{j=1}^{k} A^j x_j - \frac{1}{\theta} \sum_{j=1}^{k} v_j A^j = b,$$

which can be rearranged to give

$$\frac{1}{\theta} \sum_{j=1}^{k} A^j \left(\theta - \frac{v_j}{x_j}\right) x_j = b.$$

But, from the way we defined θ, we have $[\theta - (v_j/x_j)] \geqq 0$ for all j and zero for at least one, while $\theta \geqq 0$.

Thus the vector whose jth component is $(1/\theta)[\theta - (v_j/x_j)]\, x_j$ is a nonnegative solution to the equations $Ax = b$ and has only $(k - 1)$ nonzero components.

We can continue the reduction so long as $k > m$. The process stops at $k = m$, since at that stage the vectors become linearly independent.

It is easy to see that, if x is a basic feasible vector (that is, a basic feasible solution) it cannot be expressed as a convex combination of other feasible vectors. Suppose, to the contrary, that it could, so that we could write as some convex combination

$$x = \sum_k a_k x^k \qquad a_k > 0, \qquad \sum_k a_k = 1\,.$$

There are three classes of possibilities. If any one of the vectors x^k is nonbasic, it has more than m nonzero components, so that x has more than m nonzero components and is nonbasic. If all the vectors x^k are basic, with the same nonzero components then x is also basic, but, since all the columns from A which determine the x^k are the same, all the x^k coincide with each other and with x. If all the vectors x^k are basic, but different, then the nonzero components do not coincide and x has more than m nonzero components and is nonbasic.

Since a basic feasible solution cannot be expressed as a convex combination of other feasible vectors, it corresponds to an extreme point of the feasible set.

Part (*b*) of the equilibrium theorem, which was left unproved in the last section, is thus seen to be equivalent to the earlier theorem, that the optimum point will be an extreme point, or the set of optimum points will include an extreme point.

The set of m columns of A which correspond to the choice of nonzero variables for a basic vector is the *basis* for that vector. It is a *feasible basis* or an *optimal basis* if it corresponds to a feasible or optimal vector. The basis is a square $m \times m$ matrix which we shall write A_B.

3.6 THE BASIS THEOREM

Suppose we wish to write down the dual of the canonical problem. If the primal is a maximum, we will expect the dual to be a minimum. What happens to the constraints is not immediately obvious. Since we have assumed $m < n$, we cannot write the dual constraints as $yA = c$, since this would be a system of n equations in m variables and would have no solution since we have assumed that A is of rank m. The constraints must be *inequalities* which we take to be of the form $yA \geqq c$, since the dual is a minimum problem. If the primal had been a minimum, the dual would be a maximum and we would reverse the inequalities.

There is one more difference between forming the dual from the standard primal and forming the dual from the canonical primal. *In the canonical dual the variables y are unconstrained.*

The reason for this may be seen by considering the number of variables and constraints in the primal and dual of the standard program. The primal has n variables, m functional constraints, and n direct constraints, the dual has m variables, n functional constraints, and m direct constraints. Altogether there are $m + n$ variables and $2(m + n)$ constraints, all of which are inequalities.

In the canonical primal there are n variables, m *equality* constraints and n direct constraints. But an equality constraint is equivalent to *two* inequality constraints, so the primal has the equivalent of n variables and $2m + n$ inequality constraints. The dual adds m variables and n inequality functional constraints, so that, without direct constraints on the dual variables the whole system already contains the equivalent of $m + n$ variables and $2(m + n)$ inequality constraints, the same as the standard program.

Thus the canonical dual, for a maximum primal, is

$$\begin{aligned} &\min yb \\ \text{S.T.}\quad &yA \geqq c, \qquad y \text{ unconstrained.} \end{aligned}$$

The *equilibrium theorem* for the canonical program has the simple form:

Feasible x^*, y^* *are optimal, if and only if* $x_j^* = 0$, *whenever* $\sum a_{ij} y_i^* > c_j$.

We can now state a theorem which is important for the study of certain linear economic models.

Basis Theorem for Linear Programming

If the basis A_B *is optimal for the canonical problem* max (min) cx, S.T. $Ax = b$, $x \geqq 0$ *and is feasible for the problem* max (min) cx, S.T. $Ax = b'$, $x \geqq 0$, *it is also optimal for the latter problem.*[13]

Suppose the problem is a maximum (proof for the minimum follows similar lines). Let x^*, y^* be optimal for the original problem and its dual. From the equilibrium theorem,

$$\sum a_{ij} y_i^* > c_j \qquad \text{implies} \qquad x_j^* = 0.$$

Now let x', defined on the same basis as x^*, be feasible for the second program. We are assured such a feasible vector exists. Since x', x^* have the same basis, $x_j^* = 0$ implies $x_j' = 0$. Thus

$$\sum a_{ij} y_i^* > c_j \qquad \text{implies} \qquad x_j' = 0.$$

From the equilibrium theorem, x', y^* being feasible (note that the constraints of the dual are identical in the two problems), are also optimal.

Note that the optimal values of the dual variables are not changed by the substitution of b' for b, provided the original basis remains feasible.

We can interpret the basis theorem also in terms of a standard program. Convert the standard program

$$\begin{aligned} &\max cx \\ \text{S.T.}\quad &Ax \leqq b, \qquad x \geqq 0 \end{aligned}$$

[13] This theorem is due to Gale. (Gale [1], Chapter 9, lemma 9.3). The term "basis theorem" is not used in Gale or elsewhere for this particular theorem.

into a canonical program by using a vector, z, of slack variables. We obtain the constraints in the form:

$$[A : I]\begin{bmatrix} x \\ z \end{bmatrix} = b \qquad x, z \geqq 0.$$

A basis for $[A : I]$ specifies which components of x, z are zero. The zero and nonzero components of x are related to the effectiveness of the dual constraints, while if the ith component of z is zero it implies that the ith primal constraint is effective.

Thus we can state an alternate form of the basis theorem:

Basis Theorem for Standard Program

Define $S = \{i \mid A_i x^* = b_i\}$, *the set of constraints which are effective at the optimum of the standard program,* max cx, S.T. $Ax \leqq b$. *Now consider the program,* max cx, S.T. $Ax \leqq b'$. *If this has a feasible solution with* $A_i x = b_i'$, $i \in S$, *then it has an optimal solution with the same set of effective constraints.*

3.7 INTERPRETATION OF THE DUAL VARIABLES

Consider the variation in the feasible set caused by varying the b vector in the primal constraint. Since each constraint is of the form $\sum a_{ij}x_j \leqq b_i$ (standard form), variation in b_i is equivalent to moving the constraint boundary $\sum a_{ij}x_j = b_i$ toward or away from the origin, but maintaining the same slope, which is determined by the a_{ij}'s.

Figure 3–3 illustrates this type of constraint variation. All four cases have the same slopes for the equivalent constraint boundaries C_1, C_2, C_3, but the boundaries have been moved in or out from the origin in various ways, to give four different feasible sets (shaded).

The extreme points of the feasible set occur at intersections of constraint boundaries. The intersection of equivalent boundaries, like C_1, C_2, corresponds to one particular feasible basis, and the corresponding extreme point, V_2, is associated with the same feasible basis in all cases. We shall think of the extreme points V_2 in cases (a), (b), (c), and (d) as *equivalent*.

The basis theorem of the preceding section can then be interpreted to mean that, if constraints are varied in a parallel fashion and a given extreme point, say V_2, is optimal for one case, the equivalent extreme point (labeled V_2 also) will be optimal in every other case in which it still exists.

Note that if V_3 had been optimal in case (a) it would have been optimal

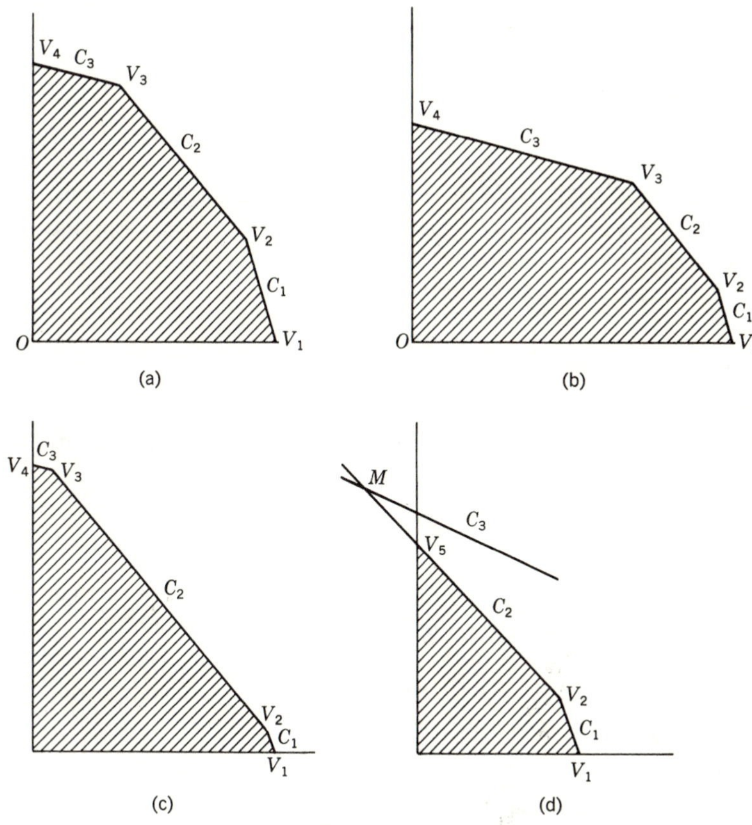

Figure 3-3. Constraint variation.

also in cases (b) and (c), but not in case (d), since it does not exist in that case. In case (d), the intersection of C_2 and C_3 which defined V_3 in the other three cases now takes place at M which does not satisfy the nonnegativity constraint and so is not a point of the feasible set, extreme or otherwise.

As can be seen from this illustration, the variations in b which can take place without destroying the feasibility of the original basis are quite wide, so the restriction included in the following important result still leaves it with powerful interpretive value:

Consider a linear program with primal constraint vector b, optimal value for the objective function V^*, and optimal dual vector y^*. Let b now be varied to $b + \Delta b$, subject to the condition that the original optimal basis remains feasible.

Then the variation ΔV^* in the optimal value of the objective function is given by

$$\Delta V^* = y^* \Delta b.$$

In particular, if only the ith component of b changes,

$$\Delta V^* = y_i^* \Delta b_i.$$

The proof follows directly from the basis and duality theorems. Let x^*, x' be the optimal vectors for the original and varied cases. Then, since the original basis remains feasible, y^* remains the optimal dual vector. Using the duality theorem, we have, for the two cases,

$$\begin{aligned} cx' &= y^*(b + \Delta b) \\ cx^* &= y^* b. \end{aligned}$$

The result then follows by subtraction.

We can now complete our interpretation of the dual variables. If only the ith constraint varies, we have

$$\frac{\Delta V^*}{\Delta b_i} = y_i^*,$$

which gives the interpretation of the ith dual variable (at optimum) as the marginal value to the program of relaxing the ith constraint. In a typical economic context, it will be the marginal social value, or marginal revenue, of increasing the quantity of the corresponding resource, thus justifying the usual interpretation of the dual variables as shadow prices.

FURTHER READING

The basic theory of linear programming has been covered in this chapter. The reader may wish to read the exposition in Gale [1] (but note the comments made earlier concerning notation) which also discusses some of the classic problems in linear programming. Dorfman, Samuelson, and Solow provide a more elementary discussion of the theory, but well spiced with examples of economic application. Allen [2] also discusses linear programming.

For the reader interested in linear programming as a computational method, it is suggested that he read the exposition of the simplex method in Gale [1], Chapter 4, then turn to one of the texts on computational linear programming, such as Gass or Hadley [2]. He should also investigate the resources of his local computer center. Gass, Chapter 9, gives a list of available computer programs as of the date of publication.

EXERCISES

1. Examine the feasibility and existence of an optimal solution for the following program and its dual:

$$\begin{aligned} &\max x_1 + 3x_2 \\ \text{S.T.}\quad &-3x_1 + 6x_2 \leqq -1 \qquad x_1, x_2 \geqq 0. \\ &x_1 - 3x_2 \leqq 2 \end{aligned}$$

2. By solving the dual problem and using the equilibrium theorem, show that (2, 1) is optimal in

$$\begin{aligned} &\max 2x_1 - x_2 \\ \text{S.T.}\quad &-4x_1 + x_2 \leqq 2 \\ &x_1 - x_2 \leqq 1 \qquad x_1, x_2 \geqq 0. \\ &2x_1 + x_2 \leqq 5 \end{aligned}$$

3. Determine (a) the basic solutions, (b) the basic feasible solutions, for the equations,

$$\begin{aligned} x_1 + x_2 + 2x_3 + x_4 &= 3, \\ 3x_1 + 2x_2 + 2x_3 + 3x_4 &= 6, \end{aligned}$$

subject to $x_1, x_2, x_3, x_4 \geqq 0$.

What is the optimal solutions of the canonical problem, $\max x_1 + 2x_2 + x_3 + x_4$ subject to the above constraints? Formulate and solve the dual.

4. Consider the standard problem with optimal solution x^*. Now replace the original objective function cx by a new objective function $c'x$ where $c'_k > c_k$, $c'_i = c_i$, $i \neq k$. If the constraints of the program are unchanged and the new optimal solution is x^{**}, show that $x_k^{**} > x_k^*$.

5. In the program of Exercise **2**, replace the b vector by the vector

$$b' = \begin{bmatrix} 1 \\ 0 \\ 4.5 \end{bmatrix},$$

and show that the basis theorem of Section 3.6 applies. Relate the constraint variation $b \to b'$ to the values of the dual variables, as in Section 3.7.

6. The Diet Problem. There are n foods each containing nonnegative amounts of each of m nutrients. Given the prices of the foods and the minimum required level of the nutrients, set up the problem of finding the minimum cost of meeting the basic nutrition level, if excess nutrients have no effect.

Examine the effect on the solution of specifying that one of the nutrients must not be consumed beyond the stated level.

7. Assuming a bounded feasible set, interpret the theory of linear programming in terms of supporting hyperplanes of convex sets (See Review R4, Sections R4.3 and R4.4).

CHAPTER

4

Classical Calculus Methods

This chapter assumes familiarity with the basic properties of continuous functions (Review R8, Sections R8.1 through R8.4). The discussion of the second-order conditions requires other material, noted at the beginning of Section 4.5, but a first reading of the chapter can be made without special study of this.

4.1 INTRODUCTION[1]

As pointed out in discussing the general structure of optimizing problems (Chapter 2), classical calculus methods can be applied to a problem with the following properties:

(a) The objective and indirect constraint functions possess suitable continuity properties. Usually they will be taken to be of class C^2, which is sufficient for all purposes.

[1] Elementary treatments of the classical method are available in a variety of texts, including Allen [1], Yamane, Henderson and Quandt. The treatment in mathematics texts (advanced calculus, mathematics for engineers and physicists) is usually cursory and confined to the first order conditions.

The fullest treatment of the subject in the economics literature is in Samuelson [1]. Other complete treatments are given in Hancock and Frisch. See, however, the footnote in Section 4.5 concerning the approach of these authors.

(b) The functional constraints are all equalities.
(c) There are no direct (nonnegativity) constraints on the variables.

The standard form of the problem will be written

$$\begin{aligned} &\max f(x) && (x \text{ a vector of order } n) \\ \text{S.T.}\quad &g^i(x) = 0, && i = 1, \ldots, m \qquad (m < n) \end{aligned}$$

Since the constraints are all effective at all times, the feasible set K contains only boundary points, so interior optima are ruled out. Unless the constraints are all linear, K will not be a convex set.

There are m constraint equations in n variables, with $m < n$. If the appropriate Jacobean is nonsingular, we can express $n - m$ of the variables in terms of the remaining m (from the *implicit function theorem*), substitute in the optimand, and reduce the problem to one of the unconstrained optimization of a function of only $n - m$ variables. This approach would seem to have all the characteristics of a good procedure: We reduce the number of variables, and come out with a problem we already know how to solve.

Direct subsitution is often used to solve simple problems when explicit solutions are desired, but the procedure is not as useful as it seems at first glance. Explicit solution of the constraint equations is possible only in a few cases, if these are nonlinear.

In the typical mathematical economics context the functional forms of f, g^i are not explicitly specified. Although some of the analytical properties of the optimal solution can be elucidated by using the relationships between the derivatives of the variables given by the *implicit function theorem*, the results are not symmetric with respect to the choice of variables as dependent and independent.

It turns out that the most useful technique for examining the properties of the optimal solution to the classical problem is a method in which, instead of the number of variables in the system being reduced, they are actually increased.

4.2 THE LAGRANGEAN FUNCTION

Deferring the justification for so doing until later, let us examine the properties of a function $L(x, \lambda)$ defined as follows:

$$L(x, \lambda) = f(x) - \sum_{i=1}^{m} \lambda_i g^i(x),$$

where the λ's are arbitrary variables. The function $L(x, \lambda)$ is the *Lagrangean function*, and the λ's are *Lagrange multipliers*. In some expositions the weighted sum of the constraint functions is added to, rather than subtracted from, the objective function. The only difference is in the signs of the λ's. The convention we have adopted is the most appropriate for the economic interpretation of the λ's that will be given later.

If we take the derivatives of $L(x, \lambda)$ with respect to the λ's, we have

$$\begin{aligned} \partial L(x, \lambda)/\partial \lambda_i &= -g^i(x) \\ &= 0 \qquad \text{for} \qquad x \in K. \end{aligned}$$

Also, for all $x \in K$,

$$L(x, \lambda) = f(x).$$

Thus, if $L(x, \lambda)$ has a maximum over the variables x, λ, at the point x^*, λ^* then, since it is unconstrained, it has a critical point there with all partial derivatives zero. Since the partial derivatives with respect to λ are zero, we have $x^* \in K$ and $L(x^*, \lambda^*) = f(x^*)$, so that x^* gives a maximum for $f(x)$ over $x \in K$ and is thus the solution to the classical constrained maximum problem. Conversely, if x^* gives a maximum of $f(x)$ for all $x \in K$, it follows that $L(x^*, \lambda) = f(x^*)$ and that $L(x^*, \lambda^*) \geqq L(x, \lambda^*)$.

Thus we can state the following, which is the justification for the introduction of the Lagrangean function:

(*a*) *If a critical point of* $L(x, \lambda)$ *occurs at* x^*, λ^*, *then* $x^* \in K$.

(*b*) *If* x^*, λ^* *maximizes* $L(x, \lambda)$, *then* x^* *maximizes* $f(x)$ *over* $x \in K$, *and if* x^* *maximizes* $f(x)$ *over* K, *then there is some* λ^* *such that* $L(x^*, \lambda^*)$ *is a maximum. The results hold for a minimum as well as a maximum.*

To use the Lagrange technique, we set up the Lagrangean and then find its critical point (or points). The partial derivatives of $L(x, \lambda)$ are divided into two sets. The partial derivatives with respect to the λ's are simply the constraint functions and equating them to zero merely ensures that $x \in K$. It is the partial derivatives with respect to the x's that play the major solution role. If these are equated to zero, we obtain n equations of the form

$$f_j - \sum_{i=1}^{m} \lambda_i g_j^i = 0 \qquad (j = 1, \ldots, n). \tag{4.2.1}$$

We can express this in vector form as

$$\nabla f - \lambda G = 0,$$

where $G = [g_j^i]$ and $\lambda = [\lambda_i]$.

These are equations both in the x's and the λ's. Since the λ's do not appear within the functions f, g^i, the above system can be regarded as locally linear in the λ's, with the matrix G as the matrix of the system and ∇f as the constant vector. Since there are n equations and only m λ's, there is a solution only if not more than m of the equations are linearly independent. The linear dependence is provided by the fact that the constraints are satisfied, so that $n - m$ of the x variables are dependent on the others, and the corresponding partial derivatives are linearly dependent on the remaining partial derivatives. Thus we can solve for the multipliers to obtain a unique set λ^*, not all of which are zero.

We can show this dependence by noting that, at the optimum, $f(x)$ is a maximum for $x \in K$. This means that $f(x)$ is stationary for small constrained variations, that is, that we can find variations in the x's, not all of which are zero, such that $df, -dg^1, \ldots, -dg^m$ are all zero. This gives an equation system of the form

$$\begin{bmatrix} f_1 & f_2 & \cdots & f_n \\ -g_1^1 & -g_2^1 & \cdots & -g_n^1 \\ \cdot & \cdot & \cdots & \cdot \\ \cdot & \cdot & \cdots & \cdot \\ -g_1^m & \cdot & \cdots & -g_n^m \end{bmatrix} \begin{bmatrix} dx_1 \\ dx_2 \\ \cdot \\ \cdot \\ dx_n \end{bmatrix} = [0];$$

In order for this to have a nonzero solution, every set of $m + 1$ columns in the system matrix must be linearly dependent. But the above matrix is simply the transpose of the system matrix in Equation (4.2.1), when those equations are considered as homogeneous equations in the $m + 1$ variables $(1, \lambda_1, \lambda_2, \ldots, \lambda_m)$.

If we write the above matrix as M, the two systems are

$$M\,dx = 0, \qquad M'\bar{\lambda} = 0 \qquad [\bar{\lambda} = (1, \lambda)],$$

so the conditions for nontrivial solution are the same in both.

Example. The production functions for each of two goods depend on the same two factors. The total quantity of each factor is fixed. If goods prices are given, under what conditions will the value of output be maximized?

Denote by x_1, x_2, the outputs of the two goods. Let x_3, x_4 be the quantities of the factors used in producing the first good with production function $x_1 = F^1(x_3, x_4)$. Similarly, we have $x_2 = F^2(x_5, x_6)$. We assume that x_3, x_5 represent the same factor, and also x_4, x_6. The problem is

$$\begin{aligned} &\max f(x) = p_1x_1 + p_2x_2; \\ \text{S.T.}\quad &x_1 - F^1(x_3, x_4) = 0, \qquad x_3 + x_5 - k_1 = 0, \\ &x_2 - F^2(x_5, x_6) = 0, \qquad x_4 + x_6 - k_2 = 0. \end{aligned}$$

The Lagrangean function is

$$L(x, \lambda) = p_1x_1 + p_2x_2 - \lambda_1[x_1 - F^1(x_3, x_4)] - \lambda_2[x_2 - F^2(x_5, x_6)] \\ - \lambda_3(x_3 + x_5 - k_1) - \lambda_4(x_4 + x_6 - k_2).$$

Equating the partial derivatives with respect to the x variables to zero, we obtain six equations,

$$\begin{aligned} p_1 &= \lambda_1 & p_2 &= \lambda_2 \\ \lambda_1(\partial F^1/\partial x_3) &= \lambda_3 & \lambda_1(\partial F^1/\partial x_4) &= \lambda_4 \\ \lambda_2(\partial F^2/\partial x_5) &= \lambda_3 & \lambda_2(\partial F^2/\partial x_6) &= \lambda_4. \end{aligned}$$

From these, we easily obtain the familiar conditions for maximizing output

$$\begin{aligned} p_1(\partial F^1/\partial x_3) &= p_2(\partial F^2/\partial x_5) = \lambda_3 \\ p_1(\partial F^1/\partial x_4) &= p_2(\partial F^2/\partial x_6) = \lambda_4, \end{aligned}$$

that is, that the value of the marginal product of both factors should be the same in each industry. Note that all the Lagrange multipliers turn out to be equal to prices, if there is competitive pricing, λ_1, λ_2 to output prices, λ_3, λ_4 to factor prices.

4.3 INTERPRETATION OF THE LAGRANGE MULTIPLIERS

In the example given in the preceding section, the Lagrange multipliers were equivalent to prices. This is reminiscent of the properties of dual variables in linear programming theory. We shall now seek the formal interpretation of the multipliers.

Consider a standard classical optimizing problem, solved by the Lagrangean technique to give solution values x^*, λ^*. Let the ith constraint be of the form $g^i(x) = b_i$.

Initially $b_i = 0$, but we wish to examine the result of a small relaxation of this constraint.

Denote the optimal value of the objective function by V^* [$V^* = f(x^*)$]. Now a small relaxation in the ith constraint will permit small changes in the optimal values of the variables, but we assume the optimum conditions remain satisfied so that the new position reached as a result of this relaxation is also optimal. The effect on the optimal value of the objective function will be given by

$$\frac{\partial V^*}{\partial b_i} = \sum_j \frac{\partial f(x^*)}{\partial x_j} \frac{\partial x_j}{\partial b_i}. \tag{4.3.1}$$

From the constraints we have

$$\begin{aligned} \sum_j \frac{\partial g^k(x^*)}{\partial x_j} \frac{\partial x_j}{\partial b_i} &= 0 \qquad k \neq i \\ &= 1 \qquad k = i. \end{aligned} \tag{4.3.2}$$

If we multiply the kth equation in (4.3.2) by λ_k^*, and sum over k, we obtain

$$\sum_k \sum_j \lambda_k^* \frac{\partial g^k(x^*)}{\partial x_j} \frac{\partial x_j}{\partial b_i} = \lambda_i^*.$$

Subtracting this from the right-hand side of (4.3.1) we obtain after minor rearrangement,

$$\frac{\partial V^*}{\partial b_i} = \lambda_i^* + \sum_j \left[\frac{\partial f(x^*)}{\partial x_j} - \sum_k \lambda_k \frac{\partial g^k(x^*)}{\partial x_j}\right] \frac{\partial x_j}{\partial b_i}.$$

The bracketed expressions on the right are the optimum conditions and are zero, so that we have

$$\partial V^*/\partial b_i = \lambda_i^*.$$

Thus λ_i^* corresponds to the marginal rate of change of the optimal value of the objective function with respect to a small relaxation of the ith constraint, other constraints being unchanged. This interpretation is directly analogous to the interpretation of the dual variables in linear programming theory (see Section 3.7 of the previous chapter).

In typical economic applications, the constraints might be resource limitations and the objective function some index of social welfare. The optimal Lagrange multipliers would then correspond to the marginal social valuation of the resources. In the example given in the previous section, the multipliers corresponded to goods and factor prices. The factor prices correspond to marginal valuations of the fixed factor supplies. What of the goods prices? These are seen to correspond to the marginal valuation of the production functions as constraints or the efficiency parameters in the production functions.

The relationship between Lagrange multipliers and the dual variables of linear programming will be further developed in the next chapter (Section 5.5).

4.4 A GEOMETRICAL NOTE

The simplest constrained optimum problem of a classical kind, and one that appears often in economics, is one with a single constraint. The first order conditions have the simple form,

$$f_j - \lambda g_j = 0 \qquad j = 1, \ldots, n.$$
$$g = 0$$

We can eliminate λ immediately by taking any variable, say the nth, as reference or "numeraire." Solving for λ in the nth equation and substituting in the remainder, we obtain

$$f_j/f_n = g_j/g_n \qquad j = 1, \ldots, n.$$

If $n = 2$, this reduces to the single equation $f_1/f_2 = g_1/g_2$, the condition that the slope of the contour curve $f = c$ be the same as the slope of the constraint function at the optimal point. The result for general n is analogous and is equivalent to the familiar condition of tangency between the objective function contour and the constraint curve at the optimum.

A simple illustration is given in Figure 4–1. The objective function is some monotonic transformation of $x_1^2 + x_2^2$ and the constraint is also circular but with its center in the positive quadrant. There are two points of tangency: A corresponding to a constrained maximum and B to a constrained minimum. If we took the negative of the original objective function, the maximum and minimum positions would be reversed.

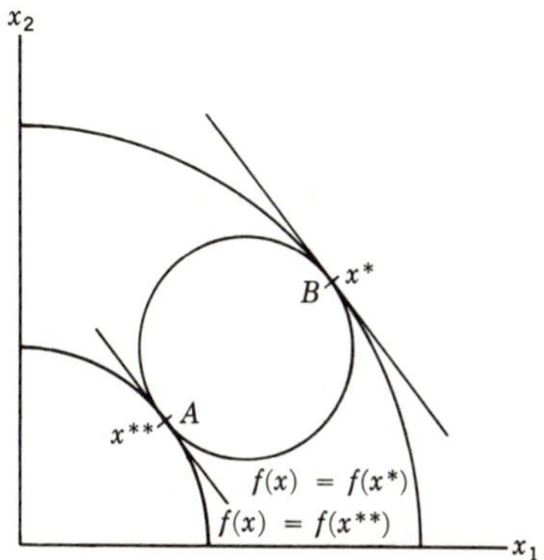

Figure 4-1. Geometry of the classical case.

The point of interest in this very simple example is that the common tangent *separates* the constraint and objective function contours in case A, but both contours lie on the same side of the tangent in case B. In the classical optimum analysis, the second order conditions which are discussed in the section following can handle both points A and B, since they depend on the relationship between the properties of the constraint and objective functions. The sufficient conditions for determining a maximum or minimum in the *extension* of the classical case (treated in the following chapter) are based on the sufficient conditions for a global optimum as set out in Chapter 2 (Section

2.8). Points like B cannot be handled by these conditions. This provides the reason why the conditions of the next section may seem complex.

4.5 SECOND ORDER CONDITIONS FOR THE CLASSICAL CASE

The analysis of this section requires a knowledge of the properties of quadratic forms subject to linear constraints, as set out in Section R6.3 of Review R6. Some readers may wish merely to familiarize themselves with the results for the most common economic problems, which are given in the latter part of the section.

So far we have been concerned with the *necessary* or *first order* conditions for $f(x)$ to have a maximum or minimum at the point x^*. These conditions give the critical points of $L(x, \lambda)$ but no information as to whether the critical point represents a maximum or minimum for $f(x)$.

We are interested here in the *second order* or so-called "*sufficient*" conditions that x^* give a proper (local) maximum or minimum. Strictly speaking, the term "sufficient" is inappropriate. The first order conditions are necessary, and the first and second order conditions together are necessary and sufficient, but the second order conditions by themselves are certainly not sufficient. Nevertheless the term is common, on the implicit assumption that the first order conditions are already satisfied.

Full discussions of the second order conditions for the classical constrained optimum problem are not easy to find[2] except for the single constraint case. In view of the importance of this problem in mathematical economics, an attempt is made here to discuss the general case.

The second order conditions for the maximum or minimum of an unconstrained objective function are well-known and discussed in Review R8 (Section R8.5). They are obtained by making a Taylor series expansion of the function around the optimum point, then studying the condition that the second order term in the expansion is always positive (for a minimum) or always negative (for a maximum). These conditions, based on the properties

[2] Most texts on advanced calculus give the second order conditions only for the single constraint case, if that. The classic reference is Hancock, who does not develop the conditions in the form most suitable for economic analysis. Frisch gives the second order conditions in terms of the characteristic roots of the bordered Hessian, truly necessary and sufficient but useless for most analytical purposes. Samuelson [1] (Appendix A) derives the determinantal conditions which are generally given. He gives as *necessary and sufficient* the conditions given here as *sufficient* only, implicitly assuming no singularity in the relevant matrices. For further discussion on this matter, see Review R6, Section R6.3.

of quadratic forms, are usually expressed in terms of the Hessian, the matrix of second order partial derivatives, and its principal minors.

We seek conditions of a similar kind for the constrained optimum problem. Unfortunately, although the results we end up with are similar in many ways to the results for the unconstrained problem, the route to these is much less straightforward.

Since $L(x, \lambda)$ and $f(x)$ have their maximum or minimum at the same values of x, it might seem that we can simply treat $L(x, \lambda)$ as an unconstrained objective function and use the ordinary methods. We cannot do this for two reasons. In the first place, $L(x, \lambda)$ is not a proper maximum or minimum with respect to the variables λ, since, for optimal x^*, $L(x^*, \lambda) = f(x^*)$ and is stationary for all variations in λ^*, even to the second order. In the second place, the variations in x^* are not free variations as in the unconstrained case but must satisfy the constraints.

Thus we shall analyze the second order conditions for $L(x^*, \lambda^*)$ considered as a function of x only, and taking into account only constrained variations in x.

If we expand $L(x^*, \lambda^*)$ as a function of the x variables only by means of a Taylor series, the second order term is given by

$$Q(dx) = (1/2!) \sum_j \sum_k (\partial^2 L/\partial x_j\, \partial x_k)\, dx_j\, dx_k,$$

where the variations dx must satisfy the constraints

$$dg^i = \sum_j g_j^i\, dx_j = 0 \qquad i = 1, \ldots, n.$$

Thus we have a minimum if $Q(dx) > 0$ for all dx satisfying the constraints and a maximum, if $Q(dx) < 0$ for all dx satisfying the constraints. The constraints, being in differential form, are linear. Thus we seek the conditions that a quadratic form is positive or negative for variables satisfying a set of linear constraints.

This problem is discussed in detail in Section R6.3 of Review R6, and we shall simply draw on those results here.

Denote by L the symmetric matrix of order $n \times n$,

$$\left[\frac{\partial^2 L(x^*, \lambda^*)}{\partial x_j\, \partial x_k}\right] = \left[f_{jk} - \sum_i \lambda^*{}_i g^i_{jk}\right],$$

and denote by G the $m \times n$ matrix of constraint derivatives

$$[g_j^i].$$

Then the second order conditions (with y written in place of dx) become

$$y'Ly > 0, \qquad \text{subject to} \quad Gy = 0 \qquad \text{(minimum)},$$
$$y'Ly < 0, \qquad \text{subject to} \quad Gy = 0 \qquad \text{(maximum)},$$

in which form we can directly use the results of Section R6.3.

To use those results, we form the bordered matrix

$$\hat{L} = \begin{bmatrix} 0 & G \\ G' & L \end{bmatrix}.$$

$\hat{L}$ is of order $(n + m) \times (n + m)$, G is of order $m \times n$, G' of order $n \times m$, and the northwest zero submatrix is of order $m \times m$.

We then have *sufficient* conditions for the second order conditions as follows:

(*a*) *For a minimum, the determinant of* $\hat{L}$ *and all its principal minors of order greater than* $m + 1$ *should have the sign of* $(-1)^m$, *where* m *is the number of constraints.*

(*b*) *For a maximum, the determinant of* $\hat{L}$ *should have the sign of* $(-1)^n$, *where* n *is the number of variables, the principal minor of order* $(m + n - 1)$ (*the largest principal minor*) *should have a sign opposite to this, and successively smaller principal minors should alternate in sign, down to the principal minor of order* $m + 1$.

Note that these are sufficient for the second order conditions to be satisfied, but not necessary. The second order conditions *may* be satisfied even if the above determinantal conditions are not.

The matrix $\hat{L}$ contains Lagrange multipliers. These can be eliminated by using the first order conditions to obtain a matrix containing only first and second order partial derivatives of the objective and constraint functions. The second order conditions depend on the *combination* of the properties of the constraint and objective functions.

If there is only a single constraint, the structure of the matrix $\hat{L}$ is simplified, as is the elimination of λ. For two variables and a single constraint, we have

$$\hat{L} = \begin{bmatrix} 0 & g_1 & g_2 \\ g_1 & f_{11} - \lambda g_{11} & f_{12} - \lambda g_{12} \\ g_2 & f_{12} - \lambda g_{12} & f_{22} - \lambda g_{22} \end{bmatrix}.$$

To eliminate λ, either of the substitutions $\lambda = f_1/g_1$ or $\lambda = f_2/g_2$ can be made.

The conditions are made even more simple if (as is the case in some of the most important pieces of economic analysis utilizing the classical optimum conditions), either the constraint function or the objective function is linear. If the constraint function is linear, so that $g(x)$ has the form $\sum a_j x_j - b = 0$, then $g_j = a_j$ and $g_{jk} = 0$. We can substitute $g_i = kf_i$, where $k = 1/\lambda$, from the first order conditions to obtain $\hat{L}$ entirely in terms of derivatives of f,

$$\hat{L} = \begin{bmatrix} 0 & kf_1 & kf_2 & \cdots & kf_n \\ kf_1 & f_{11} & f_{12} & \cdots & f_{1n} \\ \cdot & \cdot & \cdot & \cdots & \cdot \\ kf_n & f_{1n} & \cdot & \cdots & f_{nn} \end{bmatrix} = \begin{bmatrix} 0 & k\nabla f \\ k(\nabla f)' & H_f \end{bmatrix}.$$

$\hat{L}$ differs from the bordered Hessian of f, $\hat{F}$, where

$$\hat{F} = \begin{bmatrix} 0 & f_1 & \cdots & f_n \\ f_1 & f_{11} & \cdots & f_{1n} \\ \cdot & \cdot & \cdots & \cdot \\ f_n & f_{1n} & \cdots & f_{nn} \end{bmatrix} = \begin{bmatrix} 0 & \nabla f \\ (\nabla f)' & H_f \end{bmatrix},$$

only in having the first row and first column multiplied by the factor k. From ordinary properties of determinants, it follows that $\det \hat{L} = k^2 \det \hat{F}$, and the same relationship holds for corresponding principal minors of the two determinants. It is obvious that the signs of the determinants and the principal minors will be the same for both $\hat{L}$ and $\hat{F}$, so we will usually express the second order conditions in terms of $\hat{F}$.

If the constraint function is nonlinear, but the objective function is linear, we have $f_{ij} = 0$. The submatrix L of $\hat{L}$ now has typical element $-\lambda g_{ij}$, while the borders are simply first derivatives of g. Thus we have a matrix of the following kind:

$$\hat{L} = \begin{bmatrix} 0 & \nabla g \\ (\nabla g)' & -\lambda H_g \end{bmatrix}.$$

Using the notation $\hat{F}$ now to refer to the bordered Hessian of g, and multiplying every row of $\hat{L}$ by $-(1/\lambda)$, we obtain a new matrix, $\hat{L}^*$, such that

$$\det \hat{L}^* = (-1/\lambda)^{n+1} \det \hat{L} = \det \begin{bmatrix} 0 & kg_1 & \cdot & \cdot & \cdot & \cdot \\ kg_1 & g_{11} & \cdot & \cdot & \cdot & \cdot \\ \cdot & \cdot & \cdot & \cdot & \cdot & \cdot \\ kg_n & \cdot & \cdot & \cdot & \cdot & g_{nn} \end{bmatrix},$$

where $k = -(1/\lambda)$. The relationship of $\det \hat{L}^*$ and its principal minors to $\det \hat{F}$ and its principal minors is the same as in the previous case. But here we see that $\det \hat{L} = (-\lambda)^{n+1} \det \hat{L}^*$. To obtain the principal minor of order n (the largest principal minor) in $\det \hat{L}$ from that in $\det \hat{L}^*$, we have to multiply only n rows by $(-\lambda)$, and so on for smaller principal minors. Thus the principal minors of order s will be related by the factor $(-\lambda)^s$.

Thus, in the linear objective function case, if $\lambda > 0$ (which will be the case in typical economic contexts), the *ratio* of successive principal minors of det $\hat{L}$ will have opposite sign to the ratio of principal minors of det $\hat{L}^*$ and therefore to the ratio of principal minors of the bordered Hessian $\hat{F}$.

We can now summarize the rules for the important economic case in which: (a) there is only one constraint, and (b) either the constraint or objective function is linear.

In a classical optimizing problem with a single constraint, a sufficient condition that a critical point represent a local maximum or minimum, when either the objective or constraint function (but not both) is linear, is that the sequence consisting of the determinant of the bordered Hessian of the nonlinear function, followed by its principal minors of successively lower order, ending with the principal minor of order 2, *should*:

(*a*) *alternate in sign for a maximum when the constraint function is linear, or have the same sign for a maximum when the objective function is linear*;

(*b*) *have the same sign for a minimum when the constraint function is linear, or alternate in sign for a minimum when the objective function is linear.*

The part of these conditions most often used is the sign of the ratio of the first two terms in the sequence, the determinant and its largest principal minor. It should be emphasized once again that the above conditions are sufficient for a proper maximum or minimum but not necessary.[3] Strictly speaking, the argument that a local optimum exists, therefore the above conditions are satisfied, is not legitimate.

4.6 THE SUBSTITUTION EFFECT OF NEOCLASSICAL DEMAND THEORY[4]

The classic and best example of the use of second order conditions as a tool of economic analysis is provided by the Slutsky-Hicks analysis of consumer

[3] The conditions are sufficient only because they do not apply in certain unusual circumstances, such as in the case of objective functions which are "flat" in some direction that coincides with a constraint. Other degenerate cases may occur when one of the constraints has the same shape as the objective function contour in the neighborhood of the optimum.

[4] The formulation given here, minimizing income for a given level of utility, is inverse to the usual formulation, maximizing utility for a given income. It is the simplest approach for deriving the pure substitution effect.

The more usual formulation is given in Chapter 8, Section 8.2, where the neoclassical demand theory is discussed in full. The reader will find references to the literature in the later section.

behavior, with respect to the change in the price of a single good within a competitive market framework. We shall examine this analysis more fully in Chapter 8, but here we shall examine the pure substitution effect alone. We formulate the problem in a somewhat different way from Slutsky or Hicks.

The consumer has a continuous utility function $u(x)$ on n goods x_i. He faces fixed prices p_i, and we are interested in the smallest money income, M, that will enable him to attain a given utility level. The problem is

$$\begin{aligned} &\min M = px \\ \text{S.T.}\quad &u(x) - u^0 = 0. \end{aligned}$$

This is a simple classical problem with a single constraint and a linear objective function. The first order conditions are

$$\begin{aligned} u(x) - u^0 &= 0 \\ p_i - \lambda u_i &= 0. \end{aligned} \tag{4.6.1}$$

Obviously $\lambda > 0$, and we can use the simplified form of the second order conditions given at the end of the preceding section.

Now consider the effect of a small change in a single price (other prices unchanged) on the optimal values of the variables x, the utility level u^0 being unchanged. Without loss of generality, we can take the nth price to be the price that changes.

The change in p_n will change the optimal values x^*, λ^* of the variables but the constraint and first order conditions still hold, so we can differentiate within the Equations (4.6.1) to obtain

$$\begin{aligned} \sum_j u_j \frac{\partial x_j}{\partial p_n} &= 0 \\ -u_i \frac{\partial \lambda}{\partial p_n} - \lambda^* \sum_j u_{ij} \frac{\partial x_j}{\partial p_n} &= 0 \qquad i \neq n. \\ 1 - u_n \frac{\partial \lambda}{\partial p_n} - \lambda^* \sum_j u_{nj} \frac{\partial x_j}{\partial p_n} &= 0 \end{aligned} \tag{4.6.2}$$

Dividing all but the first equation in (4.6.2) by $1/\lambda^*$ and rearranging, we obtain

$$\begin{aligned} \sum_j u_j \frac{\partial x_j}{\partial p_n} &= 0 \\ u_i \frac{1}{\lambda^*} \frac{\partial \lambda}{\partial p_n} + \sum_j u_{ij} \frac{\partial x_i}{\partial p_n} &= 0 \\ u_n \frac{1}{\lambda^*} \frac{\partial \lambda}{\partial p_n} + \sum_j u_{nj} \frac{\partial x_j}{\partial p_n} &= \frac{1}{\lambda^*}. \end{aligned} \tag{4.6.3}$$

The system can be written in matrix-vector form,

$$\begin{bmatrix} 0 & u_1 & u_2 & \cdot & \cdot & \cdot & u_n \\ u_1 & u_{11} & u_{12} & \cdot & \cdot & \cdot & u_{1n} \\ \cdot & \cdot & \cdot & & & & \cdot \\ \cdot & \cdot & \cdot & & & & \cdot \\ u_n & u_{1n} & & \cdot & \cdot & \cdot & u_{nn} \end{bmatrix} \begin{bmatrix} \dfrac{1}{\lambda^*}\dfrac{\partial \lambda}{\partial p_n} \\ \dfrac{\partial x_1}{\partial p_n} \\ \dfrac{\partial x_2}{\partial p_n} \\ \cdot \\ \dfrac{\partial x_n}{\partial p_n} \end{bmatrix} = \begin{bmatrix} 0 \\ 0 \\ 0 \\ \cdot \\ 1/\lambda^* \end{bmatrix}. \quad (4.6.4)$$

If we solve for $\partial x_n/\partial p_n$ by Cramer's rule, we insert the constant term vector in place of the last column of the left-hand matrix. Since this vector contains only one nonzero term, $1/\lambda^*$, we have

$$\frac{\partial x_n}{\partial p_n} = \frac{1}{\lambda^*}\frac{U_{nn}}{\det U},$$

where U is the matrix of the equation system (4.6.4) and U_{nn} is the cofactor of u_{nn} in det U. The principal minor $|U_{nn}|$ is related to the cofactor U_{nn} by

$$\begin{aligned} U_{nn} &= (-1)^{(n+1)+(n+1)}\,|U_{nn}| \\ &= |U_{nn}|. \end{aligned}$$

Now this is a minimum problem with a single constraint and an objective function that is linear. U is the bordered Hessian of the constraint function and $\lambda^* > 0$. Thus, if the sufficient conditions for a regular minimum given at the end of the preceding section are satisfied, we have

$$|U_{nn}|/(\det U) < 0,$$

so that

$$\left(\frac{\partial x_n}{\partial p_n}\right)_{u=u^0} < 0.$$

This is the proof of the negativity of the pure substitution effect of neoclassical demand theory.

We repeat the caution already given, that this proof depends on assuming that, if a consumer has a regular minimum at his chosen x^*, the determinantal conditions are necessarily satisfied. This is not true in general, although it can be assumed to be true for well-behaved utility functions, since the constraint is linear.

4.7 GLOBAL OPTIMUM CONDITIONS IN THE CLASSICAL PROBLEM

Since the m constraints in the classical problem are all equations and since $m < n$, we can, in principle, solve the constraint equations to give a single constraint $G(x) = 0$ in $n - m$ independent variables. The only form of equation in which the solutions form a convex set is a linear equation, so we have a convex feasible set only if $G(x)$ is linear. It is sufficient for $G(x)$ to be linear that all the $g^i(x)$ be linear. It is not necessary, since nonlinear surfaces (such as ruled surfaces) may intersect linearly.

In general, however, only the linear constraint case will satisfy the general conditions for a global optimum among the classical problems. This does, in fact, cover a large class of the simpler economic cases.

The consumer choice problem, as set up in the preceding section, does not satisfy the global optimum conditions, since the feasible set defined by $u(x) = u^0$ is not convex. In this case we know that we can set up the same choice situation as that of maximizing $u(x)$, subject to a linear budget constraint. Since this does satisfy the global optimum conditions, we can argue that the problem formulated in the opposite way must also have a global optimum.

Actually, in a case like that of the consumer choice formulation, we can satisfy the global optimum conditions by making the constraints inequalities. The set $u(x) \geqq u^0$ *is* convex. We no longer have a classical problem, although we will see in the next chapter that, in a case like this, the solution with inequality constraints will be exactly the same as the classical solution, with the added advantage of the assurance of global optimality.

On the other hand, although the classical second order conditions give only the assurance of a local maximum or minimum, they give this assurance in certain conditions (as mentioned in Section 4.4) under which the more general conditions of the next chapter break down.

EXERCISES

1. If $f(x) = x_1^2 + 2bx_1x_2 + x_2^2$, what values of b give:

   (a) a local maximum for $f(x)$ S.T. $x_1 + x_2 = 1$?
   (b) a local minimum for $f(x)$ S.T. $x_1 + x_2 = 1$?

2. By considering the directional derivative $D_v f(x)$ at x as a function of v given by $\phi(v) = \nabla f(x) \cdot v$, subject to $v'v = 1$, show that the directional derivative has a maximum in the direction of the gradient vector.

3. The welfare function for a two-good, two-person economy is $W = u_1^a u_2^{1-a}$. The individual utility functions are $u_1 = x_{11}^b x_{21}^{1-b}$, $u_2 x_{12}^c x_{22}^{1-c}$, where x_{ij} is the amount of the ith good consumed by the jth individual.

If the total amounts of the two goods are fixed, what is the optimum allocation between the two individuals? Interpret the dual variables. (Assume $0 < a, b, c < 1$.)

4. Discuss the second order conditions for the problem of Exercise **1** when $b = 1$.

5. Second best problem. Discuss the relationship between the solutions to the following two problems for different values of k.

$$\text{(a)}\ \max f(x) \quad \text{S.T.} \quad g(x) = 0$$
$$\text{(b)}\ \max f(x) \quad \text{S.T.} \quad g(x) = 0, \qquad h(x) = 0,$$

where $h(x) = f_1/f_2 - k(g_1/g_2)$. (Assume x is a vector of three components.)

6. Envious consumers. In the economy of Exercise **3**, assume that each individual's utility depends on the ratio of his own consumption of a good to that of the other individual's. That is, we have

$$u_1 = r_1^b r_2^{1-b}, \qquad u_2 = r_1^{-c} r_2^{c-1},$$

where $r_i = x_{i1}/x_{i2}$.

Determine, as before, the optimum allocation and interpret the dual variables.

CHAPTER

5

Advanced Optimizing Theory

This chapter follows on from Chapters 2, 3, and 4. Sections 5.1 through 5.5 require no additional mathematical background. Sections 5.6 and 5.7 (which may be omitted on first reading) require some familiarity with mappings and fixed point theorems, provided by Reviews R8 (Section R8.7) and R9.

5.1 INTRODUCTION[1]

The strict classical method, discussed in the preceding chapter, assumes both equality in the functional constraints and the absence of direct constraints, which can typically be considered to be nonnegativity constraints, on the variables. Although this method is widely employed in economic analysis, the fact is that most economic problems have implicit, if not explicit, properties that do not entirely fit the classical case. Nonnegativity constraints on at least some variables are almost always implicit, and the functional constraints may be more accurately described by inequalities than equalities. Traditional economic analysis has been based on the faith (often justified) that the functional constraints will always be effective and the direct constraints always ineffective in the region of the optimum. Problems have often arisen,

[1] A good basic treatment of the direct extension of the classical method is given in Hadley [3], Chapter 3.

because implicit constraints have been neglected in circumstances in which they need to be taken into account. As economists develop models of greater and greater complexity, the unmodified classical method becomes less applicable.

Consider the following simple example. Find the equilibrium consumption of a consumer whose utility function on two goods x_1, x_2 is $u = (1 + x_1)(1 + x_2)$, who has unit income, and who faces fixed prices 4, 1 for x_1, x_2, respectively. This is a simple optimizing problem,

$$\max u = (1 + x_1)(1 + x_2)$$
$$\text{S.T.} \quad 4x_1 + x_2 = 1.$$

If we solve the problem, using traditional methods, we obtain the optimal values for x as $x^* = (-\frac{1}{4}, 2)$, with the optimal level of utility, u^*, as 9/4. Thus the solution gives a negative value for one of the variables, although the utility function is an entirely acceptable one.

Faced with the numerical solution, the economist would reject it with the reply, "I forgot to tell you that we cannot have negative quantities of goods." In an analytical solution, however, the possibility that the nonnegativity constraints might be effective would not be so apparent.

Suppose now we put $x_1 = 0$, the closest we can come to its optimal value without the nonnegativity constraint. Then $x_2 = 1$, and $u(0, 1) = 2$, less than u^* (as expected). At (0, 1) the ratio u_1/u_2 is 2, whereas the ratio p_1/p_2 is 4, so that the classical first order condition, $u_1/u_2 = p_1/p_2$, is not satisfied. Direct calculation shows, however, that (0, 1) is optimal, given the nonnegativity condition.

A similar situation to the above arises in a simple model of a two-factor, two-good economy with fixed factor endowments. For production functions whose isoquants intersect the axes (as the indifference curves of the above utility function do), there will be factor endowments for which the strict classical conditions for optimal production—tangency between the two isoquant sets—imply negative quantities of one factor. Production functions leading to this situation are quite permissible, and we obviously need to be able to handle such cases. Whereas production with capital alone or labor alone may seem a very special case in a simple model, production with a zero quantity of one or more specialized types of labor or of capital in a complex model is a likely situation.

5.2 NONNEGATIVE VARIABLES

The most straightforward extension of the classical calculus method is to the case in which some or all of the variables are subject to direct constraints, which we can take to be nonnegativity constraints. Assuming the problem is a maximizing one of the form,

$$\begin{aligned} &\max f(x) \\ \text{S.T.}\quad &g^i(x) = 0, \qquad i = 1, \ldots, m \\ &x \geqq 0, \end{aligned}$$

we can define the function $L(x, \lambda)$ in the usual way. Three situations are possible:

(a) $L(x, \lambda)$ has a regular local maximum at a critical point x^*, λ^*, with $x^* > 0$, and the problem satisfies the strong global optimum conditions (the feasible set is convex, $f(x)$ is concave, and either the convexity or concavity is strict).
(b) $L(x, \lambda)$ has a regular local maximum at a critical point, with $x^* > 0$, but the strong global optimum conditions are not satisfied.
(c) $L(x, \lambda)$ does not have a critical point with $x > 0$ which is also a local maximum.

In the first case the nonnegativity constraints are ineffective in the neighborhood of the global optimum and can be neglected. It is this case which is presumed to occur in traditional economic analysis.

In the third case the global optimum *must* be at a point at which some nonnegativity constraint is operative. In the second case it *might* be at such a point, and we will usually need to check.

Consider the properties of $L(x, \lambda), f(x)$ at a point at which $x_k = 0$ for at least one k. Since the functional constraints $g^i = 0$ are still equalities, the maxima of $L(x, \lambda)$ and $f(x)$, $x \in K$ still occur at the same point x^*. Also, since $\partial L/\partial \lambda_i = g^i$, we must still have $\partial L/\partial \lambda_i = 0$, so that some of the first order conditions for a maximum of L remain satisfied.

What of the remaining first order conditions $\partial L/\partial x_j = 0$? By coincidence, they might still be satisfied, but in general they will not all be. What conditions replace these?

Divide the indices of the components of x into two groups, so that $j \in S$ if $x_j = 0, j \notin S$, if $x_j > 0$. Then we can argue in the following way:

If $j \notin S$, small variations in x_j are possible in both positive and negative directions, so that x cannot be optimal unless $\partial L/\partial x_j = 0$, by the usual marginal reasoning.

If $j \in S$, small variations in x_j are possible in the positive direction, so that x cannot be optimal if $\partial L/\partial x_j > 0$, because we could increase L by a small and permissible increase in x_j. But small variations are not possible in the negative direction, so we cannot rule out $\partial L/\partial x_j < 0$ as nonoptimal.

Summarizing the above discussion:

The optimal point x^* *of the problem* $\max f(x)$ S.T. $g^i(x) = 0,\ x \geqq 0$, *satisfies the following conditions*:

(*a*) $\dfrac{\partial L(x^*, \lambda^*)}{\partial x_j^*} = f_j - \sum \lambda_i g_j^i \leqq 0,$ *and either* $\dfrac{\partial L(x^*, \lambda^*)}{\partial x_j^*} = 0$ *or* $x_j^* = 0;$

(*b*) $\dfrac{\partial L(x^*, \lambda^*)}{\partial \lambda_i^*} = -g^i(x) = 0$

If the problem is a minimizing one, the inequality in (1) *is reversed.*

The last point follows from an analogous argument to that used for the maximum.

These conditions can be stated rather more succinctly if we use the following notation, which we shall make use of in the next section also. Write $\hat{L}_x$ for the vector of the partial derivatives of L with respect to the x variables, and $\hat{L}_\lambda$ for the vector of the partial derivations of L with respect to the λ variables. For reasons that will be apparent in later use, we shall regard $\hat{L}_x$ as a column vector and $\hat{L}_\lambda$ as a row vector. Then the optimal conditions become (for the maximum problem):

$$\text{(1)}\quad \hat{L}_x \leqq 0, \qquad \hat{L}'_x \cdot x = 0, \qquad x \geqq 0,$$
$$\text{(2)}\quad \hat{L}_\lambda = 0,$$

where the prime on $\hat{L}_x$ transposes it to a row vector.

Example. Let us return to the earlier example, now properly stated as $\max u = (1 + x_1)(1 + x_2)$; S.T. $4x_1 + x_2 = 1$; $x_1, x_2 \geqq 0$.

We already know that there is no critical point of L with $x_1, x_2 > 0$. We need to try putting x_1, then x_2, to zero. Given the budget constraints, this gives the points $(0, 1)$, $(\frac{1}{4}, 0)$.

At $(0, 1)$ we have

$$\partial L/\partial x_2 = 1 - \lambda = 0, \quad \text{giving} \quad \lambda = 1$$
$$\partial L/\partial x_1 = -2 \quad \text{for} \quad \lambda = 1$$
$$\partial L/\partial \lambda = 0.$$

At $(\frac{1}{4}, 0)$ we have

$$\partial L/\partial x_1 = 1 - 4\lambda = 0, \quad \text{giving} \quad \lambda = \tfrac{1}{4}$$
$$\partial L/\partial x_2 = 1 \quad \text{for} \quad \lambda = \tfrac{1}{4}$$
$$\partial L/\partial \lambda = 0.$$

The point $(0, 1)$ satisfies the optimal conditions, the point $(\frac{1}{4}, 0)$ does not, as we would expect from the original investigation of the problem.

There is, unfortunately, no universal rule for determining which variables, when put to zero, are likely to lead to an optimal solution. In principle, we may have to try putting one variable at a time to zero, then two at a time, three at a time, and so on, and then compare the results of all cases that satisfy the optimal conditions. In general, not more than $n - m$ variables can take on zero values at the same time, since the functional constraints already determine $n - m$ of the variables in terms of the remainder. Thus there will be, in general, at least m of the conditions $\partial L/\partial x_j \leqq 0$ for which the equality holds, enough to determine the values of the m Lagrange multipliers unambiguously.

Because of the astronomical number of ways of putting the variables to zero in groups of k variables, when k varies from 1 to $n - m$, these optimum conditions are not, of themselves, a solution method for complex problems.

In economic analysis, however, we are usually interested in what happens to the optimal conditions when we *do* have a solution on the nonnegative boundary, and to this we have the answer. Furthermore boundary problems of this kind in economics frequently occur as the result of a movement to the boundary from an interior point as some parameter is changed, so that the zero variables are specified for us. Economic interpretation of the optimal conditions at zero values for the variables presents no particular difficulties, but depends on the particular problem. For a utility function on n goods with a linear budget constraint, the conditions would be interpreted as requiring that the ratio of marginal utility to price be the same for all goods actually consumed and be no greater (typically less) for goods not consumed. If there is more than one functional constraint, the interpretation is made more difficult.

5.3 INEQUALITY CONSTRAINTS

We are now in a position to consider the general optimum problem, with the two restrictions of the classical calculus method removed,

$$\begin{aligned} &\max f(x) \\ \text{S.T.}\quad &g^i(x) \leqq 0, \qquad x \geqq 0,\ i = 1, \ldots, n \end{aligned}$$

This can be converted into the case discussed in the preceding section, with equality functional constraints but with nonnegativity constraints on the variables, by adding slack variables z_i to give the ith constraint as

$$g^i(x) + z_i = 0, \qquad z_i \geqq 0.$$

There are now $n + m$ variables in the problem, an n-vector of x variables and an m vector of z variables, with nonnegativity constraints on all.

We form the Lagrangean

$$\begin{aligned} L(x, z, \lambda) &= f(x) - \sum \lambda_i(g^i(x) + z_i) \\ &= f(x) - \sum \lambda_i g^i(x) - \sum \lambda_i z_i. \end{aligned}$$

The optimal conditions with respect to x_j are as before. The optimal conditions with respect to z_i are

$$\frac{\partial L(x, z, \lambda)}{\partial z_i} = -\lambda_i \leqq 0; \qquad \text{either} \qquad \lambda_i = 0 \qquad \text{or} \qquad z_i = 0.$$

These conditions impose no direct constraints on x. Their content is entirely represented by the equivalent statement

$$g^i(x) \leqq 0, \qquad i = 1, \ldots, m, \qquad \lambda \geqq 0, \qquad \sum \lambda_i g^i(x) = 0.$$

Thus if we form the Lagrangean, ignoring the inequalities in the functional constraints, as

$$L(x, \lambda) = f(x) - \sum \lambda_i g^i(x),$$

adding the nonnegativity constraint $\lambda \geqq 0$, all points which are optimal for z in $L(x, z, \lambda)$ satisfy

$$L(x, z, \lambda) = L(x, \lambda).$$

Consider $\partial L(x, \lambda)/\partial \lambda_i = -g^i(x)$. From the constraints we have $g^i(x) \leqq 0$, and from the optimal conditions with respect to z in $L(x, z, \lambda)$ we have $\lambda \geqq 0$ and either $g^i(x) = 0$ or $\lambda_i = 0$. These together imply

$$\frac{\partial L(x, \lambda)}{\partial \lambda_i} \geqq 0; \qquad \text{either} \qquad \frac{\partial L(x, \lambda)}{\partial \lambda_i} = 0 \qquad \text{or} \qquad \lambda_i = 0 \qquad \text{(or both).}$$

The above conditions can be recognized as the conditions for a *minimum* of $L(x, \lambda)$ with respect to λ, given the nonnegativity constraint $\lambda \geqq 0$.

How does it come about that, although we are seeking a *maximum* for $f(x)$ subject to the constraints, and although we were able to treat $L(x, \lambda)$ as having a *maximum* with respect to λ in the strict classical case, we now seek a *minimum* with respect to λ of the expression which is analogous to the strict classical Lagrangean?

First we note that, in the strict classical case, we have $g^i(x) = 0$ for all i. Thus, if the optimal point is x^*, λ^*, and we replace λ^* by some λ' we have

$$L(x^*, \lambda^*) = L(x^*, \lambda') = f(x^*).$$

That is $L(x, \lambda)$ is neutral with respect to changes in λ over the feasible set. *We could have regarded $L(x, \lambda)$ as having a minimum with respect to λ just as well as having a maximum, in the strict classical case.* It was more convenient, at the time, to consider it a maximum.

Second, we note that, in the present case, at any optimal point of $L(x, \lambda)$, whether a maximum or a minimum with respect to λ, we have either $\lambda_i = 0$ or $g^i(x) = 0$, so that at all optimal points $L(x, \lambda) = f(x)$. Whether a maximum or a minimum is involved, with respect to λ, is related to the constraints, not to the objective function.

Consider the effect of a small variation in λ from its optimal value λ^* in the present case, taking note of the nonnegativity restriction. If λ_i^* is positive, $g^i(x^*) = 0$, so that $L(x^*, \lambda^*)$ is unchanged, as in the classical case. If, however, $\lambda_i^* = 0$ we may have $g^i(x^*) < 0$. Because of the nonnegativity constraint on λ, the only permissible variation is to some small positive value ε. In this case, the term $-\lambda_i g^i(x^*)$ in the expression for $L(x^*, \lambda)$ will take on a positive value and we will have $L(x^*, \lambda') > L(x^*, \lambda^*)$. Thus the optimal point gives a *minimum* of $L(x, \lambda)$ with respect to λ. The minimum property will only be apparent when at least one constraint is ineffective.

We can now summarize the properties of an optimum point x^*, λ^* for the general maximizing problem, $\max f(x)$ S.T. $g^i(x) \leqq 0$, $i = 1, \ldots, m$, $x \geqq 0$.

(1) $f_j - \sum \lambda_i g_j^i \leqq 0$; either $f_j - \sum \lambda_i g_j^i = 0$ or $x_j = 0$.
(2) $\lambda \geqq 0$, $g^i \leqq 0$; either $g^i = 0$ or $\lambda_i = 0$.[2]

If we are not trying to discover whether a certain point is optimal or not, but are merely interested in the properties of a point already known to be optimal (as is often the case in economic analysis), conditions (2) state that *we can ignore ineffective constraints at the optimum.*

If the problem is one of minimizing, the direction of the inequalities in (1) is reversed.

5.4 SADDLE POINTS AND DUALITY

The optimum conditions given above for the general maximum problem are usually referred to as the *Kuhn-Tucker conditions.*[3] Normally these would be expressed in terms of the Lagrangean function,

$$L(x, \lambda) = f(x) - \sum \lambda_i g^i(x),$$

[2] "Or" is used in the logical sense to include "both" as one possibility.

[3] The original conditions, which include some additional to those given here, are given in Kuhn and Tucker [1]. See Hadley [3] for a complete exposition.

and written in the vector notation developed previously. In this notation they take the form,

$$(1)\ \hat{L}_x \leqq 0, \qquad \hat{L}'_x x = 0, \qquad x \geqq 0;$$
$$(2)\ \hat{L}_\lambda \geqq 0, \qquad \lambda \hat{L}_\lambda = 0, \qquad \lambda \geqq 0.$$

In the original formulation by Kuhn and Tucker, and in most expositions of Kuhn-Tucker theory, the constraints appear in the form $g^i \geqq 0$, but the Lagrangean is written in the form $L(x, \lambda) = f(x) + \sum \lambda_i g^i(x)$. Reversals in the constraint inequalities and the λ_i's cancel each other, so that the form of the above conditions is unchanged.

We have already noted that $L(x, \lambda)$, a function of two sets of variables, has a maximum with respect to x and a minimum with respect to λ at the optimum. A point which gives a maximum of a function with respect to some variables, and a minimum with respect to others, is called a *saddle point*, a term descriptive of the shape of the function in three dimensions.

The solution of the general optimum problem may therefore be regarded as finding a saddle point of the Lagrangean, or alternatively as *maximinimizing* or *minimaximizing* the Lagrangean. The latter terminology follows from the fact that we find a maximum over x and a minimum over λ. It will be shown in Section 5.6 that, if x^*, λ^* is the optimal point of L,

$$L(x^*, \lambda^*) = \max_x \min_\lambda L(x, \lambda) = \min_\lambda \max_x L(x, \lambda).$$

We are now in a position accurately to relate the Lagrangean multipliers to the dual variables of linear programming, a relationship that was sketched in Chapter 4.

Consider the standard linear program with its dual

$$\begin{array}{llll} & \max cx & & \min yb \\ \text{S.T.} & Ax \leqq b, \quad x \geqq 0 & \text{S.T.} & yA \geqq c, \quad y \geqq 0. \end{array}$$

The Langrangean function for the primal program is

$$L(x, \lambda) = cx - \lambda Ax + \lambda b,$$

and we have $\hat{L}_x = c - \lambda A$, $\hat{L}_\lambda = -Ax + b$, giving the Kuhn-Tucker conditions in the form,

$$(1)\ \lambda A \geqq c, \qquad (c - \lambda A)\,x = 0, \qquad x \geqq 0;$$
$$(2)\ Ax \leqq b, \qquad y(b - Ax) = 0, \qquad \lambda \geqq 0.$$

If we put $\lambda = y$, these are seen to be the constraints and equilibrium conditions for the primal and dual. Also, if we set up the Lagrangean for the dual, we find that the Lagrangeans for both the primal and dual have the same value $cx^* = y^*b$, satisfying the duality theorem of linear programming.

The interpretation is therefore complete.

The Lagrange multipliers of the general optimum problem are analogous to the dual variables in linear programming and reduce to them if the general optimum problem is linear.

In economic problems, therefore, we can give exactly the same kind of interpretation to the Lagrange multipliers (usually as some implicit or shadow price) as we give to the dual variables.

At this point, it is clear that the Kuhn-Tucker conditions are a true generalization of the optimum conditions in all special types of optimum problem. As a theoretical framework, we could proceed from these more general conditions to the special cases of the linear programming and strict classical problems.

5.5 THE DUAL VARIABLES

Let us suppose that we have a general maximum problem for which the solution x^* is known, and we wish to find the optimal values of the dual variables λ. This kind of problem is of considerable importance in economic analysis, where we assume the individual economic unit (usually the firm or the consumer) has located the optimum by some means or other (trial and error, perhaps). We are interested for analytical purposes in the nature of the dual variables associated with this optimum, which will usually be some kind of shadow price.

Since the optimal value, x^*, is assumed to be given, the values of $f(x^*)$, $g^i(x^*)$ and of their derivatives can be treated as constants. To simplify notation, we shall write $A = [a_{ij}]$ for the matrix $[g^i_j]$, c for the vector $[f_j]$, and b for the vector $[-g^i]$. The matrix A is of order $m \times n$, c is a column vector of order n, and b a row vector of order m.

If the problem were in strict classical form, with the constraints entirely as equalities and no nonnegativity constraints on x (consequently none on λ), the optimal conditions with respect to x would be written, in the above notation,

$$c - \lambda A = 0.$$

This is a set of linear equations in λ. There are n equations and only m variables, but the nature of the optimal solution guarantees that only m are linearly independent. We can take a basis A_B of any m rows of A, and take the corresponding elements of c, then determine λ as

$$\lambda = A_B^{-1} c_B.$$

In the general maximum problem we have inequalities in the equivalent optimum conditions and also a nonnegativity constraint on λ. Taking the part of the Kuhn-Tucker conditions relevant to the determination of x^*, we have

$$\lambda A \geqq c, \qquad \lambda \geqq 0.$$

These look like the constraints for a linear programming problem. We search for an appropriate objective function. Now $L(x, \lambda)$ which has a minimum with respect to λ at x^*, λ^*, is linear in λ with coefficients $(-g^i)$. $f(x^*)$ appears as a constant term and can be ignored during the optimization. Thus the minimization of $L(x^*, \lambda)$ with respect to λ is equivalent to the minimization of $-\lambda b$, where $-b$ has the values given above.

Thus we have established that the optimal values of λ, given x^*, are determined in the general maximum problem by the linear program

$$\min -\lambda b,$$
$$\text{S.T.} \quad \lambda A \geqq c, \qquad \lambda \geqq 0,$$

where $a_{ij} = g^i_j(x^*)$, $\qquad b_i = g^i(x^*)$, $\qquad c_j = f_j(x^*)$.

In Chapter 4 we showed that, in the strict classical case, the multipliers satisfied the relationship

$$\partial V^*/\partial b_i = \lambda_i^*,$$

where $V^* = f(x^*)$ and $b_i = g^i(x^*)$. This led to the interpretation of the optimal value of the ith multiplier as the marginal value of relaxation of the ith constraint.

Do the multipliers have the same interpretation in the general case? They do, but with an important modification.

If we proceed as in the classical case, putting $g^i(x^*) = b_i$, we note that, if $b_i < 0$, the fact that the constraint is ineffective implies that a small relaxation will have no effect on the optimal values of x and will therefore have no effect on $f(x^*)$. Thus we will have $\partial V^*/\partial b_i = 0$ if $b_i < 0$. But the Kuhn-Tucker conditions imply that $\lambda_i^* = 0$, if $b_i < 0$, so the same relationship holds as in the classical case when the constraints are ineffective.

For small variations around the optimum, the ineffective constraints remain ineffective and can be ignored. Thus we can treat the problem as one with only equality constraints. If the direct constraints on x are all ineffective, the analysis is the same as in the strict classical case when we ignore the ineffective constraints.

It is the nonnegativity constraints on x, rather than inequalities in the functional constraints, that cause trouble. If we proceed as suggested above, ignoring the ineffective functional constraints and performing the same analysis as in the classical case, we obtain, as we did then (Chapter 4 Section 4.3), the relationship:

$$\frac{\partial V^*}{\partial b_i} = \lambda_i^* + \sum_j \left(f_j - \sum_{k \in S} \lambda_k^* g_j^k\right) \frac{\partial x_j^*}{\partial b_i},$$

where S is the set of effective constraints and all derivatives are taken at optimal values.

In the classical case, the optimum conditions make the bracketed expressions all zero and the equality of $\partial V^*/\partial b_i$ and λ_i^* is established. The same condition holds in the present case if x^* is strictly positive.

If, however, $x_j^* = 0$ for some j, the bracketed expression may be negative and the Kuhn-Tucker conditions are still satisfied. Taking the change in b_i to be essentially positive (that is, the constraint is relaxed, not tightened), the nonnegativity constraints on x require $\partial x_j^*/\partial b_i$ to be nonnegative if $x_j^* = 0$. Thus we have

$$\left(f_j - \sum_{k \in S} \lambda_k g_j^k\right) \frac{\partial x_j^*}{\partial b_i} \leqq 0.$$

For the general case, therefore, we have the following relationship:

$$\frac{\partial V^*}{\partial b_i} \leqq \lambda_i; \quad \text{either} \quad \frac{\partial V^*}{\partial b_i} = \lambda_i \quad \text{or} \quad x_j^* = 0, \quad \text{some } j;$$
$$= 0 \quad \text{if} \quad b_i < 0.$$

The economic interpretation is quite clear. λ_i represents the marginal value of relaxing the ith contraint (which is zero if the constraint is ineffective), *provided there are no effective direct constraints on the adjustment of* x^*. If there is an effective direct constraint on x^*, this limits its adjustment and tends to reduce the value of relaxing the functional constraint.

The nonnegativity constraints on λ do not complicate the situation. A relaxation of constraints cannot decrease the optimal value of V, so that $\partial V^*/\partial b_i$ is essentially nonnegative.

5.6 THE MINIMAX THEOREM[4]

It was asserted earlier in the chapter (Section 5.4) that, for a problem giving a proper constrained maximum, the Lagrangean would satisfy

$$\max_x \min_\lambda L(x, \lambda) = \min_\lambda \max_x L(x, \lambda) = L(x^*, \lambda^*),$$

where x^*, λ^* gives a saddle point of L.

[4] This section is more advanced than those preceding and requires use of the Kakutani fixed point theorem. The following Reviews, which have not been required to this point, provide the background: R8 (Section R8.7) and R9.

In this section we shall give two results of importance for the fundamental theory of optimization. First we shall show that minimax $F(x, y) =$ maximin $F(x, y)$, if and only if $F(x, y)$ has a saddle point. Then we shall give conditions on $F(x, y)$ that guarantee the existence of a saddle point.

Consider a continuous function $F(x, y)$ defined over compact sets X, Y. We can then state the following:

Theorem I

$\text{Min}_y \max_x F(x, y) \geqq \max_x \min_y F(x, y)$. *The equality holds, if and only if $F(x, y)$ possesses a saddle point at x^*, y^*, in which case*

$$\max_x \min_y F(x, y) = \min_y \max_x F(x, y) = F(x^*, y^*).^5$$

First we note that, whether $F(x, y)$ has a saddle point or not, $\max_x F(x, y) \geqq F(x, y)$ for all x, any y, by definition. Then

$$\min_y [\max_x F(x, y)] \geqq \min_y F(x, y)$$

for all x, giving the first part of the theorem. This proposition (minimax $\geqq$ maximin) can be regarded as a *fundamental lemma.* Its similarity to the fundamental lemma of linear programming is obvious.

Now suppose $F(x, y)$ has a saddle point at x^*, y^*. Define $s = F(x^*, y^*)$. We have, by definition of a saddle point,

$$F(x^*, y) \geqq s \qquad \text{all } y;$$

hence

$$\min_y F(x^*, y) \geqq s,$$

and

$$\max_x \min_y F(x, y) \geqq \min_y F(x^*, y) \geqq s.$$

In a similar way, we argue from the definitional inequality,

$$F(x, y^*) \leqq s \qquad \text{all } x,$$

that

$$\min_y \max_x F(x, y) \leqq s.$$

Combined with the previous inequality $\max\limits_x \min\limits_y F(x, y) \geqq s$ and the inequality of the fundamental lemma, this implies

$$\max_x \min_y F(x, y) = \min_y \max_x F(x, y) = s = F(x^*, y^*),$$

[5] This theorem underlies all game theory, and is related to Von Neumann's original theorem (1928). The proof here is similar to that given in Gale [1].

so that the existence of a saddle point is sufficient to give the equality of the theorem.

To prove necessity, choose x^* such that

$$\min_y F(x^*, y) = \max_x \min_y F(x, y) = s_1,$$

and y^* such that

$$\max_x F(x, y^*) = \min_y \max_x F(x, y) = s_2.$$

Since $s_1 = s_2 = s$, we have

$$F(x, y^*) \leqq s \leqq F(x^*, y),$$

so that $s = F(x^*, y^*)$ is identified as a saddle point, completing proof of the theorem.

We shall now prove an important existence theorem:

Theorem II

If the compact sets X, Y are also convex and $F(x, y)$ is convex in y for every $x \in X$ and concave in x for every $y \in Y$, then $F(x, y)$ has a saddle point.[6]

Define the following sets:

$$Y(x) = \{y \mid F(x, y) = \min_{y \in Y} F(x, y); \text{ each } x \in X\};$$
$$X(y) = \{x \mid F(x, y) = \max_{x \in X} F(x, y); \text{ each } y \in Y\}.$$

Since $F(x, y)$ is convex in y and concave in x, the sets $Y(x)$, $X(y)$ are compact and convex. From the continuity properties of optimal solutions (Review R9, Section R9.4), the mappings $x \to Y(x)$ and $y \to X(y)$ are upper semicontinuous. Thus the mapping $(x \times y) \to [Y(x) \times X(y)]$ is an upper semicontinuous mapping of $X \times Y$ into compact convex subsets of itself. The conditions of the Kakutani fixed point theorem (Review R9, Section R9.5) are satisfied. From the theorem, there is some x^*, y^* such that $x^* \in X(y^*)$ and $y^* \in Y(x^*)$. That is, x^* maximizes $F(x, y^*)$ and y^* minimizes $F(x^*, y)$, so that

$$F(x, y^*) \leqq F(x^*, y^*) \leqq F(x^*, y),$$

which proves the theorem.

[6] The proof given here is due to Kakutani. See Kakutani, or the exposition in Karlin [1], Chapter 1. There are many variations of the minimax problem, and various proofs of the basic theorems. For an alternative proof of the above theorem, not involving a fixed point theorem, see Karlin [1], Chapter 1.

5.7 EXISTENCE OF OPTIMAL SOLUTIONS

We can use the theorems of the previous section to establish conditions for the existence of optimal solutions. We have already seen that the general optimizing problem has an optimal solution, if the Lagrangean has a saddle point, and we have now established sufficient conditions for the existence of a saddle point.

Consider the Lagrangean written in the form

$$L(x, \lambda) = f(x) + \sum \lambda_i[-g^i(x)],$$

with the constraints $x \geqq 0$, $\lambda \geqq 0$, $g^i(x) \leqq 0$.

Since $L(x, \lambda)$ is linear in λ, it can be taken as convex in λ. Also $L(x, \lambda)$ as a function of x is a positive linear combination of the functions $f(x)$, $[-g^i(x)]$ and will certainly be concave if these functions are concave. But $-g^i(x)$ is concave if $g^i(x)$ is convex, so that $L(x, \lambda)$ is a concave function of x if $f(x)$ is concave and each $g^i(x)$ is convex.

Now consider the sets over which x, λ are confined. The only constraint on λ is the nonnegativity constraint, so that λ is defined on a convex set (the nonnegative orthant). If $g^i(x)$ is convex, the set $\{x \mid g^i(x) \leqq 0\}$ is convex (see Review R8, Section R8.6). Thus the feasible set is the intersection of convex sets and is convex.

Thus we have shown x, λ to be defined on convex sets. We also need the sets to be compact. The nature of the constraint inequalities ensures that they are closed, so it remains to consider boundedness. Since we seek a minimum for λ, and it is bounded below by the nonnegativity constraint, we can impose an arbitrary upper bound so large it does not affect the optimal solution. The feasible set for x presents the difficulties. We cannot avoid adding the special assumption that the feasible set is bounded.

Thus if $f(x)$ is concave, every $g^i(x)$ is convex and the feasible set is bounded (and nonempty), the Lagrangean satisfies the conditions of Theorem II of the previous section, so that it possesses a saddle point and the general optimizing problem possesses a solution. Furthermore, the conditions for a global optimum (Chapter 2, Section 2.7) are satisfied by the same convexity-concavity conditions. Thus we can state:

Existence Theorem

The general maximum problem $\max f(x)$ S.T. $g^i(x) \leqq 0$, $i = 1, \ldots, m$, $x \geqq 0$ *always possesses a solution if*: (*a*) $f(x)$ *is concave and every* $g^i(x)$ *is*

convex; (*b*) *the feasible set* $K = \{x \mid g^i(x) \leqq 0,\ i = 1, \ldots, n;\ x \geqq 0\}$ *is bounded and nonempty.*
Under these conditions the Lagrangean $L(x, \lambda) = f(x) - \sum \lambda_i g^i(x)$ *possesses a saddle point* x^*, λ^*, *where* x^* *is optimal in the maximum problem and* $\lambda^* \geqq 0$. *Furthermore the values* x^*, λ^* *satisfy the Kuhn-Tucker conditions* (*Section* 5.4), *which are then sufficient for a global optimum. If* $f(x)$ *is strictly concave, the point* x^* *is unique.*

By the same arguments as used in Chapter 2, Section 2.6, the theorem can be extended to cover cases in which $f(x)$, $g^i(x)$ are *positive monotonic* transformations of concave, convex functions.

Note that a linear programming problem satisfies the concavity-convexity conditions. If the primal program has an optimal solution its feasible set must be nonempty and can be considered bounded. Since the Lagrangean must then have a saddle point, and since the λ^* of the Lagrangean is the optimal solution y^* of the dual in this case, we can complete the existence theorem for linear programming (See Chapter 3, Section 3.3) by stating that *if the primal has an optimal solution, then so does the dual.*[7]

In the usual formulation of the conditions for the existence theorem, the functions $g^i(x)$ are required to be *concave*. This is because the constraints are put in the form $g^i(x) \geqq 0$, and the Lagrangean formed by addition. The conditions are obviously equivalent to those above.

The conditions given in the theorem are rather strict, requiring individual concavity or convexity of the objective and constraint functions. As pointed out in Chapter 4 (Section 4.4), the second order conditions in the classical case operate on the *combined* convexity-concavity of the functions taken together, which is a much weaker condition. On the other hand, the classical conditions do not guarantee a global optimum.

We can often have the best of both worlds, in a sense. If the number of constraints is less than the number of variables and the nonnegativity constraints on x are ineffective at the optimum, we will probably have all the functional constraints effective. Thus we can insert inequalities in the functional constraints and use the global optimum conditions for the general case, while the Kuhn-Tucker conditions will reduce to the ordinary first order conditions for the classical problem. Thus we can show, for example, that the solution to the classical consumer problem $\min px$ S.T. $u(x) = 0$ gives a global optimum if $x \gg 0$.

[7] The existence proof for linear programming can be completed without direct reference to the minimax theorem by using Farkas' Lemma. See Gale [1] (Theorem 3.1), for a proof based on the results set out here in Review R3, Sections R3.6 through R3.7.

FURTHER READING

For a more complete discussion of the Kuhn-Tucker conditions, see Hadley [3], Chapter 6. For a different advanced discussion of this topic, see Karlin [1], Chapter 7.

Discussions of the relationships between linear programming and game theory are widely available. See, for example, Dorfman, Samuelson, and Solow, Gale [1], Allen [2] and, for a more advanced treatment, Karlin [1].

EXERCISES

1. In Exercises (1) and (4) of Chapter 4, put the constraints in the inequality form $x_1 + x_2 \leqq 1$, and impose nonnegativity constraints on the variables. Then investigate the problems by the methods of this chapter.

For what values of b are the global optimum conditions fulfilled?

Examine the relationship between the global optimum conditions above, and the second order conditions of the original problem.

2. In Exercise (5) of Chapter 4, investigate the change in the solution when the constraint $h(x) = 0$ is replaced by $h(x) \leqq 0$, and the additional constraint $x \geqq 0$ is imposed.

3. Discuss the nature of the optimum solution for different values of a in the problem:

$$\begin{aligned} &\max 10 - (x_1 - 2)^2 - (x_2 - 2)^2 \\ \text{S.T.}\quad &x_1^2 + x_2^2 \leqq 1 \qquad x_1, x_2 \geqq 0. \\ &x_1 + x_2 \leqq a \end{aligned}$$

4. Maximize $ax_1^3 + x_2^2 + x_3 + x_4^2$, subject to the constraints of Exercise (3), Chapter 3. How does the optimum point vary with the value of a? (Assume $a \geqq 0$.)

5. In the two-good, two-factor transformation curve problem,

$$\begin{aligned} &\max p_1x_1 + p_2x_2 \\ \text{S.T.}\quad &x_1 = f^1(v_{11}, v_{21}), \qquad x_2 = f^2(v_{12}, v_{22}) \\ &v_{11} + v_{12} = v_1 \text{ (constant)} \\ &v_{21} + v_{22} = V_2 \text{ (constant)}, \end{aligned}$$

it is generally assumed that we have an interior equilibrium.

Assume that the isoquants of *one* of the production functions f are straight lines of constant slope. By setting up the problem with nonnegative constraints, investigate the nature of the boundary optima.

Relate the optimum conditions to the profit maximizing conditions for a competitive economy.

PART

II

Static Economic Models

CHAPTER

6

Input-Output and Related Models

This chapter assumes familiarity with the properties of square matrices and characteristic roots (Review R5), and draws extensively on the properties of semi-positive matrices (Review R7, Sections R7.1 through R7.3). Except for Section 6.7 (the substitution theorem), this chapter does not depend on optimizing theory. Readers who wish to pass to examples in the use of optimizing theory may care to give a preliminary reading of this Chapter without first studying Reviews R5 and R7, then pass on to Chapter 7.

6.1 INPUT-OUTPUT MODELS[1]

Consider a simple linear fixed coefficient model of production with several production processes or activities, each producing a single output. Unit production of some output, say the jth, requires some fixed quantity, a_{ij}, of the ith input. Since the model is linear, production of an amount x_j of the

[1] Static input-output models are discussed in Baumol [2], Allen [2] (both briefly), and Dorfman, Samuelson, and Solow (at length). Variations on the basic theme are developed in Gale [1], and Karlin [1].

Input-Output originated as an empirical and applied approach and there is a considerable literature on this aspect. The best general picture is obtained by reading the essays in Leontief [3].

jth output requires an amount $a_{ij}x_j$ of the ith input. Since there are fixed production coefficients, there is no substitution between inputs, so that production of x_j requires $a_{ij}x_j$ of the jth input *and* $a_{kj}x_j$ of the kth input. It is to be considered essential to the model that some, at least, of the required inputs are themselves current outputs of other processes in the system.

Such a model is an *input-output model* and the a_{ij}'s are the *input coefficients* of the model. The matrix $[a_{ij}]$ is the *input matrix.*

If the set of commodities which appear as inputs at least once in the system is identical with the set of commodities which appear as outputs, and if there is no source of inputs other than current production and no use for outputs except as inputs, then we refer to the model as a *closed model.* Otherwise we have an *open model.*

An input-output model which is a representation of a complete economic system is usually called a *Leontief model.* It was Leontief's pioneer work on an input-output model of the American economy that originally promoted interest in models of this kind.[2]

6.2 THE CLOSED MODEL

Since the set of inputs is the same as the set of outputs, we can assemble the input coefficients into a square matrix $\bar{A}$. It is obvious that the input coefficients are essentially nonnegative, that every output requires at least one input and that every input gives rise to at least one output, so that $\bar{A}$ is a *semipositive matrix*. The analysis of the model depends crucially on this fact.

The diagonal coefficients a_{ii} require consideration. a_{ii} represents the quantity of the ith commodity used up in its own production. The generation of electricity itself uses electricity for generator exciting, fuel pumping, and so on. If we subtract this use from total electricity generated, we obtain net output of electricity. We shall suppose, unless it is otherwise stated, that outputs are measured net in this sense, so that $a_{ii} = 0$ for all i. (This is a common convention, but the outcome of the analysis is unchanged if it is not adopted.)

Let x be a vector of outputs produced by the system. Then $\sum a_{ij}x_j$ is the quantity of the ith input necessary to produce this output bundle. Thus $y = \bar{A}x$ is the vector of inputs. Since the outputs, x, provide the only source of inputs, the system cannot operate and produce x unless $y \leqq x$. Production processes are assumed to be irreversible so that x is essentially nonnegative.

[2] Leontief's first statement appeared in 1936 (Leontief [1]), but the topic really commenced to develop with the publication of his first complete study in 1941 (Leontief [2]).

We say that the model is *viable* if there exists some x such that

$$\bar{A}x \leqq x, \qquad x \geqq 0.$$

Such a solution, if it exists, is a *viable solution.*

If the model is viable, we search for some solution that we are willing to consider as an equilibrium. Let us investigate the existence of the strongest kind of equilibrium, an *interior equilibrium.* We shall define this as some x such that

$$\bar{A}x = x, \qquad x \gg 0.$$

This clearly conforms to a strong equilibrium idea, since every commodity is produced and the production just equals the requirement for every commodity.

By definition of a characteristic root, a necessary condition for the existence of an x which satisfies the above equilibrium conditions is that $\bar{A}$ has one of its characteristic roots equal to unity. Equilibrium x is then the associated characteristic vector, which must be strictly positive for interior equilibrium.

Now $\bar{A}$ is semipositive. Assume it is also *indecomposable.* Then we can use the results of Review R7 (Section R7.3). It is clearly sufficient that the unit characteristic root be λ^*, the dominant root, which is certainly possible since λ^* is known to be real and positive [Result (a.1)]. If this is so, then the associated characteristic vector x^* is strictly positive [Result (c.2)], so that x^* gives an interior equilibrium. But x^* is also the only nonnegative vector satisfying the inequality $\bar{A}x \leqq x$ [Result (g)], so it is the only viable solution.

Thus we can enunciate a very strong result if it can be shown that $\lambda^* = 1$.

If $\bar{A}$ is indecomposable and its dominant root is unity, the closed model has a unique[3] *interior equilibrium which is also the only viable solution.*

It is clear that if $\lambda^* > 1$, the model is not viable. If $\lambda^* < 1$ the model is viable but has no interior equilibrium. Does it have a viable solution that would meet some other acceptable equilibrium condition?

If $\lambda^* < 1$, $\bar{A}x^* = \lambda^* x^* \ll x^*$, giving a surplus of every commodity. Clearly, we are not willing to regard this as an equilibrium. We certainly cannot close the surplus gap in all commodities, since $\bar{A}x \geqq \lambda^* x$ has no nonnegative solution other than x^* [Result (g)]. Further, we cannot remove the surplus even in one commodity, since $\bar{A}x \leqq x$ with at least one equality is possible only if $\lambda^* = 1$ [also Result (g)]. Thus we must conclude that, if $\lambda^* < 1$, the model has no equilibrium in any acceptable sense.

[3] Unique here means unique to a scalar multiple, since kx is a solution if x is.

If $\bar{A}$ is an actual numerical matrix, we can determine whether or not $\lambda^* = 1$ by direct computation. Our interest in the model is chiefly theoretical, however, and we really wish to find some economically meaningful condition that will guarantee that $\lambda^* = 1$. The following condition fulfils this requirement:

If there exists any set of semipositive prices which gives zero profit in at least one industry and EITHER no losses in all other industries OR no profits in all other industries, then the matrix of the model has a unit dominant root with consequences already outlined.[4]

If p_i is the price of the ith commodity, $p_i a_{ij}$ is the value of the ith commodity incorporated in unit output of the jth. $\sum p_i a_{ij}$ is the total value of inputs incorporated in unit output of the jth commodity and is therefore the unit cost of production.

The profit per unit output of the jth industry is $p_j - \sum p_i a_{ij}$, so that the condition set out above is that there is some nonzero vector p ($\geqq 0$) such that either $p\bar{A} \leqq p$, with at least one equality, or $p\bar{A} \geqq p$, with at least one equality.

Taking the first condition, we can transpose to obtain $\bar{A}'p' \leqq p'$, with at least one equality. But $\bar{A}'$ is semipositive indecomposable if $\bar{A}$ is, and has the same characteristic roots. From Review R7, the above inequalities imply $\lambda^* = 1$, [Result (g)] and the same conclusion if the second condition is taken.

There are two corollaries related to the above condition.

(*a*) *If there exists ANY nonnegative price vector giving losses in all industries, the model is not viable.*

(*b*) *If there exists ANY nonnegative price vector giving profits in all industries, the model is viable but has no equilibrium.*

If we take the first, this implies some nonnegative p such that

$$pA \gg p.$$

But, since p is *strictly* less than pA, we can find some $k > 1$ such that

$$pA \geqq kp, \qquad \text{with at least one equality.}$$

From the same type of argument as before, this implies $\lambda^* = k > 1$ and the model is not viable.

[4] This is weaker than the usual condition that there has to be a strictly positive price vector giving zero profits everywhere.

The second corollary is proved in a similar way, leading to the proof that $\lambda^* < 1$ and then using the earlier results for this case. Notice that introduction of prices emphasizes the lack of equilibrium here, since positive profits are made in all industries although there is a surplus of all products.

6.3 THE LEONTIEF OPEN MODEL

A model may be open in a large number of ways. We shall confine our attention here to the *Leontief open model.* This is an input-output model of a complete economy containing a production sector of n outputs which are also inputs within the sector, an extra input which is not an output of any production process, and a demand for outputs over and above their use as inputs. The extra input is most often identified as labor, but this identification is not essential.

If we consider only the production sector, it has an $n \times n$ input matrix which is semipositive and which will be assumed indecomposable. We shall write this input matrix A, the use of $\bar{A}$ in the preceding sections being designed to leave this notation open.[5]

Neglecting the extra input for the time being, if x is the physical output vector, then Ax is the vector of input requirements from these outputs within the production sector. Then

$$x - Ax = (I - A)\,x$$

is the vector of *net outputs,* that is, of quantities available for disposal outside the production sector itself.

The fundamental problem of the open model analysis is whether the economy can supply any arbitrary list of net outputs, referred to as *final demand* or *the bill of goods,* at least up to a scalar multiple.

If we write the final demand as a column vector c (which is essentially nonnegative), then the fundamental problem is whether there is a feasible x such that $y = c$, that is, whether there is an x such that

$$(I - A)\,x = c, \qquad x \geqq 0,$$

for all $c \geqq 0$.

If $I - A$ is nonsingular, we can always solve for x directly as

$$x = (I - A)^{-1}c,$$

[5] In some expositions the input matrix is written heavy type a, and the matrix $I - a$ is written A, identified as the *technology matrix.*

but this does not guarantee that $x \geqq 0$. If, however, we can show that $(I - A)^{-1}$ (sometimes referred to as the *Leontief Inverse of A*) is a positive matrix, then $(I - A)^{-1}c$ is always nonnegative if c is nonnegative and the problem is solved.

Let us consider the effect of the existence of an extra input, outside the production sector, in the open model. If we denote the amount of this input required for unit output of the jth industry by a_{oj}, and the vector of the a_{oj}'s by a_o, then the model must satisfy some subsidiary condition

$$a_o x \leqq l_o,$$

if the total quantity of this input is limited to an amount l_o. It is obvious that, for any x we can find some scalar multiple kx that satisfies the condition.

The real substance of the open model problem is whether final demand can be met *in any proportion.* If net outputs can be produced in all proportions, then scale can always be adjusted so that a subsidiary condition such as that above can then be satisfied. In the discussion, we shall suppose that c can take on any proportions, but is of such scale that the subsidiary condition can be met.

To return to the central problem, we are interested in conditions that guarantee that $(I - A)^{-1}$ is positive, which is a necessary and sufficient condition that final demand in all proportions can be met. From Result (d.2) of Section R7.3, a sufficient condition for this is that the dominant root, λ^*, of A be less than unity. Direct computation would show this to be true or not, given the input matrix. However, as always, we seek economically meaningful conditions that guarantee this. There are two given below, one based on quantity considerations, the other on price considerations.

(*a*) *If ANY nonzero final demand can be met with all industries producing, then final demands in all proportions can be met.*

(*b*) *If there is ANY set of positive prices for which every industry can at least cover the cost of its industrial inputs, and at least one industry can more than cover them, then final demands in all proportions can be met.*

If condition (*a*) is satisfied, then

$$(I - A)\,x \geqq 0, \qquad \text{with} > \text{for some } i, \qquad x \gg 0;$$

that is,

$$Ax \leqq x, \qquad \text{with} < \text{for some } i, \qquad x \gg 0.$$

Now if we write outputs as a diagonal matrix X, instead of as a vector, then $x = X[1]$, where [1] is the unit column vector. We can write the above inequality system as

$$AX[1] \leqq X[1], \qquad \text{with} < \text{for some } i.$$

Premultiplying both sides by X^{-1} (since $x \gg 0$, X is nonsingular),

$$X^{-1}AX[1] \leqq [1], \qquad \text{with} < \text{for some } i.$$

But $X^{-1}AX[1]$ is the vector of row sums of the matrix $X^{-1}AX$ which has the same roots as A. Each of these row sums is less than or equal to unity with at least one less than unity. If s, S are the smallest and largest of the row sums we have $s < 1$, $S \leqq 1$. Then $\lambda^* < 1$ from Result (f) of Section R7.3 and the remaining conclusions follow.

If condition (*b*) is satisfied, then

$$pA \leqq p, \qquad \text{with} < \text{for some } j, \qquad p \gg 0,$$

and the argument follows the same lines as those for condition (*a*).

These seem to be the weakest economically meaningful conditions that guarantee solution of the open model problem, and are less restrictive that those usually given.

6.4 DIRECT AND INDIRECT INPUT REQUIREMENTS

The input coefficient a_{ij} represents the amount of input i required for unit output of commodity j, if all inputs are supplied from outside the system, and only commodity j is produced. We can consider a_{ij} to represent the *direct requirement* of input i in industry j.

But if industry j is considered part of the complete system, then unit output of this industry requires the use of some of the inputs $k = 1, 2, \ldots, i - 1, i + 1, \ldots, n$ as well as input i. To produce these other inputs, other industries have to operate and these will, in general, also require the use of input i. Furthermore, these other industries will also require j as an input, so that industry j will have to produce more than a unit amount in order to supply the industries that are themselves supplying inputs to industry j. The requirement of input i which arises from these sources constitute an *indirect requirement* of input i in industry j.

The *total requirement* of input i for unit operation of industry j is the sum of the direct and indirect requirements.

We can compute the total requirement of all the inputs for the jth industry in the following way. Assume that the model is indecomposable and productive, so that any bill of goods can be produced. For a bill of goods c, we have

$$x = (I - A)^{-1}c.$$

Since we shall be using the Leontief inverse matrix $(I - A)^{-1}$ a good deal in the succeeding analysis, let us simplify notation by writing

$$A^* = (I - A)^{-1},$$

so that we have $x = A^*c$.

Now let us choose a particular bill of goods c' in which $c_j' = 1$, $c_i' = 0$, $i \neq j$.

The solution x' of $x' = A^*c'$ gives the levels at which all industries must operate to produce the bill of goods c'. But the only *net* output of the industrial sector is a single unit of commodity j. Any production of commodity i which is taking place is simply to supply input needs of that commodity. Thus x_i' gives the total requirement of input i to produce unit (net) output of commodity j.

Now in the matrix-vector product A^*c', all components of c' other than c_j' are zero, so that only the jth column of A^* plays a role in the product. Denote this column by A_j^*. Then

$$x' = A_j^*, \qquad \text{since} \qquad c_j' = 1.$$

Thus the various input requirements can be read off from the matrices A, A^*.

a_{ij} *is the direct requirement of i in j*;
a_{ij}^* *is the total requirement of i in j*;
$a_{ij}^* - a_{ij}$ *is the indirect requirement of i in j.*

We can now state and then prove an important theorem:

For EVERY industry in an indecomposable productive system, the total requirement of EVERY input exceeds its direct requirement.

To prove the theorem we note that, if A is indecomposable and productive ($\lambda^* < 1$), $A^* = (I - A)^{-1}$ is *strictly* positive [Result (d.2) of Section R7.3]. Now expand A^* in the algebraic series (permissible since every root has modulus less than one).

$$A^* = (I - A)^{-1} = I + A + A^2 + A^3 + \cdots.$$

Consider this series. A^2 is semipositive indecomposable if A is, so that, for any i, j there is a positive element in the (i, j)th place in A^{2r} for all i, j

and some r [Result (i) of Section R7.3]. Thus the strictly positive character of A^* cannot depend on A which we can remove from the series and still leave it strictly positive. Then we have

$$\begin{aligned} A^* - A &= I + A^2 + A^3 + \cdots \\ &\gg 0. \end{aligned}$$

Thus $a^*_{ij} > a_{ij}$ for all i, j and the theorem is proved. Note that this result depends crucially on the indecomposability of A.

A point of some interest is that, although we have adopted the convention that $a_{jj} = 0$, we have $a^*_{jj} > 0$. This is because output j is required as an input in the industries which supply j with its other inputs. Although we net out the electricity used directly in electricity generation from the input matrix, electricity is still required to produce and transport coal for use in electricity generation.

A Simple Numerical Example. The various points that have been made concerning the open Leontief model can be illustrated by a very simple numerical example.

Let the input matrix, A, be

$$\begin{bmatrix} 0 & 2 \\ \frac{1}{3} & 0 \end{bmatrix}.$$

By direct calculation, the roots are found to be $\sqrt{\frac{2}{3}}$, $-\sqrt{\frac{2}{3}}$, so that $\lambda^* = \sqrt{\frac{2}{3}} < 1$. The associated characteristic vector x^* has proportions $(1, \frac{1}{2}\sqrt{\frac{2}{3}})$. The other characteristic vector, $(1, -\frac{1}{2}\sqrt{\frac{2}{3}})$ is not a nonnegative vector.

The technology matrix, $I - A$, is

$$\begin{bmatrix} 1 & -2 \\ -\frac{1}{3} & 1 \end{bmatrix},$$

and its inverse can be found by simple calculation to be given by

$$A^* = (I - A)^{-1} = \begin{bmatrix} 3 & 6 \\ 1 & 3 \end{bmatrix}.$$

A^* is strictly positive, with $A^* \gg A$, so that total input requirements exceed direct input requirements everywhere.

We shall continue this example later in the chapter to illustrate other aspects of input-output model analysis.

6.5 FACTOR INTENSITY IN THE LEONTIEF MODEL

An important corollary of the above analysis arises from applying the idea of direct and indirect input requirements to the extra input, which we shall usually identify with labor.

If a_0 is the vector of input coefficients for labor (not a part of the basic input matrix A), then a_{0j} is clearly the *direct labor input coefficient* for industry j. The *total labor input coefficient* is obtained, as in the preceding section, by operating the whole system so as to produce a net output of one unit of j and zero net outputs elsewhere. If x' is the vector of gross outputs corresponding to this operation of the system, then

$$x' = A_j^* \qquad \text{(as before)},$$

and the total labor requirement, which we shall label a_{oj}^*, is given by

$$\begin{aligned} a_{oj}^* &= a_o x' \\ &= a_o A_j^*. \end{aligned}$$

Now $a_o A_j^*$, written out in full, is $\sum_i a_{oi} a_{ij}^*$.

The relationship of this sum to a_{oj} is not immediately apparent. Consider, once again, the series expansion of A^*.

$$A^* = I + A + A^2 + A^3 + \cdots.$$

The diagonal terms of A^* consist of 1's (from I) plus the sum of diagonal terms from A^r, $r = 1, 2, \ldots$. The diagonal terms of A^r are certainly nonnegative and are, in fact, positive for some r [Result (i) of Section R7.3], so that the diagonal terms of A^* are greater than unity.

Now in the expansion of $a_o A_j^*$, one of the terms is $a_{oj} a_{jj}^*$, and the remainder are certainly nonnegative. Thus, since $a_{jj}^* > 1$, the sum is greater than a_{oj} provided a_o contains at least one nonzero element. Thus we can state the following:

In an indecomposable system, the total labor requirement exceeds the direct labor requirement in all industries, even in an industry which is the only direct user of labor. Provided at least one industry uses direct labor, even industries using no direct labor require indirect labor.

An important property of the a_{oj}^*'s is that they are additive. That is, the actual labor requirement for an arbitrary bill of goods can be calculated by multiplying each total labor requirement coefficient by the quantity of that particular commodity appearing in the bill of goods, then adding over commodities.

If c is an arbitrary bill of goods, then the total labor required to produce it is

$$\begin{aligned} a_o A^* c &= \sum_j a_o A_j^* c_j \\ &= \sum_j a_{o\,j}^* c_j, \end{aligned}$$

which establishes the additive property.

We can compare the labor content of two bills of goods by computing the total labor requirement of each and then comparing them. If the two bills of goods are, in some sense, "equivalent," we can the one which requires the greatest total labor to be more *labor intensive* than the other.

The most usual sense of "equivalence" in economics is that the two bills of goods have the same value. Given a set of prices and two arbitrary bills of goods, we can always find a scalar such that, when one of the bills of goods is multiplied by it, both have the same value. If the same scalar multiple is applied to the total labor, we can determine the relative labor intensities by comparing the modified total labor quantities.

An important calculation of this kind was carried out by Leontief for two bills of goods with the same total value, one representing commodities in the proportions in which they appeared in United States imports, the other representing them in the proportions in which they appeared in United States exports, using, of course, input-output data for the United States. His finding, that the export bill of goods was more labor-intensive than the import bill of goods, contrary to general expectation, is known as the *Leontief Paradox*.[6]

Relative factor intensities are correctly stated only by using total factor requirement coefficients. Direct requirements (ordinary input coefficients) may give different results. If we consider two commodities, j, k, then it is easily possible that, for some scalar μ, we have

$$a_{oj} > \mu a_{ok}$$

but

$$a^*_{oj} < \mu a^*_{ok},$$

so that, if the price ratio of the two commodities was μ we should have the more labor intensive commodity on total labor requirement appear as the less labor intensive on direct labor only.

6.6 A LABOR THEORY OF VALUE

In the open model, we have two types of inputs. There is the set of inputs which are also outputs of the industrial sector, which we refer to as *intermediate goods*. Then there are the input or inputs which are not produced within the system, or *primary inputs*.

The standard Leontief open model contains only one primary input,

[6] The original papers are printed as Essays 5 and 6 in Leontief [3]. For the impact on international trade theory, see Corden, Bhagwati, or Chipman [2].

identified with labor. As in other economic models (Ricardo, Marx) containing labor as the only primary input we expect the Leontief model to contain an implied labor theory of value. It does indeed possess this property, but the implicit value theory is more complex than in simpler models. It is probably expressed best in the following statement.

The Leontief open model contains an implicit labor theory of value, in the sense that a set of prices proportional to total labor requirement coefficients forms an equilibrium set of prices for all final demands.

By an equilibrium set of prices in this context, we mean a set of prices such that, if the wage level just enables labor to purchase the net output of the economy, profits are exactly zero in all industries.

Let p be a price vector proportional to the total labor requirements vector, so that $p = ka_o^*$ for arbitrary k. Consider an arbitrary final demand vector c, so that the value of final demand is pc. If L is the total labor used to produce this bill of goods, then $L = a_o^*c$. But, if the wage, w, is such that labor can just buy the bill of goods, we must have $wL = wa_o^*c = pc = ka_o^*c$, so that $w = k$.

The unit profit of the jth industry, π_j, is given by

$$\pi_j = p_j - \sum_i p_i a_{ij} - wa_{oj}.$$

The vector, π, of profits for all industries is given by

$$\pi = p(I - A) - ka_o \qquad \text{(putting } w = k\text{)}.$$

But $p = ka_o^* = ka_oA^* = ka_o(I - A)^{-1}$, so that

$$\begin{aligned} \pi &= ka_o(I - A)^{-1}(I - A) - ka_o \\ &= 0. \end{aligned}$$

Thus $\pi_j = 0$ for all j. This is quite independent of c, which dropped out early in the analysis, so the result is proved.

A necessary consequence of the above results is that, at the labor theory of value prices, the labor intensity of unit values of all commodities is the same. Since direct labor varies from industry to industry, this highlights the point made earlier that direct labor coefficients do not properly represent relative labor intensities.

Numerical Example. Let us continue with the numerical example given earlier in the chapter by adding a vector of direct labor inputs $a_0 = (1, 2)$.

Simple calculation gives

$$a_0A^* = [1 \quad 2]\begin{bmatrix} 3 & 6 \\ 1 & 3 \end{bmatrix} = [5 \quad 12].$$

At prices 5, 12, the net value of output of an arbitrary bill of goods (c_1, c_2) is $5c_1 + 12c_2$, just equal to the wage bill with unit wage. It is easily calculated that industry 1's unit costs are 5 ($= 12a_{21} + a_{01} = 12 \times \frac{1}{3} + 1 = 5$) and that industry 2's unit costs are 12, so that profits are exactly zero in both industries.

6.7 THE SUBSTITUTION THEOREM

This section requires familiarity with the theory of linear programming.

An essential feature of the Leontief model is the use of a single process or activity for producing each output. There is only one way of producing commodity j, and that is represented by the set of input coefficients forming the jth column of the input matrix.

We know from the analysis of the open model that, if the system is productive (that is, it can produce at least one bill of goods), it can produce any bill of goods by using the same set of processes that were used initially. But suppose there were several ways of producing some or all outputs, would it not be possible that the processes "best" in some sense for producing one bill of goods might not be "best" for producing another bill of goods?

On the face of it, it might seem quite plausible that the production activities chosen for one bill of goods might not be appropriate to a different bill of goods. Indeed, in some models this would be so. For the Leontief model, *which has only one scarce factor*, Samuelson[7] showed that only one activity would ever be used for producing each commodity, no matter how final demand changed. We can state the theorem more formally as follows:

Substitution Theorem

In the Leontief open model, with a single scarce primary factor, the set of activities optimal for producing any one bill of goods is optimal for producing any other bill of goods, where optimal means minimizing the use of the scarce factor.

We shall use a proof based on linear programming which is quite different from Samuelson's original proof.[8]

The technology is assumed to consist of various activities, each one producing a *single output*. (The no-joint-product-feature of the Leontief model is preserved here.) Each activity is defined by n input coefficients and

[7] See Samuelson [5], also Arrow.

[8] This proof is due to Gale (Gale [1], Chapter 9), who ascribes the idea to Dantzig.

by the particular commodity produced by the activity. There may be many activities that give some particular commodity as output, only one activity that has a certain other commodity as output.

Because there is no longer a one-to-one correspondence between activities and commodities, we need to set up the model in a slightly different way from the ordinary input-output formulation. Instead of defining the activity by the column of input coefficients only, as we have done so far, we define an activity by a column containing the negatives of ordinary input coefficients corresponding to commodities which are inputs in this activity, and a single unit entry in the place corresponding to the output commodity.

If we assemble these activity columns into a matrix, the row of the matrix corresponds to a commodity, the column to an activity. The (i, j)th place in the matrix contains either a unit entry, meaning that the ith commodity is the output of the jth activity, or a negative (possibly zero) coefficient meaning that the ith commodity is an input into the jth activity. Every row will contain at least one unit entry (every commodity is the output of at least one activity) and so will every column (every activity has an output).

If we set up the ordinary input-output model in this form, the matrix we would obtain would be the technology matrix $(I - A)$. In the present case, the matrix is rectangular $n \times m$ with more activities (m) than commodities (n). We shall denote this technology matrix by $\hat{A}$.

If the various activities are operated at levels given by the activity vector y (of order m), then $\sum_j \hat{a}_{ij} y_j$ gives the net output of the ith commodity with at least one of the $\hat{a}_{ij}$'s being 1 and at least one being negative. The net output vector of the system is $\hat{A}y$. Given a bill of goods c, this will be produced by activity levels y such that

$$\hat{A}y = c, \qquad y \geqq 0.$$

Since there are n equations and m $(>n)$ variables y_j, there are many solutions. We choose among the solutions by the optimality criterion, that the minimum amount of the scarce primary resource, which we can take to be labor, be used.

If $\hat{a}_{0j}$ is the direct labor coefficient for activity j, and $\hat{a}_0$ the vector of direct labor coefficients (here an m vector instead of an n vector), producing the bill of goods c with minimum labor is equivalent to solving the canonical linear programming problem

$$\begin{aligned} &\min \hat{a}_0 y \\ \text{S.T.} \quad &\hat{A}y = c, \qquad y \geqq 0. \end{aligned}$$

From linear programming theory we know that the optimal solution (which we shall assume unique for simplicity of analysis, although this is not essential to the argument) will be *basic*, that is, will contain no more than n nonzero elements of y. If the system is indecomposable all commodities need to be produced, in some amount, for any bill of goods and since each activity produces only one good, the optimal solution will contain exactly n activities operating at nonzero levels, each producing one of the n commodities.

The columns of $\hat{A}$ corresponding to zero activity levels become irrelevant to the system and the effective technology matrix is an $n \times n$ submatrix or *basis* of $\hat{A}$ which we shall write $\hat{A}_B$.

The basis $\hat{A}_B$ is, of course, just an ordinary input-output technology matrix. If we were to observe the economy producing a bill of goods c, we would observe an ordinary input-output system.

Now let the bill of goods be changed from c to c'. Will the optimal basis change?

The basis theorem of linear programming (Chapter 3, Section 3.6) assures us that, if a certain basis is optimal for a program with constraints $\hat{A}y = c$, it is also optimal for a program with constraints $\hat{A}y = c'$, *provided it is still feasible*. But we know from input-output theory that, if we can produce bill of goods c with basis $\hat{A}_B$, we can produce any other bill of goods, including c', with the same basis. Since producing a bill of goods in the input-output sense means satisfying the equation system with a nonnegative y, it therefore implies feasibility in the linear programming sense. Thus $\hat{A}_B$ is feasible for c' as well as c, and so is optimal for c' as well as c. Thus the theorem is proved.

No matter how final demand changes therefore, the set of activities originally found optimal will remain so. Thus only one set of activities will ever be used, justifying the Leontief assumption of a single activity for each commodity.

The substitution theorem depends crucially on the existence of only a single *scarce* factor. It would still apply to an economy with scarce capital and unlimited supplies of Labor (as in the Lewis and other models of less developed economies), since we would still have the single optimality criterion, in this case minimizing the use of capital. If both labor and capital were scarce, then we would expect the activity mix to change with the bill of goods.

An interesting and direct *corollary* of the substitution theorem is the following.

Consider all the input-output systems that can be constructed from the set of known activities. Every input-output system must contain exactly one activity *producing each of the n commodities, so that, if there are k_i ways of producing the ith commodity, there are $\Pi_i k_i$ systems that can be assembled. From the substitution theorem, one of these is the* optimal system, *the system that will actually be observed in operation. Associated with each system will be a vector of total labor input coefficients. Then we have the result that every total labor requirement coefficient of the optimal system will be equal to or less than the corresponding total labor requirement coefficient of any other system.*

Let a_0^* be the total labor requirements vector for the optimal system, and a_0' the total labor requirements vector for any other system. If c is any bill of goods, the labor needed to produce this is a_0^*c, $a_0'c$ in the two systems. Since a_0^* is associated with the optimal system,

$$a_0^*c \leqq a_0'c.$$

Since this must be true for all $c \geqq 0$, it follows that

$$a_0^* \leqq a_0',$$

which proves the result.

6.8 MATRIX MULTIPLIERS[9]

Since their properties are analytically similar to those of input-output models, multi-sector income-expenditure models with constant propensities to spend are discussed here.

In the typical model of this kind, an income of y_j in sector j results in expenditure $a_{ij}y_j$ on the goods of sector i, all being measured in value terms. The coefficient a_{ij} is the propensity to spend on i from j-sector income. The matrix of these coefficients is the *propensity matrix A*. The a_{ij}'s are pure numbers and there is no problem of comparability.

Income of a sector is derived from expenditure in that sector from other sectors (induced income) and from other unspecified sources (autonomous income). Usually there is a period type of analysis involving successive rounds of income and expenditure, as in the ordinary Keynesian model.

Let us commence from some initial situation and explore the effect of an autonomous increase in sector incomes, represented by the vector $y(0)$. This income generates first round expenditures, hence incomes $y(1)$ given by

$$y(1) = Ay(0).$$

[9] See Chipman [1], Metzler, Goodwin.

These induced incomes themselves induce new expenditures given by

$$y(2) = Ay(1) = A^2y(0),$$

and the process continues as in the ordinary scalar multiplier process. If the original autonomous income increases are sustained, the total sector incomes ultimately generated will be $y(\infty)$, given by

$$\begin{aligned} y(\infty) &= y(0) + y(1) + y(2) + \cdots \\ &= (I + A + A^2 + \cdots)y(0). \end{aligned}$$

The matrix series

$$I + A + A^2 + \cdots$$

converges to $(I - A)^{-1}$ if all the characteristic roots have moduli less than unity. Since A is a semipositive matrix (indecomposability is assumed) it is sufficient to show that the dominant root λ^* is less than unity.

From Result (f) of Section R7.3, λ^* is certainly less than unity for an indecomposable matrix if none of the column sums exceeds unity, and at least one is actually less. Thus the matrix series converges if no sector spends more than its income ($\sum_i a_{ij} \leqq 1$ for all j) and there is some "leakage" so that at least one sector spends less than its income inside the system ($\sum_i a_{ij} < 1$, at least one j).

If the series converges,

$$y(\infty) = (I - A)^{-1}y(0),$$

so that $(I - A)^{-1}$ is a *matrix multiplier* analogous to the ordinary scalar multiplier.

We can use the same type of analysis of direct and indirect income generation as was used for direct and indirect input requirements in the input-output model. If we denote the matrix $(I - A)^{-1}$ by A^*, as before, then a^*_{ij} gives the ultimate increase in i sector income from a unit autonomous increase in j sector income. Models of this kind can be important for government expenditure decisions since the coefficients a^*_{ij}, the ultimate effects of expenditure in sector j, may be quite different in relative ordering from the coefficients a_{ij} giving the impact effects only.

In some contexts the "sectors" are countries, the a_{ij}'s are propensities to import from i into j, with the general analysis following the same lines as above.

If the propensity matrix is decomposable the analysis is not greatly changed. We now need to have a "leakage" in *every* subsystem that does not spend on other subsystems in order to have convergence to the multiplier.

FURTHER READING

The reader who is interested in the application and empirical content of input-output methods should read the essays in Leontief [3], and follow the literature from there.

The theorist should read Dorfman, Samuelson, and Solow, Chapters 9 and 10, and the analysis in Gale [1], Chapters 8 and 9. Gale approaches the problem from a somewhat different point of view. A discussion of certain aspects of the Leontief model is also given in Karlin [1], Chapter 8. More advanced analysis, of a rather different kind from that given elsewhere, appears in Morishima [1], Chapter I.

EXERCISES

1. For the economy with input matrix

$$\begin{bmatrix} 0 & \frac{1}{4} \\ 1 & 0 \end{bmatrix}$$

and direct labor requirements vector [2, 1], calculate:

a. The indirect requirements of all inputs, including labor, for unit net output of each good;

b. The (normalized) equilibrium price vector;

c. The wage corresponding to the prices of **b**;

d. The total output of each industry, and the total labor requirement, for producing the bill of goods

$$c = \begin{bmatrix} 1 \\ 2 \end{bmatrix};$$

e. The corresponding quantities for producing the alternate bill of goods

$$c' = \begin{bmatrix} 2 \\ 1 \end{bmatrix}.$$

2. An economy can produce three goods. The first two goods can be produced in only one way, with input vectors

$$\begin{bmatrix} 0 \\ \frac{1}{6} \\ \frac{1}{3} \end{bmatrix}, \quad \begin{bmatrix} \frac{1}{3} \\ 0 \\ \frac{1}{6} \end{bmatrix},$$

respectively. For the third good, there are two possible activities, with input vectors

$$\begin{bmatrix} \frac{1}{2} \\ \frac{1}{6} \\ 0 \end{bmatrix}, \quad \begin{bmatrix} \frac{1}{6} \\ \frac{1}{3} \\ 0 \end{bmatrix}.$$

Labour input requirements are 1, 2 for the first and second goods, and 1 for both activities for the third good.

Use either linear programming or the corollary of Section 6.7 to find the optimal activity for the third industry. Check your result by showing total labor input to be less for the optimal system x, choosing an arbitrary vector of final demands.

CHAPTER

7

Linear Optimizing Models

Linear programming theory (Chapter 3) is essential for this chapter. In addition to the mathematical background required for linear programming, familiarity with convex cones (Review R4, Sections R4.5 through R4.7) is assumed.

7.1 ACTIVITY ANALYSIS OF PRODUCTION[1]

The input-output models discussed in the previous chapter were characterized by the existence of only a single method or activity for producing each output and the existence of only a single output for each activity.

We now wish to remove both these restrictions and consider a more general model in which there may be many activities with a given commodity as output, or one of several outputs, and there may be several outputs from a

[1] The classic statement of activity analysis is Koopmans [3]. An excellent discussion of the topic at a more elementary level is the first essay in Koopmans [2].

Activity analysis developed side by side with linear programming, but more or less independently of it. The original proofs of the fundamental propositions were given in terms of the properties of finite cones. Here we derive them by the use of linear programming theory.

The generalized production theory used in the next four chapters depends very much on the ideas developed in activity analysis.

single activity. The existence of alternative activities for a given output introduces substitution into the production process, and the existence of several outputs allows for joint production. It should be pointed out that the resulting production model is of a very general kind. We can have joint inputs, for a given activity, with substitution between inputs if there are many activities, We can have joint outputs for each activity, but substitution between those outputs if there are alternative activities with the same outputs.

Neoclassical production theory was almost entirely concerned with substitutability between primary inputs and had little to offer in terms of complementary inputs which are so characteristic of raw materials and intermediate goods. The input-output models are primarily concerned with input complementaries, especially among intermediate goods. The general linear model goes far in bridging the gap between the neoclassical analysis of pure substitution and the input-output analysis of pure complementarity.

The essential feature of the general linear model is the existence of a *finite* number (however large) of basic productive activities each characterized by fixed coefficients giving the inputs required for, and the outputs produced by, a unit level of operation of that activity. In addition, activities have the following properties:

(a) They are linear in the sense that, if a given activity operated at unit level produces output b_i of commodity i and uses a_r of commodity r, the activity can be operated at any level μ ($\mu \geqq 0$) and will produce μb_i of i and will use μa_r of r.

(b) They can be linearly combined so that, if unit levels of activities j, k produce b_{ij}, b_{ik} of commodity i and use a_{rj}, a_{rk} of commodity r, the simultaneous operation of the activities at levels μ_j, μ_k will be associated with commodity totals $\mu_j b_{ij} + \mu_k b_{ik}, \mu_j a_{rj} + \mu_k a_{rk}$. If commodity i was an output of activity j and an input of activity k, the associated total would be $\mu_j b_{ij} - \mu_k a_{ik}$, a net output from the combination if positive, a net input into the combination if negative.

The first property is that of constant returns to scale, the second depends on noninteraction between activities so that the operation of an activity is unaffected by the operation of other activities.

In the input-output model, each activity was associated in a one-to-one way with the output of some commodity. This provided a label for the activity (the jth activity produces the jth commodity) and also a natural unit level (the unit level of activity j being the level giving unit output of commodity j). This identification is lost in the general linear model, so we choose any arbitrary labeling of the activities and make any arbitrary choice of the unit level of each activity. Once these choices have been made, the activity is defined by the fixed input and output coefficients.

If there is a total of n commodities in the system, we can associate two n vectors with each activity. The output vector, b^j, of activity j is a nonnegative vector with components b_{ij} which are positive if commodity i is an output of that activity and zero otherwise. The input vector is a nonnegative vector a^j in which a_{ij} is positive if i is an input, zero otherwise. A commodity which is both an input and an output of the same activity is treated net, so that it appears as either a net input or as a net output.

If there are m activities, we assemble all the output vectors into an $n \times m$ matrix B, the *output matrix*, and all the input vectors into an *input matrix* A, also of order $n \times m$. In the input-output case, the input matrix was of the same kind as here, only square, while the output matrix was the identity matrix I.

The matrix $(B - A)$ is the *technology matrix* (analogous to the technology matrix $(I - A)$ in the input output case). It is a matrix containing an entry b_{ij} if commodity i is an output of the jth activity, $-a_{ij}$ if commodity i is an input of activity j, and 0 in the (i, j)th place if commodity i is neither an input nor an output of activity j.

If the various activities are operated at levels given by the m vector y, then the net outputs of the system are given by the vector x, where

$$x = (B - A)y.$$

Although y is essentially nonnegative, x is not. A given component of x may have any sign:

(a) If $x_i > 0$, i is a *final output* of the system for activity levels y.
(b) If $x_i = 0$, i is a *pure intermediate good* of the system for activity levels y.
(c) If $x_i < 0$, i is a *primary input* into the system for activity levels y.

We have classified commodities relative to the activity vector y. In some activity analysis discussions the commodities are classified *essentially* as final, intermediate or primary goods. This is unnecessarily restrictive, since the classification of commodities does not depend on the pure technology but on the demand conditions for commodities and the supply conditions for factors. Economic systems with the same technology may have different goods classifications, an intermediate good in one economy (produced within the economy) may be a primary input in another (imported from outside). The classification may even change in the same economy under different conditions.

The activity analysis of production is the study of the properties of the set of output vectors x and the set of activity vectors y, given assumptions on the technology matrix and subject to other constraints that may be placed on the system.

7.2 THE PRODUCTION SET

The *production set* is the set of all net output vectors x that can be produced from the given technology without imposing any resource constraints. It is sometimes referred to as the *producible set*, *attainable set* or *transformation set*.

Since activities can be operated at any nonnegative levels, the production set X is defined formally as

$$X = \{x \mid x = (B - A)y, \qquad y \geqq 0\},$$

and is mapped out by the linear transformation $x = (B - A)y$, as y ranges over the whole closed positive orthant.

Two related properties follow immediately from the linear nature of the transformation:

(a) *X is a closed convex set.* The closed (that is, including the boundaries) positive orthant $y \geqq 0$ is a closed convex set, and this property is retained by linear transformation.

(b) *X is a convex polyhedral cone.* Since X is a closed convex set which contains kx if it contains x, for all scalars $k \geqq 0$ it is a cone. It is a polyhedral cone since the number of activities is finite.

Figure 7–1 illustrates a two-commodity production set with three activities whose unit levels are represented by the points A_1, A_2, A_3. Activities A_1, A_2 have x_1 as input and give x_2 as output, whereas activity A_3 has x_2 as input and gives x_1 as output. The linearity property implies that all points on lines OA_1, OA_2, OA_3 extended indefinitely past A_1, A_2, A_3 are points in X. The convex combination property implies that all points on lines joining any pair of these activity lines are also in X. Thus X fills the space between OA_1 extended and OA_3 extended and is the shaded region on the diagram. All points can actually be attained by the use of activities A_1 and A_3 only: A_2 is, as we shall see in the next section, an inefficient activity.

Although the set illustrated in Figure 7–1 is typical (as typical, that is, as we can depict in two dimensions), it is not the only type which satisfies the properties (a) and (b). It is usual to rule out certain other possibilities by making the three following special assumptions on the nature of production:

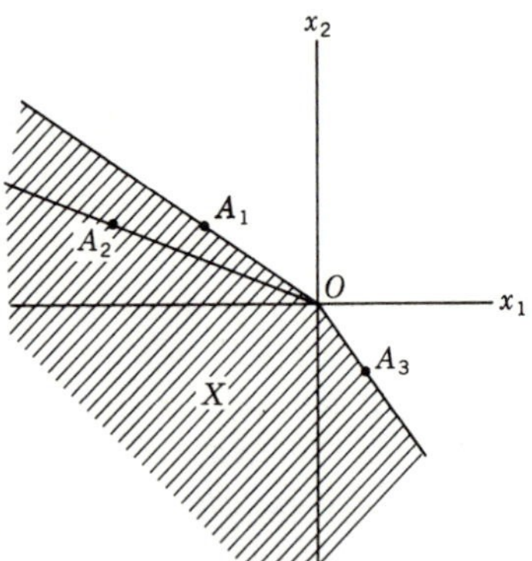

Figure 7-1. The production set.

Special Assumption I

It is not possible to produce a positive amount of any commodity without an input (negative amount) of at least one other commodity. Formally, if Ω *is the closed positive orthant,* $X \cap \Omega = 0$, *that is, the origin is the only point of the positive orthant contained in the production set.*[2]

Special Assumption II

Production processes are irreversible so that, if x is a possible production vector, $-x$ *is not. Formally,* $X \cap (-X) = 0$; *the origin is the only point common to* X *and* $(-X)$.

Special Assumption III

Wasteful production is always possible so that, if $x \in X$ *and* $x' \leqq x$, $x' \in X$. *Sometimes this same point is made by supposing the existence of a disposal activity for every commodity, that is an activity containing* -1 *in the jth place and zeros elsewhere. Wasteful production is then formally achieved by the combination of ordinary and disposal activities. A consequence of this assumption is that* $-\Omega \subset X$, *so that it is possible to dispose of any bundle of commodities.*

The reasonableness and economic implications of Assumption I are straightforward, but some discussion of the other two assumptions is called for.

Irreversibility (Assumption II) is fundamental to the *economic* theory of production, as compared with a technical production theory. Consider some

[2] Koopmans term for this postulate is more picturesque: "Impossibility of the Land of Cockaigne."

chemical process which is reversible in the chemist's sense. From the economist's point of view inputs other than molecules and energy have entered into the production process—labor is the most obvious example—which cannot be regained by reversing the chemistry. In fact, reversing the chemistry will require *more* labor input. If the production process takes time, so that the outputs are dated later than the inputs, and commodities with different dates are different commodities, then the irreversibility is clear. However we do not wish to emphasize this last point of view since most production theory is essentially static in approach.

There is a relationship between Assumptions I and II which gives added weight to Assumption II. Suppose that the activities of Figure 7–1 were reversible, so that activities $-A_1$, $-A_3$ were possible. Then, as shown in Figure 7–2, convex combinations of $-A_1$ and $-A_3$ would give points in the positive quadrant, violating Assumption I.

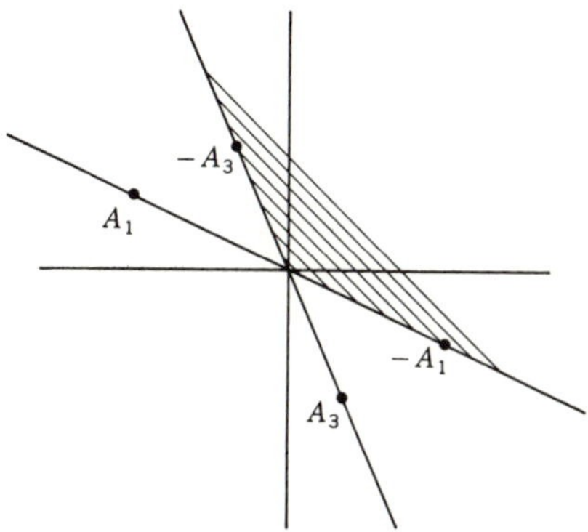

Figure 7-2. Irreversibility.

Wasteful production or free disposal (Assumption III) probably seems reasonable to the reader, but useless. This assumption is made primarily for mathematical reasons to "fill out" the production set.

The three assumptions on production sets rule out certain sets which are convex polyhedral cones. This is illustrated in Figure 7–3. The set in (a) is ruled out by Assumption II, that in (b) by Assumption I. [Actually set (b) would be extended by Assumption III to cover the whole space, as in (c).] Thus only sets of the kind (d), (e), or (f) are permissible.

The confinement of the production set by Assumptions I through III can be quite clearly defined, and leads to an important boundedness property.

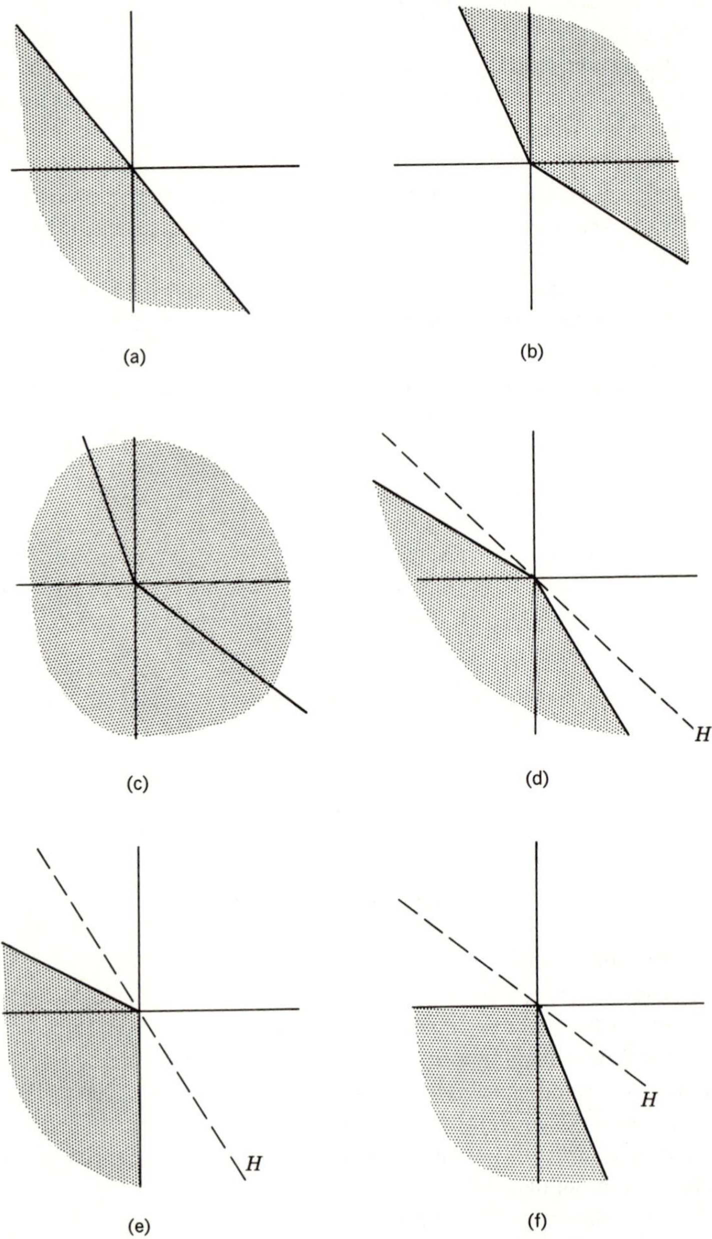

Figure 7-3. Acceptable and unacceptable production sets.

By virtue of irreversibility (II), X cannot contain a complete hyperplane. A hyperplane divides commodity space into two half-spaces so that X must be "smaller" than a half space. Thus we can find a hyperplane such that X lies entirely within one of the half-spaces so defined. By the same argument $-X$ (which can be obtained by rotating X around the origin) is contained in a half-space. since X, $-X$ intersect only at the origin, there is a hyperplane H (not necessarily unique) such that X lies in one of the half-spaces and $-X$ in the other. This is the *separating hyperplane* of X, $-X$, and it obviously passes through the origin.[3]

A hyperplane through the origin has an equation of the form $cy = 0$, in vector terms. Since the positive orthant Ω is contained in $-X$ (because $-\Omega$ is contained in X by III), and since $-X$ is contained in one of the half spaces defined by H, H itself cannot pass through Ω. Thus the equation $cy = 0$ must not permit of a nonnegative solution, which will be true if and only if all the components of c are positive. Thus $c \gg 0$. (The vector c is the *normal* to the hyperplane H.)

The two half-spaces defined by $cy = 0$ are $cx \geqq 0$ and $cx \leqq 0$. The first clearly contains Ω and so does not contain X. Thus we have the important upper boundedness property resulting from Assumptions I through III which can be expressed in four different but equivalent ways:

(*a*) *There is a strictly positive vector c such that* $cx \leqq 0$ *for all* $x \in X$.

(*b*) *There is a set* $H = \{y \mid cy = 0\}$ *such that, for every* $x \in X$, $x \leqq y$ *for some* $y \in H$.

(*c*) *The production set X lies entirely within a half-space lying on the negative side (the side containing* $-\Omega$*) of a hyperplane through the origin with a strictly positive normal.*

(*d*) *The set* $S(x^*) = \{x \mid x \geqq x^*, x \in X\}$ *is bounded or empty for all* x^*.

The first three statements have already been proved. To prove (*d*), we note that the set defined by the constraints

$$x \geqq x^*$$
$$cx \leqq 0$$

is either empty or a closed bounded convex set for every $c \gg 0$. From (*a*) it follows that $S(x^*)$ is entirely contained in such a set for every x^* and some c, so that $S(x^*)$ is bounded.

The result is easily seen in R^2 by choosing an arbitrary x^* in diagrams (d), (e), or (f) of Figure 7–3.

[3] Or we can obtain the result immediately from Assumption II and Corollary (4) of the Minkowski theorem (see Section R4.3).

7.3 EFFICIENT PRODUCTION

The most important ideas concerning the production set are those of efficiency, efficient production and efficient activities.

We define efficiency in a standard way.

A net output vector is efficient, if and only if there is no other vector which can produce more of any of the final outputs without producing less of at least one other final output or using more of at least one input, where a pure intermediate good is regarded as an input at zero level.

Since net inputs appear with a negative sign, $x'_j > x_j$ means that x' has *more* of j, if it is an output or uses *less* if it is an input, so we can state the efficiency definition more formally without distinguishing between inputs and outputs.

$x^* \in X$ is efficient if and only if, for every commodity k and every other vector $x \in X$, $x_k > x_k^*$ implies $x_j < x_j^*$ for some $j \neq k$.

If x^* is efficient, it follows from the above that $x_k^* \geqq x_k$ for all vectors x such that $x_j \geqq x_j^*, j \neq k$, so that x_k^* is the maximum for x_k among all vectors in X for which $x_j \geqq x_j^*$, $j \neq k$. The boundedness property shown in the previous section guarantees the existence of this maximum provided X satisfies the three assumptions of that section.

Thus we can set up the efficiency criterion as a maximum problem and use linear programming theory to derive an important theorem concerning efficient production:

Efficiency Theorem

Every efficient point in the production set can be attained by the use of at most n different activities, and corresponding to every efficient point are positive shadow prices such that the activities actually used make zero profits and all other activities would make losses.[4]

To simplify the terminology here, we shall write the technology matrix $(B - A)$ as T, so that $x = Ty$. Consider the submatrix of T obtained by omitting the kth row, k being an arbitrary choice. Denote this submatrix by $\hat{T}$, and denote by $\hat{x}$ the $(n - 1)$ vector obtained by omitting the kth element from x. Finally, denote by $\hat{t}$ the kth row of T, which was omitted from $\hat{T}$.

[4] It is possible that some prices are zero, rather than positive. This requires certain coincidences in the technology matrix which we shall ignore.

For any vector y of activity levels. $\hat{t}y$ gives the net output x_k of commodity k, while the vector of net outputs for all other commodities is given by $\hat{x} = \hat{T}y$. If x^* is an efficient point in X with elements $\hat{x}_j^*, j \neq k$, then $x_k^* = \hat{t}y$ for some y must be the optimal value of the linear program:

$$\begin{aligned} &\max \hat{t}y \\ \text{S.T.} \quad &-\hat{T}y \leqq -\hat{x}^*, \qquad y \geqq 0, \end{aligned}$$

which has the dual

$$\begin{aligned} &\min -\mu x^* \\ \text{S.T.} \quad &-\mu\hat{T} \geqq \hat{t}, \qquad \mu \geqq 0.^5 \end{aligned}$$

From the ordinary linear programming theory, we know that, since the above program has a solution ($x_k = \hat{t}y$ is bounded above) it has a *basic* solution with not more than n nonzero components of y, proving the first part of the theorem.

From the equilibrium theorem of linear programming, either $y_j^* = 0$, or the jth dual constraint is an equality. If $\hat{T}_j$ is the jth column of $\hat{T}$, then

$$y_j^* > 0, \qquad \text{and} \qquad \mu^*\hat{T}_j + \hat{t}_j = 0,$$

or,

$$\mu^*\hat{T}_j + \hat{t}_j < 0, \qquad \text{and} \qquad y_j^* = 0.$$

Since the constant vector in the primal constraints is x^*, and since the optimal value of the program is x_k^*, it follows that *all* the primal constraints are effective at the optimum ($\hat{T}y^* = \hat{x}^*$), so that in general all the components of μ^* will be positive.

Now, if p is a vector of prices, $\sum_{i \neq k} p_i\hat{T}_{ij} + p_k\hat{t}_j$ is the value of final outputs less the value of inputs, that is the profit, from operation of activity j at unit level. Thus, if we take the price vector p^* such that $p_i^* = \mu_i^*, i \neq k$, $p_k^* = 1$, p^* is a set of shadow prices which give zero profits for activities used and losses for other activities. Such prices exist for every efficient x and can be calculated from a program of the above kind.

It should be noted that, for any efficient vector, such a program can be set up with each commodity treated as the maximand, so there are n such programs for every efficient point. We may obtain different dual vectors μ for different programs: any of these gives a set of shadow prices which satisfy the theorem, which says nothing about *uniqueness* of the shadow prices.

[5] The primal constraints are derived from the efficiency definition requiring $\hat{x}^* \leqq \hat{x}$. This gives $\hat{T}y \geqq \hat{x}^*$, so the signs and the inequality have to be reversed to obtain the program in standard form.

There may, in fact, be a set of shadow prices associated with each efficient point, and a set of efficient points associated with given shadow prices.

Due to the association of a set of shadow prices with every efficient point such that activities not needed for attaining the point make losses, we can state the following:

Decentralization Theorem

If the production set satisfies the conditions of Section 7.2, efficient production can be attained with decentralized control of production if suitable shadow prices exist and if activity controllers avoid using activities which make losses.

The shadow prices (given from a suitable program) may have to be set by central control; they must be the same for all producers and the only prices on which decisions are based. At this stage we cannot necessarily identify these shadow prices with actual market prices.

7.4 CONSTRAINED PRODUCTION

Once we move from the study of the general properties of the production set to the study of production under constraint—typically constraint on the availability of one or more of the inputs—the number of different models is very large. The properties of any given model can be studied at the analytical level by linear programming theory and solved numerically by linear programming solution techniques. Rather than attempt any taxonomic study of the possibilities, it seems better to illustrate the kind of analysis that can be made by considering the three examples which follow. The first two are analytical, the third numerical.

(a) An example in price-quantity variation. Consider an economy, a sector, or a firm with resource vector c and technology T, facing prices p which then change to p'. We wish to study the response of optimal production to the price change from p to p'.

Optimal production under prices p will be the solution of the program:

$$(1)\ \max px = pTy, \qquad \text{S.T.} \quad -Ty \leqq -c, \qquad y \geqq 0,$$

and optimal production under prices p' the solution of:

$$(2)\ \max p'x' = p'Ty', \qquad \text{S.T.} \quad -Ty' \leqq -c, \qquad y' \geqq 0.$$

Now changing the price vector by a scalar multiple does not affect the solution of the primal program (although the dual variables will be changed in proportion), so we shall consider p' to have been adjusted by some scalar so that the optimal output $x = Ty$ in program (1) has the same value at the new prices, that is,

$$p'Ty = pTy.^{6}$$

Since programs (1) and (2) have the same constraints, the optimal vector of either program is feasible in the other. Since y is feasible in (2), we have, from the *Fundamental Lemma* of linear programming,

$$p'Ty' \geqq p'Ty.$$

But, since $p'Ty = pTy$, we have

$$p'Ty' \geqq pTy.$$

Also, since y' is feasible in (1), we have

$$pTy \geqq pTy',$$

so that

$$p'Ty' \geqq pTy',$$

or

$$(p' - p)Ty' \geqq 0.$$

But the deflation scheme for p' gives $(p' - p)\, Ty = 0$, so that, combining this with the above result, we obtain

$$(p' - p)\,(Ty' - Ty) \geqq 0,$$

or

$$(p' - p)\,(x' - x) \geqq 0.$$

This inequality summarizes the whole relationship between price and quantity changes in production, in the most general possible way. For arbitrary price changes the relationship will, in general, be a strict inequality, the equality holding only for price changes that leave the original output still optimal.

If only the jth relative price changes, so that $p_i' = \lambda p_i$, $i \neq j$, but $p_j' \neq \lambda p_j$ then, putting $p_j' = (p_j' - \lambda p_j + \lambda p_j)$ in the deflation equation we obtain

$$(p_j' - \lambda p_j)\, x_j + (\lambda - 1)\, px = 0.$$

Since $px \neq 0$, we have $\lambda - 1 > 0$ if $(p_j' - \lambda p_j)\, x_j < 0$, that is, if: (a) commodity j is an output and its price *falls* relative to the general price level, or (b) commodity j is an input and its price *rises* relative to the general price level.

Using the same substitution for p_j' in the basic price-quantity inequality, we obtain

$$(p_j' - \lambda p_j)\,(x_j' - x_j) + (\lambda - 1)\, p\,(x' - x) \geqq 0.$$

Now $p(x' - x) = pTy' - pTy \leqq 0$, as was shown before. Thus if $\lambda - 1 > 0$, we have the *Substitution Effect for Production*:

$$(p_j' - \lambda p_j)\,(x_j' - x_j) \geqq 0,$$

[6] p' can be considered to have been deflated for a change in the general price level by a Laspeyre index with weights Ty.

that is, a fall in the *relative* price of an output will lead to a reduction in the output, a rise in the *relative* price of an input will lead to reduction in the use of the input, all other prices changing in proportion. The apparent lack of symmetry in the proof, which does not consider the effect of a fall in an output price or a rise in an input price, is easily taken care of by reversing the roles of p, p' as "old" and "new" prices.

The above analysis can also be carried out by a hypothetical scalar variation in the resource vector, to give a "substitution" and "expansion" effect similar to the "substitution" and "income" effects in consumer theory.

(b) An example in resource allocation. Suppose we have two sectors of the economy, or two countries, with the same technology T, facing the same prices p, but with different resource vectors c^1, c^2. We shall investigate the relationship between the optimal behavior of the sectors and that of a consolidated economy in which the resources of both sectors are combined. The consolidated resource vector is c, where $c = c^1 + c^2$.

Optimal outputs $x^1 = Ty^1$, $x^2 = Ty^2$, $x = Ty$ for the two sectors and the consolidated economy are the solutions to the following programs:

$$\begin{array}{llll} (1)\ \max pTy^1 & \text{S.T.} & -Ty^1 \leqq -c^1, & y^1 \geqq 0, \\ (2)\ \max pTy^2 & \text{S.T.} & -Ty^2 \leqq -c^2, & y^2 \geqq 0, \\ (3)\ \max pTy & \text{S.T.} & -Ty \leqq -c, & y \geqq 0. \end{array}$$

Consider $y' = y^1 + y^2$. Since y^1 is feasible for (1), y^2 is feasible for (2), and $c = c^1 + c^2$, y' is feasible for (3). From the fundamental lemma, since y' is feasible in (3) and y is optimal in that program,

$$pTy' \leqq pTy.$$

For sufficient difference in proportions between c^1 and c^2, the optimal basis for program (3) will differ from that of at least one of the sector programs and the strong inequality will hold. Thus we can assert that:

Consolidation of production cannot reduce the total value of output over the whole economy and, for sufficient difference in resource endowments of the two sectors, will increase it.

Can we achieve the same result by the use of the market as we can by consolidation? We expect that resource trading will provide the answer, but we need appropriate shadow prices.

The duals of programs (1) through (3) are:

$$\begin{array}{llll} (1')\ \min w^1c^1 & \text{S.T.} & -w^1T \geqq pT, & w^1 \geqq 0, \\ (2')\ \min w^2c^2 & \text{S.T.} & -w^2T \geqq pT, & w^2 \geqq 0, \\ (3')\ \min wc & \text{S.T.} & -wT \geqq pT, & w \geqq 0. \end{array}$$

Since the dual constraints are the same for all programs, the dual variables of any program are feasible in any other. Since w^1, w^2 are feasible in (3′), we have, from the fundamental lemma:

$$w^1c^1 \leqq wc^1$$
$$w^2c^2 \leqq wc^2.$$

These inequalities imply that *each sector's resource vector is worth at least as much at consolidated shadow prices as at the sector's own shadow prices*, so it seems appropriate to examine the effect of permitting trade at the consolidated shadow prices. Each sector is subject to a budget constraint, that the value of resources used cannot exceed the value of the orginal resource endowment, values being measured in terms of consolidated shadow prices, w. It is obvious that the budget constraint will be satisfied by an equality, so the two sectors will now produce in accordance with the following programs:

$$(4)\ \max pTy^4 \quad \text{S.T.} \quad wTy^4 = wc^1, \quad y^4 \geqq 0,$$
$$(5)\ \max pTy^5 \quad \text{S.T.} \quad wTy^5 = wc^2, \quad y^5 \geqq 0,$$

which have duals

$$(4')\ \min \lambda_1 wc^1 \quad \text{S.T.} \quad -\lambda_1 wT \geqq pT,$$
$$(5')\ \min \lambda_2 wc^2 \quad \text{S.T.} \quad -\lambda_2 wT \geqq pT,$$

Inspection of (3′) makes it obvious that the optimal dual values are $\lambda_1 = \lambda_2 = 1$. Thus, from the duality theorem, we have, at the optimum, $pTy^4 = wc^1$, $pTy^5 = wc^2$. Also, since the dual constraints in (4′) and (5′) with equilibrium values of the λs are precisely the same as the dual constraints in the consolidated program, (3′), precisely the same activities will be used or not used in (3), (4), and (5). Finally, if we put $y^* = y^4 + y^5$, then

$$pTy^* = wc^1 + wc^2 = wc = pTy,$$

so that, *if resource trading takes place freely at consolidated shadow prices, the economy behaves in the aggregate exactly the same way it would if production was consolidated.*

Thus we have established that resource trading at consolidated shadow prices achieves the same results as consolidating the resources themselves. We have shown the incentive to trade on the supply side, since each sector's resources are worth more at consolidated shadow prices than at its own. The incentive to trade resources on the demand side follows from the gains which accrue to the sectors. For the first sector we have

$$pTy^4 = wc^1 \quad \text{[duality theorem in (4)]},$$
$$pTy^1 = w^1c^1 \quad \text{[duality theorem in (1)]},$$

and we also have $wc^1 \geqq w^1c^1$, so that $pTy^4 \geqq pTy^1$, with an equivalent result for the other sector.

(c) A simple numerical example. Consider the two-activity, three-commodity economy in which

$$T = \begin{bmatrix} 1 & 2 \\ 1 & -\frac{1}{2} \\ -2 & -1 \end{bmatrix}.$$

This is derived from input and output matrices

$$A = \begin{bmatrix} 0 & 0 \\ 0 & \frac{1}{2} \\ 2 & 1 \end{bmatrix} \qquad B = \begin{bmatrix} 1 & 2 \\ 1 & 0 \\ 0 & 0 \end{bmatrix}.$$

Consider optimum production with

$$p = [1 \quad 0 \quad 0], \qquad c = \begin{bmatrix} 0 \\ 0 \\ -1 \end{bmatrix}.$$

Only commodity 1 has direct market value, and only commodity 3 is available as a resource. The program is

$$\begin{aligned} &\max y_1 + 2y_2 \\ \text{S.T.} \quad &-y_1 - 2y_2 \leqq 0 \\ &-y_1 + \tfrac{1}{2}y_2 \leqq 0 \qquad y_1, y_2 \geqq 0. \\ &2y_1 + y_2 \leqq 1 \end{aligned}$$

It is clear that the first constraint is always ineffective, so that the first dual variable is always zero. The dual can then be written

$$\begin{aligned} &\min w_3 \\ \text{S.T.} \quad &-w_2 + 2w_3 \geqq 1 \qquad w_2, w_3 \geqq 0. \\ &\tfrac{1}{2}w_2 + w_3 \geqq 2 \end{aligned}$$

The solution to the program is easily seen to be

$$\begin{aligned} y &= (\tfrac{1}{4}, \tfrac{1}{2}) \\ x &= (1\tfrac{1}{4}, 0, -1) \\ w &= (0, 1\tfrac{1}{2}, 1\tfrac{1}{4}). \end{aligned}$$

Some features of the solution might be noted. Both activities are used because it is efficient, given the program, to produce some x_2 for use as an intermediate product. x_2 has a positive shadow price suggesting that, if a supply of it were available it would be used. Suppose the resource vector is changed to

$$c' = \begin{bmatrix} 0 \\ -1 \\ -1 \end{bmatrix}.$$

The second and third primal constraints become (the first remains ineffective)

$$\begin{aligned} -y_1 + \tfrac{1}{2}y_2 &\leqq 1 \\ 2y_1 + y_2 &\leqq 1. \end{aligned}$$

The solution to this program and its dual is

$$y = (0, 1)$$
$$x = (2, -\tfrac{1}{2}, -1)$$
$$w = (0, 0, 2).$$

Now the first activity, which was used initially to produce x_2 for use in the second activity, is no longer required, since x_2 is available as a primary resource. Actually, the system cannot use all the x_2 which is available, so it has a zero shadow price. Activity 1 is not profitable here because one of its outputs, x_2, has no value and the value of x_1 produced is less than the value of x_3 used, at the optimum shadow prices.

7.5 CONSUMPTION AS AN ACTIVITY[7]

The author has suggested elsewhere an approach to the theory of consumer behavior in which the direct objects of consumer preferences and utility are not goods as such, but the *characteristics* or properties of goods.

In the simplest version of the theory, each good is assumed to possess characteristics (assumed measurable) in fixed proportions, with the quantities of the characteristics directly proportional to the quantities of the goods. Further, the characteristics associated with a linear combination of goods are assumed to be a linear combination of the characteristics of the individual goods. These relationships between goods and characteristics are assumed to be, in principle, objective and the same for all consumers.

Thus we associate with each good a *vector of characteristics* b^j (for good j) such that b_{ij} is the amount of the ith characteristic possessed by unit quantity of the jth good. The vector b^j has all the analytical properties of an activity vector in production theory. If we assemble all the vectors b^j into a matrix B, we can refer to this as the *consumption technology* matrix. If the number of distinct characteristics is r and the number of distinct goods is n, then B is of order $r \times n$. In principle r and n may have any relationship, but we shall usually assume that $r < n$.

Denoting the vector of characteristics by z and that of goods by x, we have the following fundamental relationship

$$z = Bx.$$

In this simplest model, we assume that characteristics are associated in an essentially positive way with goods, so that B is a semipositive matrix. B is analogous to the output matrix in the production model. The chief analytical difference between this and the typical production model is that here

[7] This section is a development of ideas set out in Lancaster [3] and [4].

the typical activity has a single input (a good) and several joint outputs (characteristics), the reverse of a typical production activity.

If the consumption technology matrix is square and can be permuted into diagonal form, we have traditional consumption theory as a special case in which each good has associated with it a single characteristic which is unique to that good, so that the unique characteristic of coffee is "coffeeness."

In the typical consumer choice situation, the consumer can be considered as behaving as though maximizing a utility function subject to a budget constraint on goods. In this model, for given prices p, income k and consumption technology B, the consumer's maximizing behavior can be described as solving the following nonlinear program[8]

$$\max u(z)$$
$$\text{S.T.} \quad z = Bx, \qquad px \leqq k, \qquad x \geqq 0.$$

We can substitute for z in $u(z)$ and obtain the program, $\max u(Bx)$, S.T. $px \leqq k$, $x \geqq 0$. This can be investigated by the methods of Chapter 5. Since the utility function is private to every consumer, however, we have a different program for each individual.

Our interest here is in the use of activity analysis to investigate the set of characteristic vectors attainable for a given budget constraint. The importance of taking this approach arises from the assumed objective and universal nature of the consumption technology. Since all consumers face the same prices, the *attainable characteristics set*,

$$Z = \{z \mid z = Bx, \qquad px \leqq k, \qquad x \geqq 0\},$$

which is obviously a closed bounded convex set, is the same for all consumers except for a scalar multiple representing the size of the income k. Thus we can investigate those properties of the consumption choice which are the same for all individuals.

Since we are assuming, in the present version, that B is semipositive and x is constrained to be nonnegative, Z lies in the positive orthant. Assuming that $\partial u/\partial z_i > 0$ for all individuals and all characteristics (again, a special assumption of the simple model), no consumer will choose $z \in Z$ if there is some other $z' \in Z$ such that $z' \geqslant z$. Thus we have the idea of *efficient consumption*, analogous to efficient production.

The consumer's choice can thus be considered as having two phases. The *efficiency choice*, which will be the same for all consumers, will select efficient points of Z, then the *personal choice*, in which each consumer will choose that point in the efficient subset of Z which he, personally, prefers.

[8] This section can be regarded as an illustration of how linear programming theory can be used in a nonlinear problem.

The general nature of the set Z and the efficiency and personal choices is illustrated in Figure 7–4. There are two characteristics z_1, z_2 and five goods, the characteristics associated with the quantities of the goods that can be purchased by spending the whole budget on the good in question being given by the points $G_1 - G_5$. The set Z is the shaded area (note that, in this case, Z does not extend to the axes) and the efficient points are the heavily drawn portion of the boundary of Z. At the ruling prices, G_5 will not be purchased by any consumer. Personal choice is illustrated by indifference curves or three different consumers, giving choices A, B, C.

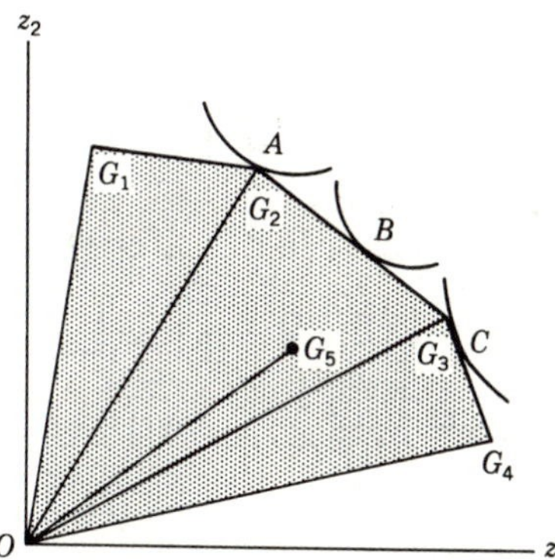

Figure 7-4. Efficiency and personal choices.

It is obvious that the goods collection which gives an efficient point z^* is the solution of the linear program

$$\begin{aligned} &\min p^*x \\ \text{S.T.} \quad &Bx = z^*, \qquad x \geqq 0, \end{aligned}$$

for which the scalar value of z^* is adjusted so that optimal x^* gives $p^*x^* = k$. (We are assuming $r < n$, so that the constraints can be satisfied as equalities.)

From ordinary linear programming theory we can immediately assert: *any efficient characteristics vector can be attained by consuming at most r different goods, where r is the number of distinct characteristics.*

Now consider the effect of a change from p^* to p^{**} on the set of efficient points and on the corresponding goods combinations. If z^* remains efficient, the new goods collection will be optimal in the program

$$\begin{aligned} &\min p^{**}x \\ \text{S.T.} \quad &Bx = z^*, \qquad x \geqq 0. \end{aligned}$$

Since the constraints in the new program are the same as in the old, we have, from the fundamental lemma of linear programming,

$$p^{**}x^{*} \geqq p^{**}x^{**}$$
$$p^{*}x^{**} \geqq p^{*}x^{*},$$

where x^{**} is the solution to the second program.

If we replace x^*in the first program by solution λx^* we obtain solution λx^*. Choose λ so that $\lambda p^{**}x^{*} = p^{**}x^{**}$. From the first of the above inequalities this implies $\lambda \leqq 1$. From the second inequality we will have $p^{*}x^{**} \geqq \lambda p^{*}x^{*}$, the direction of the inequality being unchanged since $\lambda \leqq 1$.

Now let us simplify notation by writing $x = \lambda x^*$, $x' = x^{**}$, $p = p^*$, $p' = p^{**}$.[9] From above we then have

$$p'x' = p'x$$
$$px' \geqq px,$$

with equality only if $x' = x$.

Thus the optimal goods collections x, x' can be recognized immediately as satisfying Samuelson's *Revealed Preference Axiom.*[10] If we reverse the signs in the second relationship and add to the first, we obtain the *generalized substitution effect*

$$(p' - p)(x' - x) \leqq 0,$$

and, if only the jth price changes, the *ordinary substitution effect*,

$$\Delta p_j \cdot \Delta x_j \leqq 0,$$

both for price changes compensated so that the original goods collection can just be purchased with the new income at the new prices.

The above substitution effect takes place without change in the proportions of the characteristics consumed (the proportions being given by z^*), and is a property of all efficient points in Z. This effect, the *efficiency substitution effect*, is due simply to changing the proportions of goods used to derive given characteristics in response to changes in relative goods prices, and so is *objective and universal.*

In addition to the efficiency substitution effect (which is necessarily zero in the traditional case in which there are no alternatives for obtaining given

[9] The notation x' here does not mean the transpose of x, but is merely an identifying label.

[10] For a discussion of revealed preference, see many price theory texts, for example Henderson and Quandt, or Samuelson himself, for example Samuelson [1]. The relationship above is the so-called weak axiom.

characteristics), there is the usual *personal* (or *private*) *substitution* effect due to the convexity of individual preferences. To investigate this, consider the dual of the program we have been examining

$$\max wz^*$$
$$\text{S.T.}\quad wB \leqq p^*,$$

For optimal w^* we have, from the duality theorem,

$$w^*z^* = p^*x^* = k.$$

Thus we can consider the consumer as *locally* maximizing $u(z)$ subject to the budget constraint $w^*z \leqq k$. The shadow prices w^* are actually relevant to all consumers whose optimal choices use the same basis of B.

We wish now to examine the choice of a consumer when there is no efficiency substitution effect, that is, when we constrain the consumer to the same basis. Let $\hat{B}$ be the basis for some initial choice z, subject to a budget constraint with prices p. The relevant shadow prices on characteristics will then be given by $\hat{w}\hat{B} = p$, where $\hat{w}$ is the subvector of *nonzero* shadow prices.

Now let prices change to p', with the consumer constrained to the basis $\hat{B}$. The consumer now faces *pseudo-shadow prices* (we use the term since we are preventing the attainment of a fully optimal situation) $\hat{w}'$. Let the consumer's income be adjusted so that he can just buy his original characteristics vector z even though we allow no basis change. Then, if the consumer's preferences satisfy the usual convexity assumption we will have the generalized substitution effect on characteristics

$$(\hat{w}' - \hat{w})(z' - z) \leqq 0,$$

where z' is the new characteristics choice.

Since $\hat{B}$ is a square nonsingular matrix, we have $\hat{w} = p\hat{B}^{-1}$, $\hat{w}' = p'\hat{B}^{-1}$, and also $z = \hat{B}x$, $z' = \hat{B}x'$. Substituting in the inequality above, we obtain

$$\begin{aligned}(\hat{w}' - \hat{w})(z' - z) &= (p' - p)\,\hat{B}^{-1}\hat{B}(x' - x)\\ &= (p' - p)(x' - x),\end{aligned}$$

so that

$$(p' - p)(x' - x) \leqq 0.$$

Thus convexity of preferences implies a substitution effect on goods even when no efficiency substitution effect is permitted.

Finally, we need to show that, if full adjustment is allowed, the consumer is able to buy the original characteristics at the new shadow prices after his income has been adjusted so that he can just buy the original goods at the new goods prices.

Let prices change from p to p' and income be adjusted so that $p'x = p'x'$, where x is the original goods collection and $p'x'$ the new (adjusted) income. Let w, z be the old, and w', z' the new, shadow prices and characteristics, with full adjustment permitted. Use of the duality theorem and the fundamental lemma in the relevant programs gives

$$w'z' = w'Bx' = p'x',$$
$$w'z = w'Bx \leqq p'x.$$

But $p'x' = p'x$, so that

$$w'z' \geqq w'z,$$

and the adjusted income is more than sufficient to permit the purchase of the original characteristics at the new shadow prices.

Thus we have shown the existence of two substitution effects, independent and working in the same direction, the objective and universal efficiency substitution effect and the personal substitution effect. For an uncompensated price change there will be two income effects, the efficiency income effect and the personal income effect, also working in the same direction.

The simplified version of the model which has been set out and analyzed here can be generalized to include negative (undesired) characteristics and to include the sale of personal services (labor) with associated negative characteristics. It can also be generalized to include activities requiring joint inputs of several goods (complementarity in consumption). In the more general model we separate activities (y) from goods and characteristics to obtain a two-part technology,

$$z = By,$$
$$x = Ay,$$

where A is the goods input matrix. The analogy to the general production model is obvious.

FURTHER READING

For activity analysis with a more geometric orientation, read the exposition in Koopmans [2], then the full account in Koopmans [3]. There are related papers in Koopmans [1], and also in Morgenstern. Allen [2] also discusses activity analysis in terms of its original formulation. For the material on consumption, consult Lancaster [3] and [4].

The material of this chapter is used, directly or indirectly, through Chapters 8, 9, 10, and 11.

EXERCISES

1. Show that, if there are r *essentially primary* inputs that are not produced within the system at all, an efficient point requires the use of not more than r activities. Relate this result to the substitution theorem for the Leontief model (Section 6.7 of the preceding chapter).

2. For a system with n goods, a *facet* of the production cone is a nonnegative linear combination of any set of less than n activities. Prove that, if a facet contains an interior point that is efficient (an interior point being a strictly positive linear combination of the activities defining the facet), then all points on the facet are efficient. Such a facet is referred to as an *efficient facet*.

3. Prove that an efficient facet is such that the relationships,

$$pF = 0,$$
$$pT \leqq 0,$$

(where T is the technology matrix and F the submatrix of T defining the facet) hold for some $p \gg 0$.

4. Show that the normal (at the origin) to an efficient facet is a strictly positive vector.

CHAPTER

8

Nonlinear Optimizing Models

This chapter should be read only after Chapter 4 (including a careful study of Sections 4.5 and 4.6) and Chapter 7. Essential background material which the reader may not already have studied is the theory of quadratic forms (Review R6) and of homogeneous functions (Review R8, Section R8.6).

8.1 INTRODUCTION

This chapter is concerned with the analysis of nonlinear demand and production models. Examples of the use of both neoclassical and set theory methods are given in both demand and production theory, to provide the reader with some comparison of the strengths and limitations of each type of analysis.

The neoclassical demand analysis given in Section 8.2 is a standard and simple example of this approach. The neoclassical transformation surface, the subject of Sections 8.4 through 8.6, requires deeper analysis and depends on a knowledge of quadratic forms. The simple set theory approaches to demand and production theory given in Sections 8.3 and 8.7 present no difficulty, but Section 8.7 draws on the activity analysis approach to production set out in Sections 7.1 through 7.4 of the preceding chapter.

8.2 NEOCLASSICAL DEMAND THEORY[1]

Neoclassical demand theory originated with Slutsky's[2] classic paper but owes much of its development and most of its popularization (if that can be considered the appropriate word) to the work of Hicks.[3] It is probably the greatest single achievement of strictly neoclassical mathematical economics, based on calculus and determinants.

The analysis contains three main propositions:

(a) The *Slutsky equation*, setting out the effect of an uncompensated change in a single price on the demand for each good in the system;

(b) A proof that the Slutsky equation is invariant to changes in the utility index;

(c) The *composite commodity theorem* establishing that a group of goods whose relative prices do not change can be analyzed in a similar manner to a single good.

In this analysis, the individual consumer is assumed to have a continuous (class C^2) utility function $u(x)$ defined on an n vector of goods and to be constrained to buy at given prices from a fixed money income M. The n goods of the model are assumed to comprise the consumer's universe, so that he spends all his income on those goods. His actual behavior is assumed to be as if he solved the following classical optimizing problem,

$$\begin{aligned} &\max u(x) \\ \text{S.T.}\quad &px = M. \end{aligned}$$

Using ordinary Lagrangean methods, the first-order conditions for optimum are

$$\begin{aligned} px - M &= 0, \\ u_i - \lambda p_i &= 0. \end{aligned} \tag{8.2.1}$$

It is assumed that the solution and the properties of u are such that a proper maximum is reached with $x \gg 0$, so that the determinantal conditions

[1] This is a standard analysis. See Hicks, Mosak or Samuelson [1] for equivalent expositions, or the original article by Slutsky.

[2] See Slutsky. The composite commodity theorem was due to Hicks and does not appear in Slutsky.

[3] That is, in Value and Capital, rather than the earlier article with Allen.

(discussed in Chapter 4, Section 4.5) for the appropriate second-order conditions[4] are satisfied.

We now consider the effect of a small change in one price that we take to be p_n on the consumer's optimal position. The optimal conditions are satisfied throughout the change, so we take derivatives with respect to p_n through the first-order conditions (8.2.1) to obtain, after some rearrangement

$$\begin{aligned} \lambda \sum_{j=1}^{n} p_j(\partial x_j/\partial p_n) &= -\lambda x_n \\ -p_i(\partial \lambda/\partial p_n) + \sum_{j=1}^{n} u_{ij}(\partial x_j/\partial p_n) &= 0 \qquad i = 1, \ldots, (n-1) \qquad (8.2.2) \\ &= \lambda \qquad i = n \end{aligned}$$

We can substitute $u_i = \lambda p_i$ from (8.2.1) and write the equations in matrix-vector form:

$$\begin{bmatrix} 0 & u_1 & u_2 & \cdots & u_n \\ u_1 & u_{11} & u_{12} & \cdots & u_{1n} \\ u_2 & u_{12} & u_{22} & \cdots & u_{2n} \\ \cdot & \cdot & \cdot & \cdot & \cdot \\ u_n & u_{1n} & u_{2n} & \cdots & u_{nn} \end{bmatrix} \begin{bmatrix} -(1/\lambda)(\partial \lambda/\partial p_n) \\ (\partial x_1/\partial p_n) \\ \cdot \quad \cdot \quad \cdot \quad \cdot \\ \cdot \quad \cdot \quad \cdot \quad \cdot \\ (\partial x_n/\partial p_n) \end{bmatrix} = \begin{bmatrix} -\lambda x_n \\ 0 \\ 0 \\ \cdot \\ \lambda \end{bmatrix}. \qquad (8.2.3)$$

The matrix of this equation system is simply the bordered Hessian of u, which we write $\hat{U}$.

We solve for $\partial x_r/\partial p_n$ by using Cramer's Rule (see Review R5, Section R5.2). This involves replacing the $(r + 1)$th column of $\hat{U}$ by the constant vector, then dividing the determinant of the result by det $\hat{U}$. Since the constant vector contains only two nonzero elements, the first and last, we have

$$\partial x_r/\partial p_n = \frac{1}{\det \hat{U}}\left(-\lambda x_n U_r + \lambda U_{rn}\right), \qquad (8.2.4)$$

where

U_r is the cofactor of u_r in det $\hat{U}$,
U_{rn} is the cofactor of u_{rn} in det $\hat{U}$.[5]

Now return to the first order conditions (8.2.1) and consider the effect of

[4] These conditions are inappropriately called "stability conditions" by Hicks and many of the earlier investigators.

[5] Note that U_r, U_{rn} are the cofactors of u_r, u_{rn}, not the cofactors of the elements occupying the $(1, r)$th and (r, n)th places in det $\hat{U}$, because of the bordering in $\hat{U}$.

a change in M with prices constant. Taking derivatives through these equations we obtain, in a similar way to the way we obtained (8.2.2),

$$\begin{aligned} \lambda \sum_{j=1}^{n} p_j(\partial x_j/\partial M) &= \lambda \\ -p_i(\partial\lambda/\partial M) + \sum_{j=1}^{n} u_{ij}(\partial x_j/\partial M) &= 0. \qquad i = 1,\cdots,n \end{aligned} \tag{8.2.5}$$

If we substitute $u_i = \lambda p_i$ as before, it is easy to see that the equations, in matrix-vector form, have the same structure as those in (8.2.3) except that the constant vector now contains only the single nonzero element λ, occupying the first place. Solving for $\partial x_r/\partial M$ by Cramer's Rule, we therefore obtain the simple result:

$$\partial x_r/\partial M = \frac{\lambda U_r}{\det \hat{U}}. \tag{8.2.6}$$

Write $K_{rn} = \lambda U_{rn}/(\det \hat{U})$ and substitute from (8.2.6) in (8.2.4) we obtain

The Slutsky Equation: $\quad \partial x_r/\partial p_n = -x_n(\partial x_r/\partial M) + K_{rn}. \quad$ (8.2.7)

The term $-x_n(\partial x_r/\partial M)$ is the *income term*, the term K_{rn} is often referred to as the *Slutsky term.*

K_{rn} can be shown to have various properties, including that $K_{rn} = K_{nr}$ (because $\hat{U}$ is symmetric). However, the real interest is in the case $r = n$.

Now $K_{nn} = \lambda U_{nn}/(\det \hat{U})$ and has already been identified (see Chapter 4, Section 4.6) as the *pure substitution effect*, that is, the change in x_n resulting from a change in p_n when income is so adjusted that $u(x)$ is kept constant. It is always negative, if $u(x)$ has the properties assumed of it.

Thus we can divide the effects of an uncompensated price change into *income* and *substitution effects*, illustrated by writing the demand equation (8.2.7) in the form,

$$\left(\frac{\partial x_n}{\partial p_n}\right)_{M=M_0} = -x_n\left(\frac{\partial x_n}{\partial M}\right)_{p=p_0} + \left(\frac{\partial x_n}{\partial p_n}\right)_{u=u_0} \tag{8.2.8}$$

Now we shall turn to the task of showing that the above analysis is invariant to changes in the utility index.[6] Replace the index u by the index

[6] Note we replace the utility index, not the utility function. That is, we renumber the indifference curves (the contours of u) without either changing their shape or their order.

$v(u)$, where we assume v' $(= du/dv) > 0$. We have

$$v_i = v'u_i,$$
$$v_{ij} = v'u_{ij} + v''u_iu_j.$$

Let $\hat{V}$ be the bordered Hessian of v, corresponding to $\hat{U}$. The first column of $\hat{V}$ is simply $v'\hat{U}^1$, where $\hat{U}^1$ is the first column of $\hat{U}$. The $(j+1)$th column of $\hat{V}$ is given by

$$\hat{V}^{j+1} = \begin{bmatrix} v'u_j \\ v'u_{1j} + v''u_ju_1 \\ \cdot \quad \cdot \quad \cdot \quad \cdot \\ v'u_{nj} + v''u_ju_n \end{bmatrix}$$

$$= v'\begin{bmatrix} u_j \\ u_{1j} \\ \cdot \\ u_{nj} \end{bmatrix} + v''u_j \begin{bmatrix} 0 \\ u_1 \\ \cdot \\ u_n \end{bmatrix}$$

$$= v'\hat{U}^{j+1} + v''u_j\hat{U}^1.$$

Since adding a multiple of a column of a determinant to a different column leaves the value of the determinant unchanged, and since multiplying each of $(n+1)$ columns of a determinant by v' multiplies the determinant by $(v')^{n+1}$, we have

$$\det \hat{V} = (v')^{n+1} \det \hat{U}.$$

Applying the same arguments to the cofactors V_r, V_{rn}, we find

$$V_r = (v')^n U_r, \qquad V_{rn} = (v')^n U_{rn}.$$

Now consider the first order conditions with the new utility index, with which will be associated a Lagrange multiplier μ, not necessarily the same as λ. We have

$$v_i = \mu p_i.$$

But from the first order conditions with the original utility index, we have $(1/\lambda)u_i = p_i$, and we also have $v_i = v'u_i$, so that $\mu = v'\lambda$.

Using all the information we have assembled, it follows that

$$\lambda U_r/(\det \hat{U}) = \mu V_r/(\det \hat{V})$$
$$\lambda U_{rn}/(\det \hat{U}) = \mu V_{rn}/(\det V),$$

so that the Slutsky equation is invariant with respect to a change in the utility index.

The *composite commodity theorem* can be proved by the use of the foregoing methods, but this requires some special properties of determinants. This theorem is better proved from the *generalized substitution theorem* which is demonstrated in the next section, using different methods from those used here.

The analysis can be extended very simply to cover the situation in which the consumer is endowed with some initial vector of goods x^0, rather than with a fixed money income.

The demand for inputs by a firm, given a neoclassical production function, has exactly the same analytical form as the demand for goods by a consumer. Instead of a utility function $u(x)$, we have a production function $x = f(v)$, where v is a vector of inputs. For given factor prices, the firm is assumed to maximize the output that can be obtained for any given outlay on factors. Apart from replacing the bordered Hessian $\hat{U}$ by the bordered Hessian of the production function, $\hat{F}$, and renaming the goods as factors, all the above analysis holds.

8.3 CONVEXITY PROOF OF THE SUBSTITUTION THEOREM[7]

A more efficient way of proving the negativity of the pure substitution effect in neoclassical demand theory than that set out in Chapter 4 (Section 4.6) is to use the properties of convex sets. This is simpler, more direct and gives a much more satisfactory proof of the generalized substitution effect than calculus methods.

We assume that the utility function has the following strict convexity property:

The set $S = \{x \mid u(x) \geqq u^0\}$ *is strictly convex.*[8]

This is certainly not a stronger assumption than the neoclassical assumption on $u(x)$ in the calculus method. It is, in fact, weaker, since continuity in the derivatives is not required.

We suppose that, as in the calculus proof, the consumer faces given prices p^* and chooses goods x^*, giving utility level $u(x^*) = u^0$. The prices now change to p^{**}, the consumer's income is adjusted so that his new optimal

[7] The reader may care to revise his knowledge of convex sets at this point (Review R4, Sections R4.1 through R4.4).

[8] Equivalent to assuming $u(x)$ to be a strictly *concave-contoured* function in the terminology of Review R8, Section R8.5. S is *strictly* convex if all its boundary points are also extreme points.

choice, x^{**}, just attains the original utility level u^0. The choice criterion is that x^*, x^{**} are such as to minimize $p^*x, p^{**}x$, respectively, over all $x \in S$. Denote $p^*x^*, p^{**}x^{**}$ by M^*, M^{**}.

Consider the hyperplane H defined by $p^*x = M^*$. This divides commodity space into the two half-spaces,

$$H^+ = \{x \mid p^*x \geqq M^*\},$$
$$H^- = \{x \mid p^*x \leqq M^*\}.$$

Now there can be no point of S in the *interior* of H^- since, if there were, we could find some $m < M^*$ such that there was a point of S in the set $\{x \mid p^*x \leqq m\}$, contradicting the minimum property of M^*.

Thus the only points of S in H^- are those actually in H, so that S lies wholly in H^+. That is, H is the *supporting hyperplane* of the strictly convex set S at the point x^*.

Since $x^{**} \in S$, $x^{**} \in H^+$ and either $x^{**} = x^*$, or x^{**} is in the interior of H^+. Thus we have

$$\text{either } x^{**} = x^* \qquad \text{or} \qquad p^*x^{**} > M^* = p^*x^*.$$

We can argue in precisely the same way for the hyperplane

$$H' = \{x \mid p^{**}x = M^{**}\},$$

which must be a supporting hyperplane of S at the point x^{**}, so that S lies within the positive half-space. Thus we have

$$\text{either } x^{**} = x^* \qquad \text{or} \qquad p^{**}x^* > M^{**} = p^{**}x^{**}.$$

Now a strictly convex set may have more than one supporting hyperplane (for example H, H') through one point, so the case $x^{**} = x^*$ cannot necessarily be ruled out. However, we will assume it not to be the case unless $H = H'$ (this can happen only if all prices change proportionately), an assumption not stronger than requiring continuous first derivatives for u.

The two inequalities we have can be written

$$p^{**}x^{**} - p^{**}x^* < 0$$
$$-p^*x^{**} + p^*x^* < 0.$$

Adding the inequalities, we obtain the *generalized substitution theorem*

$$(p^{**} - p^*)(x^{**} - x^*) < 0.$$

If the price of only the jth good changes, we have the ordinary own-price substitution effect

$$\Delta p_j^* \cdot \Delta x_j^* < 0.$$

We can use the generalized substitution effect to prove the *composite commodity theorem*, as was stated in the preceding section.

Define some composite good G as follows

$$G = \hat{w}\hat{x},$$

where $\hat{w}$, $\hat{x}$ are r-vectors ($r < n$). It is assumed that the prices of the r goods included in G all move together and remain proportional to the weights $\hat{w}$, so that we can write $\hat{p}^* = P_G^*\hat{w}$, $\hat{p}^{**} = P_G^{**}\hat{w}$, where P_G (a scalar) can be regarded as the price of G.

Let the price of the composite good change, with prices of all goods not in the composite good unchanged. Then all the components of $(p^{**} - p^*)$ in the generalized substitution theorem which do not correspond to goods in G are zero. Thus we have

$$\begin{aligned}(p^{**} - p^*)(x^{**} - x^*) &= (P_G^{**} - P_G^*)\,\hat{w}\,(\hat{x}^{**} - \hat{x}^*)\\ &= \Delta P_G \Delta G,\end{aligned}$$

so that the composite good has a substitution effect of precisely the same kind as a single good.

The composite commodity theorem provides the justification for the simplification of n-commodity models to two or three composite commodities.

8.4 THE NEOCLASSICAL TRANSFORMATION SURFACE[9]

The neoclassical transformation surface, that is the set of all efficient combinations of n goods that can be produced with fixed total quantities of each of m factors, the production functions being of the smooth nonlinear neoclassical kind, is a concept of great importance in welfare economics and the theory of trade, as well as in other aspects of economic theory. The shape of this surface, particularly whether it is everywhere convex outwards, has implications of great importance. If this convexity property does not hold it creates problems in welfare economics, it casts doubts on the efficacy of the price system and it leads to unusual equilibrium configurations.

We shall examine the properties of this surface here, and analyze the conditions that give rise to the convex case, which we shall consider to be the "regular" case. We are primarily interested in *strict* convexity.

[9] No complete analysis along the lines of Sections 8.4 through 8.6 seems to exist in the literature, although the results are well-known for the two-good, two factor case.

Properties of quadratic forms (Review R6) are used extensively in this and the following two sections.

The economy consists of n goods and m primary factors, with the output of any good being given by a neoclassical production function of the kind

$$x_j = f^j(v_1^j, v_2^j, \ldots, v_m^j),$$

where v_r^j is the quantity of the rth factor used in the output of the jth good. We assume that f^j is an *industry-wide* production function, so that we have a single production function for each good. There are no intermediate goods, so that goods and factors are entirely separate.

The usual assumptions on the f^j's is that each has a proper minimum cost of producing each level of output for given factor prices. As pointed out in Section 8.2, the derivation of the appropriate conditions is analytically similar to the derivation of the conditions that the consumer has a proper minimum for the cost of attaining a given utility level. The conditions themselves are expressed in terms of the properties of $\hat{F}^j$, the bordered Hessian of f^j.

Given fixed total endowments of the m factors in the economy as a whole, it is clear that an efficient output vector x (that is, a point on the transformation surface) will be such as to maximize px for some suitable $p \geqslant 0$ subject to the constraints on factor supplies.

Thus we can consider any point on the transformation surface to be given by the solution to a classical optimizing problem,

$$\max \sum_j p_j x_j = \sum_j p_j f^j(v^j)$$

$$\text{S.T.} \quad \sum_j v^j = V,$$

where v^j is the vector $(v_1^j, \ldots, v_m^j)$ and V is the vector of factor endowments. The variables of the system are the mn variables v_r^j. We assume that we can find an optimum with all these variables positive so that we do not need to consider nonnegativity constraints. If necessary, we confine our analysis to that part of the transformation surface for which this is true.

The Lagrangean is

$$L(v^1, \ldots, v^n, \mu) = \sum_j p_j f^j - \sum_r \mu_r \Big(\sum_j v_r^j - V_r\Big).$$

The first order conditions are

$$\begin{aligned} \partial L/\partial \mu_r &= \sum_j v_r^j - V_r = 0 \qquad r = 1, \ldots, m \\ \partial L/\partial v_r^j &= p_j f_r^j - \mu_r = 0 \qquad j = 1, \ldots, n, r = 1, \ldots, m. \end{aligned} \tag{8.4.1}$$

The μ_r's are shadow factor prices. If actual factor prices are proportional to them, the above conditions are the usual marginal conditions for optimal production.

Our interest here is primarily in the second order conditions. The conditions that the transformation surface be convex outward are identical with the conditions that the optimizing problem have a proper maximum.

In the present analysis it is not enough merely to reproduce standard second order sufficient conditions for a maximum. We will learn much more by considering the second order conditions at the fundamental level, as the conditions that the quadratic form based on the matrix of second order partial derivatives of L (with respect to the v variables only) be always negative for appropriate constrained variations.

These second order partial derivatives have the form,

$$\begin{aligned} \frac{\partial^2 L}{\partial v_r^j \partial v_s^j} &= 0 \qquad j \neq k, \text{ all } r, s \\ \frac{\partial^2 L}{\partial v_r^j \partial v_r^j} &= p_j f_{rs}^j. \end{aligned} \tag{8.4.2}$$

If we order the variables so that all the variables v_r^1 come first, then all the variables v_r^2, and so on, it is easily seen that the matrix A of second order derivatives has the block diagonal form,

$$A = \begin{bmatrix} p_1F^1 & 0 & 0 & \cdot & 0 \\ 0 & p_2F^2 & 0 & \cdot & 0 \\ \cdot & \cdot & \cdot & \cdot & 0 \\ 0 & 0 & 0 & 0 & p_nF^n \end{bmatrix},$$

where F^j is the *unbordered* Hessian of the second-order derivatives of f^j. If y^j denotes the vector of differentials $(dv_1^j \cdots dv_m^j)$, and y the vector $(y^1, \ldots, y^n)$, then the quadratic form in which we are interested is

$$y'Ay = \sum_j p_j(y^j)'F^jy^j. \tag{8.4.3}$$

This is subject to the constraints

$$\sum_j dv_r^j = 0, \qquad r = 1, \cdots, m$$

which we can write

$$\sum_j y^j = [0] \qquad \text{(this is a vector equation).}$$

We could also write the constraints

$$[I:I:\cdots:I]\,y = 0,$$

where the matrix is made up of n identity matrices of order m put side by side.

Thus the transformation surface is convex outward at the point corresponding to the prices p if

$$y'Ay < 0 \qquad \text{for} \qquad [I:I:\cdots:I]\,y = 0;$$

that is if

$$\sum_j p_j(y^j)'F^jy^j < 0 \qquad \text{for} \qquad \sum_j y^j = 0. \tag{8.4.4}$$

It is obviously sufficient that $(y^j)'F^jy^j$ be negative for all j and for all y^j satisfying the constraints. Since the constraints, in a large system, do not bear very heavily on any one industry, it is not too much stronger to give the following:

Condition I for regular transformation surface. It is sufficient, but not necessary, that the Hessian of every production function be negative definite.

Now $y'Ay$ must be negative for any choice of y, with at least some components nonzero, which satisfies the constraints. Consider putting the vectors y^j equal to zero. Obviously we can make only $(n - 2)$ zero and still leave at least one nonzero y^j, because of the constraints. Let all the vectors be zero except y^j, y^k. From the constraints we have $y^k = -y^j$ and so, for this particular y

$$\begin{aligned} y'Ay &= p_j(y^j)'F^jy^j + p_k(-y^j)'F^k(-y^j) \\ &= (y^j)'(p_jF^j + p_kF^k)\,y^j. \end{aligned} \tag{8.4.5}$$

The constraints have already been taken care of, so the last quadratic form is unconstrained. Thus we have

Condition II for regular transformation surface. It is necessary, but not sufficient, that the weighted sums of the Hessians of any two production functions be negative definite, where the weights are the prices appropriate to the particular point on the transformation surface.

This condition implies, among other things, that not more than one of the production functions can have a *positive definite* Hessian. Even if there is only one industry with this property, then there will be some price ratio for which the sum $p_kF^j + p_kF^k$ is not negative definite and so we will not have the regular curvature *everywhere* on the transformation surface.

The above conditions have been expressed in terms of the *unbordered* Hessians of the production functions. The standard conditions on the production functions are expressed in terms of the *bordered* Hessian $\hat{F}^j$. Now it is clear that the standard second order conditions on the production function are *necessary* for the corresponding conditions on the unbordered Hessian.

We obtained the standard conditions by investigating the circumstances under which $(y^j)'F^jy^j$ is negative for variables satisfying the constraints $\sum f_i^j y_i^j = 0$. This condition must also be satisfied for $(y^j)'F^jy^j$ to be negative for *unconstrained* variables. The standard condition is not sufficient, however. This can be illustrated by considering the two-factor production function with Hessian

$$F = \begin{bmatrix} f_{11} & f_{12} \\ f_{12} & f_{22} \end{bmatrix}.$$

The condition that F be negative definite is that $f_{11}, f_{22} < 0$ and $f_{11}f_{22} - (f_{12})^2 > 0$. The standard conditions on f for the constrained variation is that $(f_2)^2f_{11} + (f_1)^2f_{22} - 2f_1f_2f_{12}$ be negative. A large positive value of f_{12} will satisfy the second condition but deny the first.

The significance of the quadratic form based on the unbordered Hessian will be apparent from the analysis of the next section.

8.5 RETURNS TO SCALE[10]

Given a production function $f(v)$, if we take some initial input vector v^* and then compare $f(tv^*)$ with $tf(v^*)$ for all $t > 0$, we have the following definitions:

The production function is said to show decreasing, constant, or increasing, returns to scale for input proportions v^* according as $f(tv^*)$ is less than, equal to, or greater than, $tf(v^*)$ for all $t > 1$, with the inequalities reversed for $t < 1$.[11]

If the relevant condition holds for all v^*, then the function is said to show decreasing, constant or increasing returns *everywhere.*

We shall investigate the properties of returns to scale by using the properties of homogeneous functions. The function $f(v)$ is homogeneous of degree ρ if $f(tv) = t^\rho f(v)$ for all v. Clearly we have decreasing, constant or increasing returns to scale for a homogeneous function according as ρ is less than, equal to, or greater than 1.

The properties of homogeneous functions are discussed in Review R8

[10] Familiarity with the properties of homogeneous functions (Review R8, Section R8.6) is required here.

[11] That is, we have decreasing returns to scale at proportions v^*, if $f(tv^*) < tfv(^*)$ for all $t > 1$, and $f(tv^*) > tf(v^*)$ for $0 < t < 1$.

(Section R8.6). We shall make use here of Euler's Theorem and of the result, proved in Section R8.6, that, for homogeneous function $f(v)$ of degree ρ,

$$(\rho - 1)f_i = \sum_j f_{ij}v_j \qquad \text{all } i,$$

or, in matrix-vector form,

$$(\rho - 1)\nabla f = v'F,$$

where ∇f is the gradient vector and F the (unbordered) Hessian of f.

We shall assume the usual concave-contoured production function, so that the quadratic form $u'Fu < 0$ for small movements u ($u \neq 0$) along the contour $f(v) = c$, that is, for which $\nabla f \cdot v = 0$.

Now any arbitrary small change from v to $v + y$ can be considered as the sum of two parts:

(a) A small change u along a contour, satisfying the foregoing relationship;
(b) A proportional change from v to $v + \lambda v$.

Thus we have $y = u + \lambda v$. We can now consider the quadratic expression

$$\begin{aligned} y'Fy &= (u + \lambda v)'F \cdot (u + \lambda v), \\ &= u'Fu + \lambda v'Fu + u'F \cdot \lambda v + \lambda v'F \cdot \lambda v, \\ &= u'Fu + 2\lambda v'Fu + \lambda^2 v'Fv \qquad \text{(since } F \text{ symmetric)}, \\ &= u'Fu + v'F \cdot (\lambda^2 v + 2\lambda u). \end{aligned}$$

Using the result given above for homogeneous functions, we have

$$\begin{aligned} v'F \cdot (\lambda^2 v + 2\lambda u) &= (\rho - 1)\nabla f \cdot (\lambda^2 v + 2\lambda u) \\ &= (\rho - 1)[\lambda^2 \nabla f \cdot v + 2\lambda \nabla f \cdot u]. \end{aligned}$$

But, from the properties of u, we have $\nabla f \cdot u = 0$, and from Euler's theorem we have $\nabla f \cdot v = \rho f(v)$, so that

$$y'Fy = u'Fu + \lambda^2 \rho(\rho - 1) f(v).$$

Unless $u = 0$, we have $u'Fu < 0$, from the contour properties, and, since $f(v)$ is essentially positive, we have

$$\lambda^2 \rho(\rho - 1) f(v) \leqq 0,$$

for $\rho \leqq 1$, with the strict inequality for $\rho < 1$. Thus the quadratic expression $y'Fy$ is nonpositive for all u, λ. If $\rho = 1$, $y'Fy$ is zero, if and only if $y = \lambda v$ ($u = 0$). If $\rho < 1$, $y'Fy$ is negative for all u, λ (other than $u = 0$ and $\lambda = 0$).

But, by suitable choice of u subject to $\nabla f \cdot u = 0$ and of λ, any arbitrary small vector y can be expressed in the above fashion, so that F is negative semidefinite if $\rho = 1$ and is negative definite if $\rho < 1$. We can now state a result of fundamental importance for neoclassical production theory.

If $f(v)$ is a concave-contoured production function, homogeneous of degree ρ, it has a Hessian which is: (a) negative semidefinite for $\rho = 1$ (constant returns to scale; (b) negative definite for $\rho < 1$ (decreasing returns to scale).

That is, the production function is a *concave* function if it shows non-increasing returns to scale, and a *strictly concave* function, if it shows decreasing returns to scale.

From Condition I of the Section 8.4, we can now assert:

An economy consisting entirely of industries with homogeneous concave-contoured production functions showing decreasing returns to scale will have a regular (strictly convex outward) transformation surface.

Since *constant* returns to scale is the most common assumption on neo-classical production functions, we still have not yet established the full conditions for the existence of a regular transformation surface. The special additional assumptions required for the constant returns case will be taken up in the next section.

If there are increasing returns ($\rho > 1$), there is little that can be said about the transformation surface. In general, it will have a complex shape, and may well be regular over some regions.

8.6 RELATIVE FACTOR INTENSITY

In this section we shall examine the additional assumptions required in the constant returns to scale case to guarantee a transformation surface that is *strictly* convex outward.

From the preceding Section, we know that the Hessians of homogeneous concave-contoured constant returns production functions are negative semi-definite and that the quadratic expression $y'Fy$ is zero only for $y = \lambda v$, where v represents the point at which the Hessian F is evaluated.

From (8.4.3) of Section 8.4 the quadratic form for the transformation surface is thus necessarily either negative semidefinite or negative definite. If we can rule out the possibility that $y'Ay = 0$ for any choice of putting components y_r^j equal to zero, subject to the constraints, then the transformation surface will be strictly convex outward.

Now there are two ways of putting the y_r^j's equal to zero. If we put $y_r^j = 0$ but leave $y_s^j \neq 0$, then we know from the properties of F^j that $(y^j)'F^jy^j < 0$ (since this cannot be a pure scale change). If we put $y_r^j = 0$ for all r, then we remove the Hessian F^j from exercising any effect on the overall result. This can be done for $n - 2$ industries at most, because of the constraints, and those industries must then meet Condition II of Section 8.4.

Thus we need only to find conditions that guarantee the fulfilment of Condition II, that the weighted sum of the Hessians of any two production functions, $p_j F^j + p_k F^k$, be negative definite. Since F^j, F^k are both negative semidefinite, we need to show that $u'(p_j F^j + p_k F^k)\, u \neq 0$ for any u and this will guarantee that $u'(p_j F^j + p_k F^k)\, u < 0$ for all u.

Since $u'F^j u$, $u'F^k u \leqq 0$ for all u, we can have $u'(p_j F^j + p_k F^k)\, u = 0$ only if $u'F^j u = u'F^k u = 0$. The prices are irrelevant, except that they must not be negative.

From the discussion of the constant returns case in the Section 8.4, we know that $u'F^j u = 0$ only if $u = v$, the factor proportions appropriate to the point at which the Hessian is evaluated. This point is, of course, x_j, the jth component of the point on the transformation surface.

Thus the condition we need is that $v^j \neq \gamma v^k$ for a point on the transformation surface with components x_j, x_k.

From the first-order conditions (8.4.1) of Section 8.4, if x_j, x_k are components of a point on the transformation surface, then

$$p_j f_r^j = \mu_r = p_k f_r^k \qquad r = 1, \ldots, m,$$

where μ_r is the shadow price of the rth factor.

Now if two production functions are so related that $v^j = \gamma v_k$ for $f_r^j(v^j) = \beta f_r^k(v^k)$, $r = 1, \ldots, m$ that is, if the input vectors are proportional when the marginal productivity vectors are proportional (or, equivalently, when both optimize on the same shadow prices), we say they have the same *relative factor intensity*. Thus the condition is satisfied if industries j, k do *not* have the same relative factor intensities anywhere, and this must hold for all pairs of industries. We can now give our final statement on the transformation surface in the constant returns case:

It is sufficient for a regular (strictly convex) transformation surface in an economy consisting entirely of industries with constant returns to scale that: (a) Every production function be concave-contoured; and (b) no two production functions have the same relative factor intensities anywhere.

The above statement of conditions in the constant returns case is of great importance in international trade theory, where it is critical to problems of factor price equalization and the whole Heckscher-Ohlin analysis of trade.[12]

[12] For references to the relevant trade literature, and some discussion of the problems, see Bhagwati, Corden, and Chipman [2].

8.7 GENERALIZED PRODUCTION THEORY[13]

By using set-theoretic rather than calculus methods we can examine production theory at a more general level than was possible in the analysis of Sections 8.4 through 8.6. The use of these methods permits much easier proof of weak results, and proof of these at a more general level, than by the use of calculus. Strong results, like the conditions for strict convexity of the transformation surface under constant returns to scale, are not so amenable to the methods of this section.

We generalize from the picture of production given by activity analysis (Chapter 7, Sections 7.1 through 7.4). As in activity analysis, but unlike neoclassical production theory, we do not divide commodities into distinct classes of goods and factors, but assume any commodity might be either an input or an output.

Denote an n vector of commodities by y, where y_i is positive if it is an output, negative if it is an input. We change the notation from that used in Chapter 7, where the commodity vector was x and y represented the vector of activity levels. We do not use the idea of activity levels here, and we shall wish to use y for the production vector in the next chapter, where x is used for the demand vector.

An arbitrary vector y may be feasible or infeasible (equivalently attainable, unattainable; possible, impossible) under the given production technology. It is feasible if the outputs can actually be produced, in the quantities given in y, from inputs in quantities given by y.

The set Y of all feasible y's is the *production set* (directly analogous to the production set in the activity analysis model). A feasible y (that is, $y \in Y$) is referred to as a *production* or *supply*. The production set gives all the input-output combinations that are technically possible. A production y is analogous to an activity in the linear case with the following crucial differences:

(a) We do not assume that ky is necessarily feasible if y is;
(b) We do not assume that $y^j + y^k$ is necessarily feasible if y^j, y^k are both feasible.

As a consequence of (a) and (b) we cannot describe a production simply by unit level input-output proportions and an activity level, and we cannot assume that a linear combination of productions is necessarily feasible.

[13] This section builds on the activity analysis material of Chapter 7 and is a necessary preliminary to the discussion of Chapter 9.

In the linear case we were given a finite number of basic activities and were able to "fill in" the production set by taking all convex combinations of these. Here we cannot do this: every $y \in Y$ needs to be specifically given, so that we have an infinite number of these. Given the neoclassical production function $x_j = f^j(v)$, for example, we would have a different production $y = (0, 0, \ldots, f^j(v), \ldots, 0, v_1^j, \ldots, v_m^j)$ for every v. As a practical matter we would assume, of course, that the y's could be specified by a finite number of functional relationships, that is by generalized production functions.

As a consequence of the infinite number of productions, we shall assume throughout that continuity prevails, that is, that *Y is closed.*

We shall assume in all our analysis that Y satisfies the assumptions given in Section 7.2, Chapter 7, for linear production sets, namely,

I. $Y \cap \Omega = 0$ (*output without inputs is impossible, but complete inactivity* ($0 \in Y$) *is always possible*).
II. $Y \cap (-Y) = 0$ (*irreversibility*).
III. $y' \in Y$ *if* $y \in Y$ *and* $y' \leqq y$ (*wasteful production possible*).

We shall also usually assume, unless interaction between production processes is an essential feature:

IV. $y^i + y^j \in Y$ *if* $y^i, y^j \in Y$ (*additivity*).

Returns to scale can be defined for the general case as follows:

The production set Y shows weakly decreasing, constant, or weakly increasing, returns to scale according as, for nonnegative t, and all $y \in Y$.
$ty \in Y$ for $t < 1$, but $ty' \notin Y$ for some $y' \in Y$ and $t > 1$ (weakly decreasing returns);
$ty \in Y$ for all t (constant returns);
$ty \in Y$ for $t > 1$, but $ty' \notin Y$ for some $y' \in Y$ and $t < 1$ (weakly increasing returns).

We can now prove the following, which is a weaker but generalized version of the relationship between returns to scale and the convexity of the transformation surface given in Sections 8.5 and 8.6:

If the production set is additive and shows either weakly decreasing or constant returns to scale, it is convex.

For $y^i, y^j \in Y$ and t such that $0 \leqq t \leqq 1$, we have $ty^i \in Y$ and $(1 - t)y^j \in Y$ (from returns to scale, and assumption I if either t or $(1 - t)$ is zero), and we have $ty^i + (1 - t)y^j \in Y$ (additivity), proving convexity.

When we discuss general equilibrium in the Chapter 9, we will see that the assumption that Y is convex is necessary to be sure of the existence of equilibrium in a competitive economy. Nonadditivity (interactions between production processes) and increasing returns to scale are the chief enemies of convexity in the production set.

Note that the convexity proved above is *weak* convexity[14] (that is flat boundaries on the production set are possible).

Sometimes we may wish to consider a production set Y^j for a firm or industry, rather than for the whole economy. We shall usually assume that every Y^j has the properties assumed of Y, but it should be realized that Y may have stronger properties than individual Y^j's. Whereas Y is convex if every Y^j is convex, Y may be convex even if some Y^j is not. We may have increasing returns in the production of one commodity, for example, somehow swamped by decreasing returns elsewhere.

Neoclassical production functions require a change of sign convention and explicit statement of otherwise implicit assumptions to be expressed in set-theoretic terms. For example, we can express the classic Cobb-Douglas production function,

$$X = cL^a K^{1-a},$$

as a production set,

$$Y = \{y_1, y_2, y_3 \mid y_1 \leqq c(-y_2)^a(-y_3)^{1-a}; \qquad y_1, -y_2, -y_3 \geqq 0\},$$

where $y_1 = X$, $y_2 = -L$, $y_3 = -K$. The set Y then formally satisfies assumptions I, II, and III.

FURTHER READING

For deeper analysis of generalized production theory (also consumer theory in the same tradition), see Debreu [1]. For examples of neoclassic analysis see Samuelson [1] (this book is already oriented towards aspects of the "new" mathematical economics), and Samuelson's collected papers (Samuelson [2]). For neoclassical analysis in trade theory, see Samuelson [2] and Chipman [2].

[14] For the conditions for strict convexity, we cannot avoid the detailed neoclassical analysis of Sections 8.4 through 8.6, especially for the constant returns case.

CHAPTER

9

General Equilibrium

This chapter requires familiarity with the properties of point-to-set mappings (Review R9) and of the contents of the preceding chapters.

9.1 EQUILIBRIUM IN A MARKET ECONOMY[1]

The notion of economic equilibrium, which is substantive rather than merely formal, presents no special problems in a fully centralized economy. If production is fully controlled and the resulting output directly allocated to consumers without the intervention of any market or decentralized decisions, the problems are those of *feasibility* and *optimality* rather than of equilibrium.

In an economy with any degree of decentralization, especially if exchange passes through a market, there is a real problem as to whether an equilibrium, suitably defined, always exists for arbitrary initial conditions. Although the equilibrium problem is examined here for the completely decentralized competitive economy, it exists also for partly centralized economies, such as those of the Soviet type.

The notion of equilibrium in a decentralized economy encompasses two different types of relationship among the decisions which are independently

[1] For general studies on market equilibrium and competitive equilibrium generally, see Dorfman, Samuelson, and Solow (Chapter 13) and, more advanced, Debreu [1] and Karlin [1]. Other references are given later in the chapter.

made. First is *compatibility*, the decisions made by different decision makers must be compatible with each other. Second is *equilibrium* in something like the dynamic sense, that is, the decisions must be more than momentarily compatible, they must be capable of being sustained in the absence of external influences.

In analyzing the behavior of the decentralized economy, we accept the *rules* for the individual decision makers as data and assume every decision maker acts in accordance with the relevant rule. In the models examined here, we assume that there is no *direct* relationship between any decision makers so that the results of their individual acts in response to objective and universal parameters (prices) can be added.

The fundamental problems of equilibrium are thus thrown on to the market. Given a vector of prices, we aggregate the amounts that are demanded by consumers and subtract the sum of current production and available supplies, to obtain the net result or *excess demand* in each market. It is more convenient to discuss excess demand than to discuss demand and supply separately.

We assume that consumption can only come from current production, so that compatibility requires that excess demand can never be positive at equilibrium. On the other hand, if excess demand is negative at a positive price (excess supply at a positive price) we make the rudimentary dynamic assumption that there will be a downward pressure on price until either excess demand or the price is zero.

Thus, if z_i is the excess demand for the ith good, we shall define equilibrium in the ith market as

$$p_i z_i = 0, \qquad z_i \leqq 0, \qquad p_i \geqq 0,$$

which implies that either excess demand is zero or the price is zero. That is, any good with excess supply (negative excess demand) is a free good.

This definition of market equilibrium can be directly generalized to cover all markets. If p, z are the vectors of prices, excess demands, then all markets are in equilibrium, if and only if

$$pz = 0, \qquad z \leqq 0, \qquad p \geqq 0.$$

The economy will be considered to be in equilibrium if the market equilibrium conditions are satisfied and all decision makers are operating in accord with the relevant rules. (These rules, for the competitive economy, are taken to be utility or profit maximization at given prices.)

In some studies (including the Arrow-Debreu model) the market equilibrium condition is retained in the classic form $z = 0$. What is actually proved, in the first instance, is the existence of a point satisfying the above inequality definition. A "free disposal" assumption can then be used to suppose that if $z_i < 0$, it can just as well be zero. It seems simpler to adhere to the inequality definition, which is certainly necessary if there is no free disposal. Air pollution is an obvious joint product with negative excess demand and no free disposal.

Walras' original study of the problem of general equilibrium was confined to the equilibrium condition $z = 0$ and included no explicit nonnegativity constraints. Although Wald had rigorously studied the problem of the existence of equilibrium as early as 1935,[2] general equilibrium analysis in the economics profession was confined to counting equations and unknowns up into the nineteen fifties.

In this chapter we shall investigate the existence of equilibrium in two models of the market economy. The first, and simplest, is the Walras-Wald model which has simple production conditions, no profits to be distributed, and in which we shall assume the existence of well-behaved *aggregate* demand functions. The second will be a slightly simplified version of the Arrow-Debreu-McKenzie model, which is an accurate depiction of the traditional perfect competition model of the economy.

Before proceeding to examine these models, we shall investigate some of the differences between the rigorous approach to the equilibrium problem and the equation counting approach, and also assemble some mathematical analysis for later use.

The existence of equilibrium in an economy with imperfect competition is an unsolved problem of great complexity.[3] Most of the traditional models of firms in imperfect competition (including monopoly) are so rooted in partial concepts that they need to be rebuilt to fit into a general equilibrium framework. We shall not consider imperfect competition here.

9.2 WALRAS' LAW AND THE BUDGET CONSTRAINT

In the "equation counting" approach to general equilibrium, we assemble the demand and supply components of the system together into the n excess demand relationships, each a function of all n prices, to obtain the equilibrium conditions

$$z_i = F^i(p) = 0 \qquad i = 1, \ldots, n.$$

[2] The English translation was published in 1951 (Wald).

[3] Generalized game theory concepts have achieved some results in the model of universal oligopoly. See Debreu and Scarf.

There are n relationships in n variables. From the beginning it was, however, correctly realized that the count is not correct because the relationships are homogeneous of degree zero in p.[4] Assuming that the F^i are single-valued functions and not point-to-set mappings (this assumption was always made in the neoclassical analysis), they are homogeneous functions and so can be written in the form $F^i = f^i(r)$, where r is the $(n - 1)$ vector $(p_1/p_n, \ldots, p_{n-1}/p_n)$. There are now n relationships in only $n - 1$ independent variables, p_n being indeterminate. Thus it is necessary to show that only $n - 1$ of the relationships are independent, if the count is to come out right.

Walras found the dependence supplied through the budget constraint. In a pure exchange economy, if every consumer obtains his income from the sale of goods, and spends all that income on goods, it is obvious that the value of purchases equals the value of sales for all traders and so, for the economy as a whole, $pz = 0$. The same relationship holds for an economy with production, provided all profits are distributed to consumers.

The relationship $pz = 0$ is, of course, part of the equilibrium condition.

If it is assumed to hold everywhere, even when general market equilibrium has not been established, it is known as Walras' Law.

The implication of Walras' Law is that the excess demand functions $z_i = F^i(p)$ are not all independent since we can always find one excess demand, say z_n, from $z_n = -\sum_{i=1}^{n-1} F^i(p)$. This implication has been used to generate many fruitless arguments, especially in monetary theory.

Walras' Law is neither necessary nor sufficient for the purpose for which it has been used. It is not necessary that we be able to solve properly for z in terms of p except at the equilibrium point itself, for which the equilibrium condition gives $pz = 0$, and the counting method certainly is not sufficient to guarantee the existence of equilibrium, if only because of the implicit nonnegativity constraints.

Walras' Law carries important implications for dynamic behavior. It implies that, even when the markets are out of equilibrium, every economic agent acts as though the existing prices are equilibrium prices and plans sales and purchases that balance, even if all those sales and purchases cannot be made. This behavior is related to Walras' theory of *tâtonnement*, which represents a dynamic process. Walras' Law acts as a law of conservation and tends to stabilize market behavior, as we shall see in Chapter 12.

In the equilibrium analysis given in this chapter, we do not count equations and unknowns and do not interest ourselves in whether the excess

[4] Assuming no real balance or money illusion effects.

demand functions are independent or not. Indeed, we do not assume that excess demand is a single valued function so that the question of dependence is irrelevant.

We *do* use the budget constraint, in a weak form, in establishing the existence of equilibrium, but we use it in quite a different way from that of the equation counting approach.

In showing the existence of equilibrium, we shall be proceeding in the following way. We take an arbitrary set of prices, then consider a set of excess demand vectors that would be consistent with producer and consumer behavior at those prices if they were equilibrium prices. The way in which we construct this set of excess demand vectors is quite artificial and chosen for mathematical reasons. *It does not purport to represent any actual process of adjustment.* We make this construction for all feasible prices, and obtain a mapping from prices into excess demand vectors, which will be a point-to-set mapping. We then draw on a *fixed point theorem* to show that at least one of the price vectors maps into a set of excess demand vectors which includes an excess demand vector satisfying the market equilibrium conditions. Since the mapping has been chosen so that the equilibrium conditions for the individual decision makers are always satisfied, we thus prove the existence of at least one point of equilibrium in the economy.

Confining our attention to excess demand which satisfies the budget constraint at all prices is quite different from assuming that excess demand *in fact* satisfies Walras' Law. It is quite consistent with the approach here that individual's *actual* behavior out of equilibrium does *not* satisfy Walras' Law. We would still set up our artificial mapping based on the budget constraint and prove that an equilibrium *exists*, although the actual dynamic processes of the economy may not result in its being *attained.*

Finally we may note that we require only the weak budget constraint

$$pz \leqq 0,$$

whereas Walras' Law requires equality if the dependence of the excess demand functions is to be established.

9.3 THE EXCESS DEMAND THEOREM[5]

The equilibrium proofs which follow in the succeeding sections require the use of topological methods at a crucial point in the argument. Since the same

[5] This name has been chosen for simple reference. The theorem as such was proved, somewhat differently, in Gale [3], and the simpler proof provided in Kuhn [1]. An exposition of the theorem is also given in Debreu [1] and Karlin [1].

Proof depends on the Kakutani fixed point theorem (Review R9, Section R9.5).

analysis can be used for both the models which follow, it seems useful to make a package of the part of the argument which is common to both and present it here. This leaves the later discussion freer to discuss the special properties of the models themselves.

Some readers may wish merely to accept the following theorem as proved, then pass directly on to the next section. Although the theorem is mathematical and does not need to be given the particular interpretation here attached to it, we use it only in respect to excess demand, hence the name and the interpretation:

Excess Demand Theorem

If every semipositive price vector p (which we can take to be normalized so that $\sum p_j = 1$) is associated with a set $Z(p)$ of excess demand vectors such that $Z(p)$ has certain properties and the mapping $p \to Z(p)$ has appropriate continuity properties, and if the weak budget constraint $pz \leqq 0$ is satisfied for all $z \in Z(p)$, then there is some price p^* for which $z \leqq 0$ for some $z \in Z(p^*)$.

Since $pz = 0$ is a special case of $pz \leqq 0$, this theorem guarantees the existence of an equilibrium excess demand vector if the conditions of the theorem are met. Both the existence proofs given in this chapter are based on showing that the conditions of the excess demand theorem can be met with the individual decision makers operating in accordance with the appropriate rules.

The formal requirements on $Z(p)$ and the mapping $p \to Z(p)$ are:

$Z(p)$ is nonempty and convex for all p; $p \to Z(p)$ is upper semicontinuous; and the set of $Z(p)$'s is bounded.

The general sense of the theorem can be illustrated from a two-market example. We have $p_1 z_1(p) + p_2 z_2(p) \leqq 0$ everywhere. The theorem is obviously satisfied if z_1, z_2 are nonpositive everywhere. But suppose $z_1(p) > 0$ for some p. Then we must have $z_2(p) < 0$. It is obvious that, unless $z_2(p)$ is unbounded (which is ruled out) $z_1(p)$ cannot remain positive indefinitely as the ratio p_1/p_2 increases and so must be nonpositive for sufficiently large p_1/p_2. The only question is whether $z_2(p)$ may become positive. It may, indeed, do so, but not *before* $z_1(p)$ becomes negative. Thus there must be some p_1/p_2 for which both $z_1(p)$ and $z_2(p)$ are nonpositive, and thus some normalized price vector $[p_1, p_2]$ for which $z_1(p)$, $z_2(p)$ are nonpositive.

Formal proof of the theorem is a much more difficult matter. We proceed as follows. Denote by P the set of semipositive normalized prices.

This is clearly a compact convex set. Denote by Z the set of all $z(p)$ for $p \in P$ [Z is the union of the sets $Z(p)$]. If Z is not convex, we replace it by any compact convex set containing Z, which we denote by Z'.

Now define the set $S(z)$ as follows:

$$S(z) = \{p \mid pz \quad \text{is a maximum for} \quad z \in Z', \quad p \in P\}.$$

That is, we choose an arbitrary excess demand vector from the set of all excess demand vectors which are attainable at some prices, then find the price vector for which the value of this excess demand is maximized. It is important to note that the price vector is *any* price vector, not necessarily the particular p which is associated with z through the mapping $p \to Z(p)$.

Clearly $z \to S(z)$ is a mapping from Z' into a subset of P. Since Z' is convex we know this mapping to be upper semicontinuous (see *continuity properties of optimal solutions*, Review R9, Section R9.4). $S(z)$ is a convex set since it is the intersection of the hyperplane $\{y \mid yz = \max pz\}$ with P.

Consider the set $P \times Z'$, that is, the set consisting of normalized price vectors paired with excess demand vectors. If we take some point p, z in $P \times Z'$, then $Z(p)$ associates a set of excess demand vectors with p, and $S(z)$ associates a set of price vectors with z. In other words, the mapping $p, z \to Z(p), S(z)$ maps a point in $P \times Z'$ into a subset of $P \times Z'$.

We have shown the mapping $z \to S(z)$ to be upper semicontinuous, and $p \to Z(p)$ has been assumed to have the same property, so that the combined mapping is upper semicontinuous also. We have shown that $S(z)$ is convex and $Z(p)$ has been assumed convex, so that $S(z) \times Z(p)$ is convex.

Thus we have an upper semicontinuous mapping $p, z \to Z(p), S(z)$ from the set $P \times Z'$ into a convex subset of itself. These are the conditions for invoking the *Kakutani Fixed Point Theorem* (Review R9, Section R9.5). The theorem states that there exists some $p^* \in P$, $z^* \in Z'$ which is a fixed point, that is, for which $p^* \in S(z^*)$ and $z^* \in Z(p^*)$.

From the construction of $S(z)$, $p^* \in S(z^*)$ implies that, for all $p \in P$,

$$pz^* \leqq p^*z^*.$$

Using the weak budget condition it follows that, since $z^* \in Z(p^*)$,

$$p^*z^* \leqq 0.$$

Thus

$$pz^* \leqq 0, \quad \text{for all} \quad p \in P.$$

Clearly the last inequality is satisfied for all $p \in P$ only if

$$z^* \leqq 0,$$

thus proving the theorem.

We might note that a strong budget condition (Walras' Law) does not strengthen the conclusion of the theorem.

9.4 THE WALRAS-WALD MODEL

This is a simplified model of a competitive economy, based on the original Walrasian linear production model. It is similar to the model for which Wald did his pioneer study of the existence of equilibrium, although Wald's inverted demand functions (price as a function of quantities) are not used.[6] It is almost identical with the model studied by Dorfman, Samuelson, and Solow[7] and the proof is basically similar to theirs, although rearranged somewhat.

As a representation of the competitive economy, this model has been superseded by the Arrow-Debreu-McKenzie model of the next section. Since this model is simpler than the Arrow-Debreu-McKenzie model, the existence proof is rather simpler (although less so than one might, perhaps, expect). The analysis of this simpler model is chiefly justified on heuristic grounds, and on the interesting uniqueness property that can be added by a further assumption.

We define the model as follows:

Resource and Technology Limitations

There is a linear, fixed coefficient production technology with no joint products and with a single method of producing each good. There are no intermediate goods, so that the productive factors are distinct from the outputs of the system. The technology is fully defined by the input matrix A, of order $m \times n$, where m is the number of factors and n the number of goods.

It is assumed that every output requires the use of all factors ($A \gg 0$). This assumption can be modified, but we keep it for simplicity.

The total supply of resources to the economy is fixed, so that, if y denotes the output vector and v^0 denotes the fixed resource vector, the resource and technology limitations can be summed up in the production constraint.

$$(1) \qquad Ay \leqq v^0 \qquad y \geqq 0.$$

No specific theory of behavior by producers is required.

[6] For the original Wald model see Wald, also Kuhn [2].

[7] Dorfman, Samuelson, and Solow, Chapter 13.

Consumers

It is assumed that the behavior of consumers is fully expressed by *aggregate* demand functions of the following kind:

$$(2) \qquad x_j = \phi^j(p, w) \qquad x_j \geqq 0 \qquad j = 1, \ldots, n,$$

where p is the vector of goods prices, w the vector of factor prices.

The aggregate demand functions are assumed to be continuous, single valued, and homogeneous of degree zero in the combined vector (p, w).

Income is assumed to be derived from the sale of factors, all of which are assumed to be owned by consumers. Since we shall assume that consumers spend all of their income (nonsatiation) and that all consumers own valuable factors, the budget constraint is

$$(3) \qquad px = wv^0.$$

Market Equilibrium

There are two sets of markets, the goods markets and the factor markets. Define the excess demand vector for goods $z = x - y$, and the excess demand vector for resources $u = Ay - v^0$, then the market equilibrium conditions are

$$(4) \qquad pz = 0, \qquad p \geqq 0, \qquad z \leqq 0$$

$$(5) \qquad wu = 0, \qquad w \geqq 0.$$

[The condition $u \leqq 0$ is already guaranteed by (1).]

An equilibrium for the economy is given by vectors p^*, w^*, y^*, x^*, z^*, u^* *which satisfy* (1) *through* (5).

We prove the existence of an equilibrium for the economy by first proving the following:

If p^* *is the equilibrium price vector, then the equilibrium output and factor price vectors* y^*, w^* *are necessarily optimal in the program:*

$$\max p^*y \qquad \text{S.T.} \quad Ay \leqq v^0, \qquad y \geqq 0$$

with dual

$$\min w^*v^0 \qquad \text{S.T.} \quad wA \geqq p^*, \qquad w \geqq 0.$$

To prove this we note that y^* is feasible in the primal, from (1). w^* may not be feasible in the dual. However, since $A \gg 0$, λw^* is certainly feasible if λ is large enough.

Since y^*, λw^* are feasible in the primal and dual, we have, from the fundamental lemma of linear programming:

$$p^*y^* \leqq \lambda w^*Ay^* \leqq \lambda w^*v^0. \tag{9.4.1}$$

From the market equilibrium condition (4) we have

$$p^*z^* = 0,$$

that is,

$$p^*x^* = p^*y^*.$$

From the budget constraint (3) we have

$$p^*x^* = w^*v^0,$$

so that $p^*y^* = w^*v^0$ and $\lambda = 1$ in (9.4.1). Thus w^* is optimal in the program (duality theorem).

We can now proceed to the proof proper. From above, we know that, if p^* is the equilibrium price vector, the equilibrium values y^*, w^* will be optimal in the program. Thus we can confine the search for equilibrium vectors to the vectors which are optimal in the program.

Now we know that if w^* is the optimal vector associated with p^* in the program, kw^* will be the optimal vector for kp^*. Thus, since kp^* gives the vector (kp^*, kw^*) and since all the equilibrium relationships are homogeneous of degree zero in the latter, we can normalize p so that $\sum p_j = 1$. Denote the set $\{p \mid p \geqq 0, \sum p_j = 1\}$ by P.

Choose some $p \in P$. Then the linear program will give solutions $y(p)$, $w(p)$ associated with p. Since the solutions to the program may not be unique we consider the sets $Y(p)$, $W(p)$ of vectors optimal for p. Since all members of both sets are optimal, the duality theorem gives

$$py = wv^0, \qquad \text{all} \qquad y \in Y(p), \qquad \text{all} \qquad w \in W(p).$$

We shall condense the above into the statement $p \cdot Y(p) = W(p) \cdot v^0$ and use equivalent notation for similar statements elsewhere in the argument.

Since p gives $w(p)$, and since x is a function only of p, w, then p gives some $x(p)$. The demand function is single valued, but we must take account of the fact that $w(p)$ may be a set. Thus p may give rise to various $w(p)$'s and so to a set of $x(p)$'s. We shall denote this set by $X(p)$.

We can now form the set of excess demand vectors $Z(p)$ associated with p,

$$Z(p) = X(p) - Y(p).$$

The set $Z(p)$ consists of all the vectors we can obtain by taking any $y(p)$ from $Y(p)$ and subtracting it from any $x(p)$ in $X(p)$.

Since the budget constraint is satisfied for all $x \in X(p)$, and since $p \cdot Y(p) = W(p) \cdot v^0$, we have

$$\begin{aligned} p \cdot Z(p) &= p \cdot X(p) - p \cdot Y(p) \\ &= W(p) \cdot v^0 - W(p) \cdot v^0 \\ &= 0. \end{aligned} \tag{9.4.2}$$

Now the mappings $p \to Y(p), p \to W(p)$ are upper semicontinuous (*continuity properties of optimal solutions*, Review R9, Section R9.4) and the function $p, w(p) \to x_j(p)$ is continuous (it is the demand function) for all j, so that the mapping $p \to X(p)$ is upper semicontinuous. It follows that $p \to Z(p)$ is upper semicontinuous.

Thus $p \to Z(p)$ is upper semicontinuous, $p \cdot Z(p) = 0$, and the last relationship ensures that $Z(p)$ is convex.

Thus $p, Z(p)$ satisfy all the conditions for the *excess demand theorem*. Using this theorem, we know that there is some $p^* \in P$ such that $z^* \leqq 0$ for $z^* \in Z(p^*)$.

Choose this p^*, and the associated z^*. From p^* we obtain the sets $W(p^*)$, $Y(p^*)$, $X(p^*)$ by solving the linear program and using the demand functions. From $W(p^*)$, $Y(p^*)$, $X(p^*)$ choose that w^*, y^*, x^* which gives z^* (there may be several choices, any will do). Then p^*, z^*, w^*, y^*, x^* give an equilibrium of the system.

To show this, we note that the vectors satisfy conditions: (1) since it is a constraint of the program; (2) since x^* is given by this; (3) since this was built in to the solution process; and (4) since we have shown that $p^*z^* = 0$, and chosen $z^* \leqq 0$.

The last condition, (5), is satisfied by the equilibrium theorem in the linear program, which gives

$$w^*u^* = w^*(Ay^* - v^0) = 0,$$

since w^*, y^* are optimal in the program. This completes the proof.

We can now strengthen the results considerably. Since we have assumed $A \gg 0$, it follows that, if $p_j^* = 0$, $w^*A^j - p_j^* > 0$ so that, using the equilibrium theorem for the program, $y_j^* = 0$. But $x_j^* \geqq 0$ and $z_j^* \leqq 0$ so that, if $p_j^* = 0$ we have $z_j^* = 0$.

Since $z_j^* = 0$, if $p_j^* > 0$, the above implies that $z^* = 0$, that is, *all markets will be cleared and there will be no free goods in the system.*

There may, however, be free factors. If the number of factors exceeds the number of goods, then we shall expect to find the production constraints $Ay^* \leqq v^0$ are not all satisfied by equalities (unless the factor endowment happens, by chance, to be just right). From the equilibrium theorem in the program, the surplus factors will carry zero price.

If we now make the following additional assumption:

That the aggregate demand functions satisfy the revealed preference axiom; if $p^{**}x^{**} \geqq p^{**}x^{*}$, *then* $p^{*}x^{**} > p^{*}x^{*}$ *for two price vectors* p^{*}, p^{**} *and their associated demand vectors.*

Suppose that p^{**}, x^{**} and p^{*}, x^{*} were two distinct equilibria. From above, we know that $z^{*} = z^{**} = 0$, so we have $x^{*} = y^{*}, x^{**} = y^{**}$. Since the constraints of the linear program are constant, both y vectors are feasible in the program which gave the other. Thus since y^{*} is feasible in the program, $\max p^{**}y$ S.T. $Ay \leqq v^{0}, y \geqq 0$, we have

$$p^{**}y^{*} \leqq p^{**}y^{**} \qquad \text{(fundamental lemma of linear programming).}$$

Similarly, using the program relevant to p^{*}, we have

$$p^{*}y^{**} \leqq p^{*}y^{*}.$$

These two inequalities are inconsistent with the assumption on the demand function, so that there cannot be distinct equilibria.

This uniqueness proof was given in Wald's original paper. In terms of traditional demand theory, the assumption of convexity in the *aggregate* demand function is unjustified.[8] However, the "characteristics" approach to demand theory (see Chapter 7, Section 7.5) introduces a universal substitution effect which is independent of income distribution and which is additional to the traditional type of substitution effect. This makes the assumption of the aggregate revealed preference axiom more respectable. We shall see later that this same assumption gives powerful stability properties.

Note that, in this model, no assumptions were made concerning the behavior of producers, because the simplicity of the production conditions made them unnecessary. The equilibrium implies, however, that zero profits are made for goods actually produced and that goods which cannot be produced without loss will not be produced. The equilibrium is thus consistent with profit maximization.

9.5 THE ARROW-DEBREU-McKENZIE MODEL[9]

Arrow-Debreu,[10] and McKenzie[11] independently formulated somewhat different models, both good representations of the competitive economy, and

[8] Because of income redistribution, the aggregate relationship $p^{**}x^{**} \geqq p^{**}x^{*}$ may not imply that the equivalent relationship holds for every individual.

[9] This model is discussed in Debreu [1] and Karlin [1].

[10] See Arrow and Debreu.

[11] See McKenzie [1] and [2].

proved the existence of an equilibrium in these models. The best features of the original models were combined, with some additional modifications, in a later model by Debreu.[12] The model given here is a simplified version of the Debreu synthesis, closer to the original Arrow-Debreu model than to the McKenzie model.

The proof given here, which is heuristic rather than rigorous, uses the *excess demand theorem* of Section 9.3.

The specifications of the model are as follows:

Production Possibilities

Unlike the Walras-Wald model, this does not distinguish between goods and factors, allows intermediate goods and joint products, and permits any number of productive activities. Indeed, we assume the most general production conditions, which may be linear or nonlinear, of the kind discussed in Section 8.7.

The number of commodities is assumed to be some finite number n. There is assumed to be a finite number of firms, the jth firm having open to it a feasible production set Y^j with the following properties:

(1) Y^j is closed and convex;
(2) $Y^j \cap \Omega = 0$;
(3) $Y^j \cap (-Y^j) = 0$;
(4) $y \in Y^j \quad \text{if} \quad y^* \in Y^j \quad \text{and} \quad y \leqq y^*, \quad \text{all } y, y^*,$

with the above assumed true for all j.

Assumption (1) implies continuity and *nonincreasing returns to scale everywhere.* This can be weakened by requiring nonincreasing returns for the economy as a whole but not necessarily for every individual firm. Assumptions (2), (3), and (4) are usual assumptions on production sets (see Section 8.7). The irreversibility assumption (3) can be replaced by other assumptions.

Consumption Possibilities

We assume that there is a finite number of consumers, each of whom has a feasible consumption set X^i. We adopt the sign convention that commodities consumed are written with a positive sign, while commodities supplied by consumers (like labor services) are given a negative sign.

The feasible consumption set X^i should not be confused with the budget constraint. X^i merely shows consumptions that are physically pos-

[12] See Debreu [2].

sible expressing, for example, the fact that the consumer cannot provide labor services with zero consumption of goods. Specifically, we assume:

(5) X^i is closed convex and lower bounded.

The convexity assumption rules out the possibility that the consumer, viewed as a process using consumer goods to produce labor, has increasing returns to scale in this respect. Lower boundedness implies both an upper limit to the labor services the consumer is capable of supplying and a lower limit to his level of consumption (either zero or some positive subsistence level).

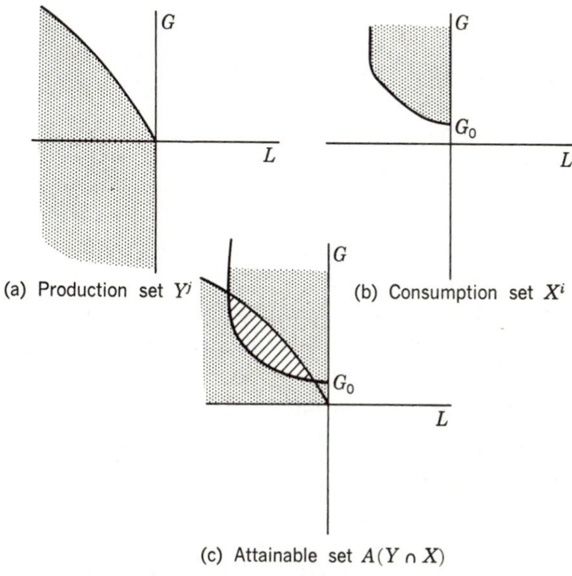

Figure 9–1 Consumption and production sets.

The general nature of the assumptions on the production and consumption sets is illustrated in Figure 9–1 (a) and (b). This gives typical feasible sets for an economy with two commodities, labor (L) and a consumption good (G). The consumption set is depicted as not passing through the origin (this is not a necessary assumption), but shows a minimum subsistence level of consumption, G_0, and also a maximum level of labor output.

The two diagrams are also used to illustrate the idea of *attainable states* of the economy. If an economy consisted of an aggregate production set like (a) and an aggregate consumption set like (b), with no initial stocks of commodities, then the only attainable states of the economy are in the intersection of the two sets, illustrated in (c).[13]

For both production and consumption we make the implicit assumption of *additivity* (no interactions between firms or between consumers) so that, if $y^j \in Y^j$ and $y^k \in Y^k$, then $y^j + y^k$ is a possible production for the economy, with a similar property for consumption. We shall denote the total production set $\sum Y^j$ by Y.

Consumer Preferences

Each consumer is assumed to have preferences defined on his consumption set, which we shall suppose to be expressed by a utility function with the following properties:

(6) $u^i(x^i)$ is a continuous differentiable function on X^i;
(7) $\{x^i \mid u^i(x^i) \geqq u^i(\bar{x}^i)\}$ is convex for all $\bar{x}^i \in X^i$;
(8) For every $\bar{x}^i \in X^i$, there is some x^i such that $u^i(x^i) > u^i(\bar{x}^i)$.

These are standard enough assumptions. The nonsatiation assumption (8) can be modified to require only that there is no satiation for any consumer in any *attainable* state, since it is obvious that what happens outside the set of attainable states is irrelevant.

The analysis can be carried out directly in terms of preference quasi-ordering without explicit reference to a utility function.[14]

Distribution of Wealth and Income

If ω is the vector of initial stocks of goods, it is assumed that these are entirely owned by consumers so that, if ω^i is the initial endowment of the ith consumer $\sum \omega^i = \omega$.

If π^j is the profit of the jth firm, it is assumed that this profit is entirely distributed to consumers. We assume that the method of distribution of profits is an initial condition of the economy, defined by fixed numbers θ_{ij} which give the share of the jth firm's profits which go to the ith consumer. We have $\theta_{ij} \geqq 0$, $\sum_i \theta_{ij} = 1$.

In this model the consumer obtains income from three separate sources

(a) From the sale of his initial holdings ($p\omega_i$);
(b) From his share of profits ($\sum_j \theta_{ij}\pi^j$);
(c) From the sale of services which are included in X^i.

[13] The set of attainable states is necessarily bounded since each X^1 is lower bounded and the set $\{y^j \mid y^j \in Y^j, \quad y^j \geqq y^*\}$ is bounded for every y^*, from Section 7.2.

[14] This is done in Debreu [1] and [2].

Due to the use of the sign convention on commodities we do not consider item (c) separately. For a consumer with no initial resources or profit shares, the budget constraint will be $px \leqq 0$, since px represents the cost of consumption goods less the value of services sold.

Items (a) and (b) together are referred to as the consumer's *wealth* in Arrow-Debreu. This is not a completely appropriate term, but we shall retain it.

It is assumed that the consumer operates within his *wealth constraint*, that is, his consumption satisfies

$$px^i \leqq p\omega^i + \sum_j \theta_{ij}\pi^j.$$

Competitive Equilibrium

The rules in accordance with which the economic agents are assumed to act are that producers maximize profits, accepting prices as given (perfect competition), and that consumers maximize utility subject to their wealth constraint.

A competitive equilibrium is a configuration of the economy in which there is market equilibrium and in which each producer and consumer is acting as above.

Formally a competitive equilibrium is a collection of n vectors $\{p^*, x_1^*, \ldots, y_1^*, \ldots\}$ in which:

(1) $p^*y_j^* = \max p^*y_j, \qquad y_j \in Y^j, \qquad \text{all } j;$

(2) $u^i(x_i^*) = \max u^i(x_i), \qquad x_i \in X^i, \qquad p^*x_i \leqq p^*\omega_i + \sum_j \theta_{ij}\pi_j^*, \qquad \text{all } i$
(where $\pi_j^* = p^*y_j^*$);

(3) $p^*z^* = 0, \qquad z^* \leqq 0, \qquad p^* \geqq 0$
(where $z^* = x^* - y^* - \omega, \qquad x^* = \sum_i x_i^*, \qquad y^* = \sum_j y_j^*$).

p is assumed to be normalized throughout.

Existence Theorem

The economy described above has a competitive equilibrium if it satisfies the following additional condition:

$$(*) \ \textit{Interior}\ [(Y + \omega_i) \cap X^i] \text{ is nonempty, } \textit{all } i.$$

Before proceeding to an outline of the proof, let us examine the nature of Condition (*). This condition is necessary, as we shall see, to eliminate a discontinuity problem which otherwise arises. It says, in effect, that there must be an attainable state of the economy which is not on the boundary of the consumption set for any consumer.

In the original Arrow-Debreu model, the additional condition was given as:

For each consumer, ω_i must be such that there is some $x_i \in X^i$ such that $x_i \ll \omega_i$.

This original condition required that every consumer had a positive initial endowment of every good in his consumption set and was intolerably strong. We can see that the earlier condition satisfies our condition here by noting that $0 \in Y$, so that $0 + \omega_i \gg x_i \in X^i$. Then $x_i + \eta \in [(Y + \omega_i) \cap X^i]$ for some $\eta \gg 0$ with $x_i \in X^i$, and $x_i + \eta$ is clearly an interior point of X^i.

The proof of the existence theorem depends on establishing the existence of a suitable mapping $p \rightarrow Z(p)$ so that we can use the *excess demand theorem* (Section 9.3). We assume that prices are normalized with P the set of non-negative normalized prices. We proceed as follows, commencing with an arbitrary $p \in P$:

Since Y^j is closed and convex, the set $Y_j(p)$ of vectors which maximize py_j for $y_j \in Y^j$ is closed and convex and the mapping $p \rightarrow Y_j(p)$ is upper semicontinuous (see Review R9, Section R9.4). Also the function $\pi_j(p)$ is continuous (also Section R9.4). Since $0 \in Y^j$, we must have $\pi_j(p) \geqq 0$.

Given p, then, we have $\pi_j(p)$, hence the wealth $w_i(p) = p\omega_i + \sum_j \theta_{ij}\pi_j(p)$ of each consumer is a function only of p.

Now, since $\theta_{ij}, \pi_j \geqq 0$, $w_i \geqq p\omega_i$. Some consumers, however, may own no profit shares or may own shares only in firms making zero profits, so that their wealth consists entirely of the initial endowment of goods. We shall now consider such consumers since it is among these that the problems which give rise to the necessity of Condition (*) will appear.

Consider the simple case of a consumer with no labor services to sell, whose consumption set consists of two commodities and whose initial endowment contains only one of the commodities. The situation is depicted in Figure 9–2, where the two goods are G_1, G_2 and the initial endowment ω consists of some amount of G_1 only. At some prices $p \gg 0$ the consumer's budget constraint will be the shaded area in the diagram and he will choose some point like A, exchanging part of his initial stock of G_1 for some amount of G_2. If we consider a sequence of (normalized) prices in which p_1 falls, the budget line will pivot counterclockwise around ω and, with continuous convex preferences, his chosen consumption will be expected to move towards the point ω. So long as $p_1 > 0$, however small it may be, he will still be constrained to a budget line lying wholly to the left of ω.

If p_1 now falls to zero, the consumer has nothing of value to sell and cannot purchase any G_2. He must consume only G_1, but now G_1 is a free good and the consumer is no longer constrained by his initial endowment.

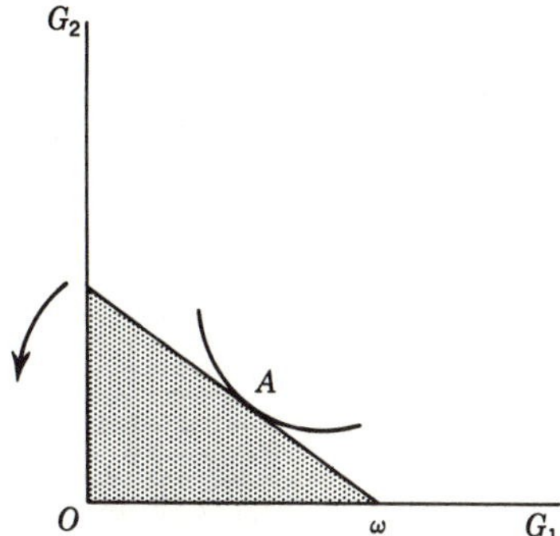

Figure 9–2 Lower boundary of the consumption set.

Assuming no satiation, the consumer will now presumably seek as much of this free good from the market as he can.

Thus, as p_1 falls from ε to 0, consumption of G_1 will increase from $\omega - \eta$, with $\eta > 0$, to an indefinitely large quantity, *giving a discontinuity in the mapping* $p \to x_i(p)$ *at the point in question.*

The additional Condition (*) is introduced to handle the problem of this discontinuity. In the original Arrow-Debreu model, the condition that ω include some G_2 eliminates the problem which can only arise at a *lower boundary point* in X^i.

Following Debreu,[15] we handle the problem in the following way. We consider a model identical with our general equilibrium model except that, whenever the situation described above arises the consumer's chosen collection is considered be the *set* $0 \leqq G_1$. This clearly removes the discontinuity by artificially smoothing the mapping into an upper semicontinuous point-to-set mapping. We now proceed to discuss the model with the smoothed mapping. If this can be shown to have an equilibrium, we define this to be a *quasi-equilibrium* for the ordinary model. Condition (*) then guarantees that there is a quasi-equilibrium point at which the smoothing operation is not actually required, and this point is therefore the equilibrium which is sought.

Formally the smoothing operation described above is specified as follows:

Given p, the set $X_i(p)$ is the ordinary preferred set $\{x_i \mid u_i(x_i)$ is maximal for $x_i \in X^i, px_i \leqq w_i\}$, provided $w_i \neq \min px_i$ for $x_i \in X^i$. If $w_i = \min px_i$ over X^i we remove the choice element and define $X_i(p) = \{x_i \mid px_i = w_i\}$.

Since $p \to w_i(p)$ is a continuous function, and because of the ordinary continuity properties of optimal solutions (Section R9.4) and the above smoothing operation, the mapping $p \to X_i(p)$ is upper semicontinuous and $X_i(p)$ is a closed bounded convex set.

[15] See Debreu [2].

Consider the sets $Y(p) = \sum Y_j(p)$, $X(p) = \sum X_i(p)$ and $Z(p) = X(p) - Y(p) - \omega$. Since all the mappings $p \to Y_j(p)$, $p \to X_i(p)$ are upper semicontinuous mappings into closed bounded convex sets, the mapping $p \to Z(p)$ has the same property.

For any $z \in Z(p)$ we have:

$$\begin{aligned} pz &= \sum px_i - \sum py_j - p\omega, \\ &= \sum_i px_i - \sum_i \sum_j \theta_{ij}\pi_j - \sum_i p\omega_i, \\ &= \sum_i (px_i - w_i), \\ &\leqq 0 \qquad \text{(wealth constraint).} \end{aligned}$$

Further, since there is no satiation within the attainable set for any consumer, we will have $px_i = w_i$, all i, so that

$$pz = 0.$$

Thus the conditions for invoking the *excess demand theorem* (Section 9.3) are established. From this theorem, there exists some p^* such that $z^* \leqq 0$ for $z^* \in Z(p^*)$. Then we have $p^*z^* = 0$, $p^* \geqq 0$, $z^* \leqq 0$, so that the market equilibrium conditions are satisfied. The behavior conditions for firms and consumers have been satisfied, by construction, everywhere in $Z(p^*)$ except where smoothing has been used, so we have established the existence of a *quasi-equilibrium.*

Now consider the effect of Condition (*). This implies that, if ω_i is on the lower boundary of X^i it is an interior point of the production set Y. But profit maximization will always result in production being on a boundary point of Y. Thus the quasi-equilibrium cannot occur at a point at which $w_i(p^*) = \min p^*x_i$ for $x_i \in X^i$, and i, and hence the *quasi-equilibrium is an equilibrium.*

This completes the existence proof.

FURTHER READING

The reader should study the exposition of the Walras-Wald model given in Dorfman, Samuelson, and Solow, Chapter 13, and also the analysis in Karlin [1], as well as the interesting simple model in Gale [1], Chapter 3.

For the Arrow-Debreu-McKenzie Model, the expositions in both Debreu [1] and Karlin [1] should be studied. Both discuss the relationship between competitive equilibrium and pareto-optimality, a topic not taken up in this book. Karlin's discussion of the welfare problem is especially interesting, but requires a study of his material on advanced optimizing theory.

PART

III

Dynamic Economic Models

CHAPTER

10

Balanced Growth

All but the last section follow directly from the work of Chapters 6 and 7. The final section requires some additional mathematical background, given in Review R8, Section R8.8.

10.1 INTRODUCTION[1]

Balanced growth, in which all the *proportions* of the economy remain stationary over time, but the scale of production increases, provides the link between the static and dynamic analysis of multisector economic models.

In many ways, the analysis of balanced growth is more similar to static equilibrium analysis than to the analysis of more complex growth modes. In models with homogeneous production and consumption conditions, such as those with which we shall be concerned in this and the next chapter, scalar expansion involves no real adjustment process over time.

The questions we shall seek to answer in this chapter are simply whether certain economic models possess a mode of balanced growth, whether it is unique, and what the characteristics of that mode are.

[1] A discussion of balanced growth with linear technology is given in Dorfman, Samuelson, and Solow, Chapter 11, and Gale [1], Chapter 9. More advanced expositions are given in Karlin [1] and Morishima [1]. Excellent brief discussions of matters covered in this and the following chapter are given in Hahn and Matthews and Koopmans [4].

We shall see, in the next chapter, that the properties of balanced growth paths are central to the description of other modes of growth. In this chapter we shall do no more than describe these properties; questions of whether balanced growth is optimal on some criterion will be taken up in the following chapter.

10.2 A LEONTIEF-TYPE MODEL[2]

Let us extend the static input-output model in a simple way by supposing that, because of time lags between production of a commodity and its availability as an input, or for other reasons of a similar kind, the economy requires to hold *stocks* of all its inputs. We shall make the simplest possible assumption, that the economy as a whole requires a stock of the ith commodity equal to at least k_i times the total use of that commodity as an input during unit time period. The unit time period may be short (a "week") or long (a "year")—the analysis is not affected by the length of the period, but the k_i's could usually be expected to be smaller for the longer time period.

In Leontief's own dynamic model, and other dynamic input-output models, the stocks *for each industry* are taken to be proportional to that industry's use of inputs. It is simpler and probably more realistic to assume that stocks are not specific to the industry but may be moved from one industry to another as the need arises, so that only the total stocks held by the economy need be considered.

We shall express the set of stock requirement coefficients $k_i, i = 1, 2, \ldots, n$ as a diagonal matrix K. The vector giving total input use of commodities is Ax, so that the stock requirements necessary for the economy to produce gross outputs x are given by the vector

$$KAx.$$

Thus, if output $x(t)$ is to be produced in time period t, stocks at the commencement of the time period, $s(t)$, must be sufficient to support this level of output. We must have

$$KAx(t) \leqq s(t).$$

If c is an arbitrary bill of goods, it requires gross outputs given, as in the ordinary open model, by

$$x = (I - A)^{-1} c.$$

[2] Leontief's own model is given in Leontief [4]. The model discussed here is closely related to that in Dorfman, Samuelson, and Solow.

Thus the particular bill of goods $c(t)$ can only be produced in time period t if

$$KA(I - A)^{-1} c(t) \leqq s(t).$$

This is the *fundamental constraint* of the input-output model with stocks.

What of the labor requirement? We shall now reverse the approach of Chapter 6 and suppose that labor is never the limiting factor in the present model, but the scarce resource is always the level of stocks. If we wish to include labor, then we can include it as one of the n commodities in the central technology and require that a stock of labor, proportional to its use in the system, is required.

Since stocks are limiting, we obtain *growth* in this model only if stocks increase, and this increase must come from production. We can consider the bill of goods at time t to consist of two parts, $c'(t)$ the vector of goods currently consumed (if labor appears explicitly in the system, consumption may be regarded as input into a labor-producing activity), and $\Delta s(t)$ the vector of additions to the stocks $s(t)$.

This gives two relationships

$$\begin{aligned} c(t) &= c'(t) + \Delta s(t) \\ s(t+1) &= s(t) + \Delta s(t). \end{aligned}$$

If current consumption $c'(t)$ is given, the first set of equations relates stock increases to the technology and to current stocks, the second provides the link between stocks in one period and those in the next. The materials for a dynamic model are present.

For our current purposes we shall make a drastic simplification. *We shall suppose that the consumption of every commodity is a constant fraction of its net output.*[3]

If γ_i is this fraction for the ith commodity $(0 < \gamma_i < 1)$, we shall form the diagonal matrix Γ of these propensities to consume.

Then we have

$$\begin{aligned} c'(t) &= \Gamma c(t) \\ \Delta s(t) &= (I - \Gamma)\, c(t) \\ c(t) &= (I - \Gamma)^{-1}\, \Delta s(t) \end{aligned}$$

Since $0 < \gamma_i < 1$ for all i, $(I - \Gamma)$ is a diagonal matrix with strictly positive diagonal, so that $(I - \Gamma)^{-1}$ exists and is semipositive. Its diagonal elements are $1/(1 - \gamma_i)$.

[3] Assumptions giving equivalent results are made through this and the next chapter, in relation to various different models.

It is usual in setting up this simple model to assume that the propensity to consume is the same for all commodities, so that $c(t)$, $\Delta s(t)$ are related by a *scalar* multiple $(1 - \gamma)^{-1}$. This, however, decreases the generality of the model with no gain in analytical simplicity.

If we insert the relationship between stock increases and net outputs into the fundamental constraint, we obtain

$$KA(I - A)^{-1}(I - \Gamma)^{-1}\Delta s(t) \leqq s(t).$$

To simplify the notation we shall write this in the form

$$K^*\Delta s(t) \leqq s(t),$$

where

$$K^* = KA(I - A)^{-1}(I - \Gamma)^{-1}.$$

We shall assume that the basic technology gives a productive open model, so that we will take it that A is semipositive indecomposable with a dominant root less than unity and that, as a consequence, $(I - A)^{-1}$ is a strictly positive matrix.

It follows immediately that, since A is semipositive indecomposable, so is $A(I - A)^{-1}$. Premultiplication by a diagonal matrix of positive diagonal elements K, and postmultiplication by another diagonal matrix of positive diagonal elements $(I - \Gamma)^{-1}$ do not affect this property. *Hence K^* is semipositive indecomposable.*

Let us consider the conditions for *balanced equilibrium growth*, that is, growth such that the ratio $\mu = \Delta s_i(t)/s_i(t)$ is the same for all commodities and at least one stock is required in full (at least one equality in the fundamental constraint). μ is the growth rate of the system.

Thus we seek a solution to the special relationship

$$K^*\Delta s(t) \leqq \frac{1}{\mu}\Delta s(t), \qquad \text{at least one equality, } \Delta s(t) \geqq 0.$$[4]

From Result (g) of Review R7, Section R7.3, the only solution of the above is $1/\mu^* = \lambda^*$, $\Delta s(t) = x^*$, where λ^*, x^* are the dominant root and associated characteristic vector of the semipositive indecomposable matrix K^*. Thus, although we require equilibrium for only one commodity, we shall obtain it for all since the solution values give equalities throughout the fundamental constraint.

[4] Stock decumulation, $\Delta s_i(t) < 0$ for some i, t, is not inherently ruled out of the general model. Here we seek balanced growth with stock increments proportional to existing stocks. It is the nonnegativity of stocks, rather than of stock increments, that is reflected in the above nonnegativity constraint.

It is essential to note that the above simple solution gives $\Delta s(t)$ for all t, and hence $s(t)$ for all t. We have, in fact,

$$\begin{aligned} s(t) &= \Delta s(t-1) + s(t-1) \\ &= (1+\mu^*)\, s(t-1) \\ &= (1+\mu^*)^2\, s(t-2) \\ &= (1+\mu^*)^t s(0), \end{aligned}$$

where $s(0)$ is the initial level of stocks. Thus we obtain the balanced growth path, *only if initial stocks are in the same proportion as* x^* *the relevant characteristic vector.*

If initial stocks are not in the appropriate proportions, the growth path is quite different from the simple balanced growth path.

Since K^* has characteristic vectors other than λ^*, there are other balanced growth paths with equalities in the constraints. Since λ^* is the root with *largest* modulus, the growth rate, which is the inverse of the root, is *smallest* for λ^*. No other growth paths of this kind, however, have associated nonnegative stock vectors [Result (e) of Section R7.3].

It is possible, of course, to have balanced growth without equilibrium, that is with surplus stocks. Unless there are surplus stocks of *all* commodities, the above analysis holds. If there are surpluses everywhere we have

$$K^*\Delta s(t) \ll (1/\mu')\Delta s(t).$$

Since $\Delta s(t) \geqq 0$, we can find some strictly positive matrix E such that

$$(K^* + E)\,\Delta s(t) \leqq (1/\mu')\,\Delta s(t), \qquad \text{at least one equality.}$$

The solution to this system is $1/\mu' = \lambda^*_{(K^*+E)}$. But K^*, $K^* + E$ are both semipositive indecomposable, with $K^* + E \gg K^*$, so that $\lambda^*_{(K^*+E)} > \lambda^*_{K^*}$ [Result (h) of Section R7.3] and $\mu' < \mu$. Thus the rate of balanced growth with surplus stocks is less than with equilibrium.

The balanced equilibrium growth rate μ^*, like many other equilibrium quantities in economics, has a kind of minimax or saddle-point quality. It is the smallest balanced growth rate with market equilibrium but the largest balanced growth rate without stock decumulation.

Numerical Example. We can dynamize the simple numerical example given in Chapter 6 by choosing

$$K = \begin{bmatrix} 4 & 0 \\ 0 & 3 \end{bmatrix} \qquad \Gamma = \begin{bmatrix} \frac{1}{2} & 0 \\ 0 & \frac{3}{4} \end{bmatrix} \qquad (I - \Gamma)^{-1} = \begin{bmatrix} 2 & 0 \\ 0 & 4 \end{bmatrix}.$$

The matrix K^* is then given by

$$K^* = \begin{bmatrix} 4 & 0 \\ 0 & 3 \end{bmatrix} \begin{bmatrix} 0 & 2 \\ \frac{1}{3} & 0 \end{bmatrix} \begin{bmatrix} 3 & 6 \\ 1 & 3 \end{bmatrix} \begin{bmatrix} 2 & 0 \\ 0 & 4 \end{bmatrix} = \begin{bmatrix} 16 & 96 \\ 6 & 24 \end{bmatrix}.$$

Direct calculation gives $\lambda^* = 45.1$, $x^* = (3.30, 1)$, so that the system has a growth rate of just over 2% per period along a balanced path with stocks in the ratio of 3.30 to 1. The other root is negative and its associated characteristic vector negative. The growth path associated with this would be an alternation, running down one stock in one period to build up the other, then reversing the roles of the two stocks in the next period.

10.3 THE VON NEUMANN GROWTH MODEL[5]

As we did with the dynamic input-output model of the previous section, we shall examine the simple balanced growth properties of a generalized linear model without investigating its full dynamics.

The Von Neumann growth model is defined by a technology of the standard activity analysis kind, with an output matrix B and an input matrix A. It is assumed that every good in the system is the output of some activity and that every activity requires some good in the system as an input. It can be viewed as a closed model in which even labor is produced by an activity, using consumer goods as inputs, or as an open model in which labor is never scarce and no activity has labor as its only input.

We shall make use of the idea of indecomposability as applied to this system. In Review R7 (Section R7.2) indecomposability is discussed as applied to a single matrix. Here we have a pair of matrices, but the general idea is the same. The system (B, A) is said to be indecomposable, if there is no subset of goods which can be produced without using at least one input not in the subset. The decomposability of the system depends on the relationship between the zeros in the B and A matrices. Von Neumann's original assumption was that the place (i, j) was occupied by a nonzero in either the B or the A matrix, a very strong assumption that rules out many interesting economic models.[6]

We shall construct a simple growth model by assuming that the activities $y(t)$ require the whole of the time period to produce their output, so that the inputs have to be in existence at the commencement of the period.

[5] Von Neumann's original paper was published in 1937. An English translation was published in 1946 (Von Neumann). The analysis given here is based on Gale [4] and [1].

[6] The original Von Neumann conditions were first relaxed in Kemeny, Morgenstern, and Thompson.

Because of the possibility of joint outputs, the Von Neumann model is particularly well adapted to a true capital model with depreciation by use, as opposed to inventory or eternal capital models. Consider, for example, a simple model with two capital goods, both used in the production of either. We suppose that either type of capital is part worn out after one use, completely worn out after two uses. Define four goods in the system as follows:

x_1; new type I capital,
x_2; once-used type I capital,
x_3: new type II capital,
x_4: once-used type II capital.

Then an activity using new capital will give an output of used capital of the same kind jointly with its new product. If once-used capital is an input, it does not reappear.

Such a model would have a technology of the following kind, where the first four activities, producing new type I capital, are shown:

$$B = \begin{bmatrix} b_{11} & b_{12} & b_{13} & b_{14} & \cdot \\ 1 & 1 & 0 & 0 & \cdot \\ 0 & 0 & 0 & 0 & \cdot \\ 1 & 0 & 1 & 0 & \cdot \end{bmatrix}, \qquad A = \begin{bmatrix} 1 & 1 & 0 & 0 & \cdot \\ 0 & 0 & a_{23} & a_{24} & \cdot \\ 1 & 0 & 1 & 0 & \cdot \\ 0 & a_{42} & 0 & a_{44} & \cdot \end{bmatrix}.$$

This kind of capital model cannot be constructed unless joint products are permitted.

The basic constraints of the Von Neumann model are, therefore,

$$Ay(t) \leqq By(t-1) \qquad y(t), y(t-1) \geqq 0.$$

We are interested in the balanced growth mode of the model, with $\alpha y(t-1) = y(t)$. We drop the dating of y and investigate the following problem, the *technological expansion problem*:

Find a positive α which is a maximum, subject to the constraints

$$\alpha Ay \leqq By, \qquad y \geqq 0.$$

This is a nonlinear problem since α, y, both unknowns, appear multiplied together in the constraints. To show that we can find a positive α and that this has a maximum, write the constraints in the form

$$(B - \alpha A)y \geqq 0, \qquad y \geqq 0.$$

Since $By \gg 0$ for some $y \geqq 0$ (because every good is the output of some activity), we can find some positive α small enough to satisfy the constraints.

On the other hand, α is bounded and so has a maximum since, by making it large enough, we can make $(B - \alpha A)y$ contain at least one negative element for every $y \geqslant 0$.

We shall denote the maximum value of α by α^*, and the associated vector y by y^*.

The real insights into the working of this model are obtained by setting up a dual problem, the interpretation of which will be apparent later.

Find β^*, p^* such that β^* is the *minimum* of all β's which satisfy

$$\beta pA \geqq pB, \qquad p \geqq 0.$$[7]

We can use the same kind of arguments as those used for the primal problem to convince ourselves that β has a minimum which is positive.

We shall now show that $\beta^* \leqq \alpha^*$.

It is clear that$(B - \alpha^* A)y \gg 0$ has no nonnegative solution since, if it had, we could increase α^*, contradicting its maximal property. Using a property of linear inequalities (Review R3, Section R3.7) we can therefore assert that $p(B - \alpha^* A) \leqq 0$ has a semipositive solution. It follows that, since β^* is the minimum of all β's satisfying $p(B - \beta A) \leqq 0$ for $p \geqq 0$, we have $\beta^* \leqq \alpha^*$.

Using the materials we have assembled, we are now ready to state and prove the following fundamental theorem:

Von Neumann Theorem

For the model defined by the technology (B, A) *in which every good is the output of some activity and every activity requires some good as input, there exist semipositive vectors* y, p *and a positive scalar* γ *such that: (a)* $\gamma Ay \leqq By$*; (b)* $\gamma pA \geqq pB$*; (c)* $p(\gamma A - B)y = 0$*; and (d) If the technology* (B, A) *is indecomposable,* γ *has a unique value* $\gamma^* = \alpha^* = \beta^*$*, where* α^* *and* β^* *are as defined earlier.*

Result (c), which implies that if $\gamma A^i y < B^i y$ then $p_i = 0$ or if $\gamma p A^j > pB^j$ then $y_j = 0$, has a resemblance to the equilibrium theorem of linear programming which should be noted.

To prove the theorem, put $\alpha^* \geqq \gamma \geqq \beta^*$, which is possible since $\alpha^* \geqq \beta^*$. Then γ, along with the vectors y^*, p^* associated with α^*, β^* certainly satisfy (a) and (b).

[7] This is often referred to as the *economic expansion problem*.

To prove (c), we have $pB \leqq \gamma pA$ and, since $y \geqslant 0$, this implies $pBy \leqq \gamma pAy$. In the same way, $By \geqq \gamma Ay$ implies $pBy \geqq \gamma pAy$, so that we must have $pBy = \gamma pAy$ or $p(\gamma A - B)y = 0$.

In the general case any value of γ between α^* and β^* is a solution, and there will usually be vectors p, y other than p^* and y^*, so the solution is not unique. To prove uniqueness in the indecomposable case, we first note that since p^*, y^* are semi-positive,

$$\alpha^* p^* A y^* \leqq p^* B y^* \leqq \beta^* p^* A y^*,$$

so that

$$(\beta^* - \alpha^*)\, p^* A y^* \geqq 0.$$

If we can show that $p^*Ay^* > 0$, then we will have shown that $\beta^* \geqq \alpha^*$. But we have already shown that $\beta^* \leqq \alpha^*$, so this will imply that $\beta^* = \alpha^*$.

It is obvious that $B^i y^* \geqq 0$ for every i and that $B^i y^* > 0$ for at least one i since y^* is semipositive and every good appears as the output of some activity. Suppose that we had $B^i y^* = 0$ for some i. Since $B^i y^* \geqq \alpha^* A^i y^*$ for all i, we could only have $B^i y^* = 0$, if we also had $A^i y^* = 0$. This would imply that the activity vector y^* neither produced nor used good i and therefore that all goods other than i formed an independent subset. Thus, if the system is indecomposable, we must have $By^* \gg 0$ and, since $p^* \geqslant 0$, $p^*By^* > 0$. But $\beta^* p^* A y^* \geqq p^* B y^*$ and $\beta^* > 0$ so that $p^*Ay^* > 0$.

Thus, if the system is indecomposable $\beta^* = \alpha^*$ so that Result (d) of the theorem is proved.

We can interpret the theorem, for the indecomposable case, in the following way:

From (a), the output of each good is at least equal to γ^* times the input of that good, which is assumed not greater than its output in the previous period. Thus the output of every good is growing at a rate at least equal to $\gamma^* - 1$.[8] From (c) any good whose output is growing at a rate greater than $\gamma^* - 1$ will have a zero price. All goods with positive prices will have the same growth rate, $\gamma^* - 1$.

From (b), the proceeds from operating any activity at unit level are less than or equal to γ^* times the cost of the inputs used, at the equilibrium prices p^*. Since the inputs can be regarded as having been invested in the process one period earlier, $\gamma^* - 1$ can be regarded as a "shadow" interest rate or rate of return on investment. From (c), no activity which, at the equilibrium

[8] We will assume the input-output relationships are such as to give a positive rate of growth. This implies some quantitative relationship between input and output coefficients analogous to the conditions that the simple input-output system be productive. The purely structural relationships required for the Von Neumann theorem merely guarantee that γ^* is positive, not that it is greater than unity.

prices, fails to give the shadow rate of return will be used. All activities actually used will give the same rate of return, $\gamma^* - 1$.

Although we have examined only the dynamically simple balanced growth mode, the general ideas which emerge from analysis of the Von Neumann model are fundamental to the general theory of growth in a multisector economy.

A simple input-output technology, with $B = I$ and A square, could be treated as a special case of the Von Neumann model. In this case we would have $\gamma^* = 1/\lambda^*$, where λ^* is the dominant root of the semipositive square matrix A, while y^*, p^* would be the associated column and row characteristic vectors (sometimes called the right- and left-hand eigenvectors). Otherwise the analysis would be basically the same.

Note that the dynamic input-output model actually discussed in the preceding section was not this special case of the Von Neumann model, but was a model with a different dynamic specification.

10.4 THE VON NEUMANN-LEONTIEF MODEL[9]

Suppose we have a model which is similar to the Von Neumann model except that there are no joint products. Each activity has but a single output, although there may be many activities producing that same output. We can regard such a model as a simplified Von Neumann model or a generalized Leontief model. In spirit it is probably best described as a generalized Leontief model, since the existence of joint outputs gives the ordinary Von Neumann model its special character as a capital expansion model. Whichever way we regard it, this model forms a bridge between the type of analysis used in the Von Neumann model and the analysis, based on the theory of semipositive square matrices, used in discussing Leontief models.

The B matrix, in this case, will consist of columns each containing a single nonzero entry, 1, at the place corresponding to the good which is the output of that activity. We shall suppose the activities to have been numbered so that the first set of activities consists of those that have good 1 as output, the next set of those having good 2 as output, and so on.

The B matrix will then have the general form:

$$\begin{bmatrix} 1 & 1 & 1 & \cdot & 0 & 0 & \cdot & 0 & \cdots \\ 0 & 0 & 0 & \cdot & 1 & 1 & \cdot & 0 & \cdots \\ 0 & 0 & 0 & \cdot & 0 & 0 & \cdot & 1 & \cdots \\ \cdot & \cdot & \cdot & \cdot & \cdot & \cdot & \cdot & \cdot & \cdots \end{bmatrix}.$$

[9] The analysis given here is based on the first part of the Morishima turnpike theorem. See Morishima [1] and [2].

The columns of A, which will not be different in form from those of the A matrix in the regular Von Neumann model, will have the corresponding arrangement. Now let us make all the subtechnologies of the model possible by taking exactly one activity for producing each good and denote the input matrix of the typical subtechnology by A_r. A_r is, of course, a square $n \times n$ submatrix of A. If the number of activities which have good j as output is m_j, the number of such matrices A_r will be $\Pi_j m_j$.

Each matrix A_r is a square semipositive matrix, which we take to be indecomposable, and so has a positive dominant characteristic root associated with strictly positive characteristic row and column vectors, from the results of Review R7.

Let us choose, from all the A_r, that whose *dominant* characteristic root is the *least*. Denote this matrix by A^*, its dominant characteristic root by λ^*, and the characteristic row and column vector associated with this root by p^*, x^*, respectively. We have $p^*, x^* \gg 0$.

Now define an activity vector y^* in the following way: If the jth activity is absent from A^*, $y_j^* = 0$; among the n nonzero components which remain, we put each one equal to the component of x^* which is the output of that particular activity. Thus y^* operates the activities which appear in A^* so as to produce x^* and has all other activities at zero level. We have $By^* = x^*$.

Then we can state the following:

The Von Neumann-Leontief model has a unique expansion factor $\gamma^* = 1/\lambda^*$ and unique price and activity vectors p^*, y^*, where λ^*, p^*, y^* are as defined above. Outputs of all goods grow in accordance with the relationship $\gamma^* A^* x^* = x^*$, so that there are no surpluses. All activities in A^* break even at interest rate $\gamma^* - 1$ and all other activities are unprofitable.

These results are much stronger than for the general model. Although indecomposability guarantees a unique γ^* for the general case, we cannot guarantee uniqueness of the price or activity vectors, nor can we guarantee that surpluses will not exist.

The above result is an analog of the *substitution theorem* for a growth version of input-output. So long as the economy grows along the unique Von Neumann path x^*, the activities not in A^* can be treated as though they do not exist.

Note that, in this case, we have conditions that the growth will actually be an expansion already available. For expansion to occur, we must have $\gamma^* > 1$, that is, $\lambda^* < 1$. But this is simply the ordinary condition that the input-output model with input matrix A^* be productive and was discussed in the preceding chapter.

To prove the theorem, we first prove the following *lemma*:

Any activity not in A^* makes a loss at prices p^* and interest rate $\gamma^* - 1$.

Let A^j be such an activity producing good i. Then we wish to show that

$$\lambda^* p_i^* < p^* A^j. \tag{10.4.1}$$

Suppose, to the contrary, that $\lambda^* p_i^* \geqq p^* A^j$. Replace the appropriate column (that giving output i) in A^* by A^j, and call the new matrix A^{**}. Then it is obvious that

$$\lambda^* p^* \geqq p^* A^{**}.$$

If λ^{**} is the dominant characteristic root and x^{**} the associated characteristic vector of A^{**}, we have, since $x^{**} \gg 0$,

$$\lambda^* p^* x^{**} \geqq p^* A^{**} x^{**} = \lambda^{**} p^* x^{**} > 0,$$

giving $\lambda^{**} \leqq \lambda^*$, in contradiction of the minimal property of λ^*, thus proving the lemma.

To prove the theorem, we have, from the lemma

$$\lambda^* p_i^* \leqq p^* A^j, \tag{10.4.2}$$

where A^j is *any* column of A (whether in A^* or not) and i the output of A^j. From this we have

$$\lambda^* p^* B \leqq p^* A. \tag{10.4.3}$$

Also, from the definition of λ^*, y^*, we have

$$\lambda^* B y^* = A y^*, \tag{10.4.4}$$

which also gives

$$\lambda^* p^* B y^* = p^* A y^*. \tag{10.4.5}$$

If we put $\gamma^* = 1/\lambda^*$, we see that (a), (b), and (c) of the Von Neumann theorem are satisfied by (10.4.4), (10.4.3) and (10.4.5), respectively. Thus $1/\lambda^*$, p^*, y^* are solutions to the Von Neumann model.

To prove uniqueness, we need to show that any solution to the model is equivalent to this. Let γ, p, y be a solution, and write $\mu = 1/\gamma$. Then, from the Von Neumann theorem

$$\mu B y \geqq A y.$$

But $p^* \gg 0$, so that

$$\mu p^* B y \geqq p^* A y. \tag{10.4.6}$$

From (10.4.3) we have $\lambda^* p^* B \leqq p^* A$. Since $y \geqq 0$, this gives

$$\lambda^* p^* By \leqq p^* Ay. \tag{10.4.7}$$

From (10.4.6) and (10.4.7) together we have

$$\mu p^* By \geqq p^* Ay \geqq \lambda^* p^* By. \tag{10.4.8}$$

Since $p^* By > 0$, this implies

$$\mu \geqq \lambda^*. \tag{10.4.9}$$

But, from (10.4.4), $\lambda^* By^* = Ay^*$, hence

$$\lambda^* pBy^* = pAy^*. \tag{10.4.10}$$

From the Von Neumann theorem, $\mu pB \leqq pA$, so that

$$\mu pBy^* \leqq pAy^*. \tag{10.4.11}$$

Thus (10.4.10) and (10.4.11) together imply

$$\lambda^* pBy^* = pAy^* \geqq \mu pBy^*; \tag{10.4.12}$$

hence that

$$\mu \leqq \lambda^*. \tag{10.4.13}$$

From (10.4.9) and (10.4.13) we have $\mu = \lambda^*$.

This equality, together with (10.4.12) gives $\lambda^* pBy^* = pAy^*$. But $By^* = x^*$, $Ay^* = A^* x^*$, so that we have

$$\lambda^* px^* = pA^* x^*.$$

Since $x^* \gg 0$, this implies

$$\lambda^* p = pA^*,$$

so that p is the characteristic row vector of A^* associated with λ^* and thus is equal (or proportional to) p^*.

We can complete the proof very easily by showing that $y = y^*$, using similar arguments to those used for $p = p^*$.

10.5 GENERAL BALANCED GROWTH MODELS[10]

The three balanced growth models discussed so far have all been characterized by linear technologies, with or without joint products. In this section

[10] This section is closely related to the original analysis of Solow and Samuelson. The material of Review R8, Section R8.8, is necessary background. The discussions of both this section and Section R8.8 can be carried considerably further. See Morishima [1], Appendix.

we shall generalize the results somewhat to cover cases in which the technology may be nonlinear, provided it shows constant returns to scale.

Consider an expanding model of the kind discussed in the Von Neumann and Von Neumann-Leontief cases, but in which each good has a linearly homogeneous production function of the form

$$y_i = f^i(x_{i1}, x_{i2}, \ldots, x_{in}) \qquad i = 1, \ldots, n. \tag{10.5.1}$$

We place no further restrictions on the functions f^i except that they are assumed to be positive nondecreasing functions of the inputs, and increasing functions of at least some inputs. We assume that total inputs are fixed, that is,

$$\sum_i x_{ij} = x_j \qquad j = 1, \ldots, n. \tag{10.5.2}$$

The relationships (10.5.1) and (10.5.2) define a transformation from the input vector x into an output vector y. This transformation represents a *point-to set* mapping, since x can be allocated in an infinite number of ways among the individual industries subject to (10.5.2), and each allocation may give rise to a different output vector y.

Let us suppose we are given a *homogeneous allocation rule*, that is, a rule which associates a unique set of x_{ij}'s with each x, subject to the requirement that if x^* gives the allocation x_{ij}^*, then kx^* gives the allocation kx_{ij}^*, for all i, j. It is obvious that such a rule is one that defines the *proportions* in which the goods are to be allocated. Since the rule gives a unique allocation of inputs for each x, it will result in a unique vector of outputs y. Due to the homogeneity of the allocation rule and the production functions, if x^* gives y^*, then kx^* gives ky^*. Thus we can state the following:

Given n linearly homogeneous production functions in n inputs, a homogeneous allocation rule will give a point-to-point mapping defined by the linearly homogeneous vector-valued function $y = F(x)$.[11]

It is clear that the functions F are of the kind studied in Review R8, Section R8.8. If we assume the simple dynamic relationship

$$x(t) = y(t - 1), \tag{10.5.3}$$

we can draw on the results of Section R8.8 and state immediately:

An economy of constant returns to scale production functions has at least one mode of balanced growth associated with every homogeneous allocation rule.

[11] Note that, whereas a given allocation rule gives a unique function F, the inverse relationship does not hold. A given F may be derived from a set of different allocation rules, since it may be possible to produce each good with a variety of different input combinations.

If the economy is indecomposable (there is no subset of goods that can be produced without the use of goods from outside the subset), there is a unique mode of balanced growth associated with each allocation rule.

Among the various allocation rules, clearly one (or more) lead to balanced growth at the highest rate. Such a mode we shall refer to as the *Von Neumann path*, generalizing from the original Von Neumann model.

In the next chapter we shall use these results in relation to neoclassical transformation functions. We can note here that, for the Von Neumann-Leontief model of the previous section, each subtechnology defines a function F of the kind discussed above. For the Von Neumann model itself, the situation is more complex. If we take any n activities, these define a square submatrix $\hat{A}$ of A and $\hat{B}$ of B. If $\hat{A}$ is nonsingular, we can write

$$y(t) = \hat{A}^{-1}\hat{B}y(t-1).$$

The matrix $\hat{A}^{-1}\hat{B}$ then defines a function F of the appropriate kind if and only if $\hat{A}^{-1}\hat{B} \geqq 0$. There will be a mode of balanced growth associated with every set of n activities which satisfy the above conditions.

FURTHER READING

The reader may care to follow up references given throughout the chapter, and to read the expositions in Dorfman, Samuelson, and Solow, Chapter 11, and Gale [1], Chapter 9. Otherwise it is suggested that Chapter 11 be read before branching out into the other literature.

CHAPTER

11

Efficient and Optimal Growth

This chapter follows on from the analysis of Chapter 10. No additional mathematical techniques are required for Sections 11.1 through 11.5. The simplest calculus of variations techniques are required for Section 11.6, and the relevant material is covered in Review R11, Section R11.2.

11.1 EFFICIENCY AND OPTIMALITY IN DYNAMIC MODELS[1]

In the previous chapter we investigated, in a purely descriptive fashion, the balanced growth mode in various dynamic models. Here we shall investigate optimal properties of growth paths.

Consider a simple dynamic model of an economy with three goods, a consumption good (C) and two capital goods (K_1, K_2). Production is assumed to be some function of the capital stocks only so that combinations of the consumer good and increments in the capital stocks are related by a production function:

$$T[C(t), \Delta K_1(t), \Delta K_2(t); K_1(t), K_2(t)] = 0.$$

[1] A slightly more elementary account of many of the topics discussed here is given in Dorfman, Samuelson, and Solow, Chapter 12. This chapter has been seminal in much recent growth theory. The reader is also referred to the survey in Hahn and Matthews, and to Koopmans [4].

The growth of the capital stocks is given by the dynamic relationships

$$K_1(t+1) = K_1(t) + \Delta K_1(t),$$
$$K_2(t+1) = K_2(t) + \Delta K_2(t).$$

It is obvious that, given arbitrary initial stocks $K_1(0)$, $K_2(0)$, the economy has open to it a doubly infinite set of growth paths in the sense that the time paths of any two of the variables $C(t)$, $K_1(t)$, $K_2(t)$ can be arbitrarily set within a certain range. We are interested in criteria for choosing among these paths.

If there are only a finite number of time periods[2] we can, in principle, optimize some objective function of the variables of the general kind,

$$\phi[C(0), K_1(1), K_2(1), C(1), \ldots, K_1(N), K_2(N), C(N)],$$

subject to the production constraints and the initial stocks $K_1(0)$, $K_2(0)$.

Most work on optimal dynamic models involves some simplification of the objective function ϕ. Two important cases can be distinguished.

(a) *Terminal Optimality*. Here C(n) is given for all n [in the simplest models $C(n) = 0$, all n], and the objective function $\phi[K_1(N), K_2(N)]$ is *a function of the terminal capital stocks only*. The constraints are the path $C(n)$, the production constraints, and the initial stocks.

(b) *Continuous Optimality*. Here the values of $C(n)$ are the most important ingredients in the objective function. This may have the form $\phi[C(0), \ldots, C(N)]$, *a function of the time path of consumption only*, subject to constraints on the terminal stocks and to the usual production and initial constraints. By making N large enough, the effect of the terminal stocks may be almost erased.[3]

In terms of traditional welfare ideas, continuous optimality should be the problem to which we should give our attention, since only consumption can generate individual utility or aggregate welfare. Considerable difficulties arise, however, in specifying the problem in this case because, in spite of some progress in the area in recent years,[4] there is no well-accepted theory for specifying the nature of a dynamic utility function. The simplest version of

[2] If the number of time periods is not finite (as in a continuous time model), we must use other methods. *Calculus of variations* is especially appropriate in continuous time formulations and is so used in Section 11.6.

[3] A careful distinction should be made between a continuous time model, with an infinite number of periods of infinitesimal length but a *finite* length of time until the terminal point, and the case in which $N \to \infty$, in which the time horizon itself recedes to an infinite distance.

[4] See, for example, Koopmans [5], Koopmans, Diamond, and Williamson, and Chakravarty.

the continuous optimality problem, the *optimum savings problem*, was investigated by Ramsey in 1927[5] and recent versions have been discussed by various authors.[6] The objective function is usually assumed to have one or other of the simple forms, $\phi = \sum C(n)$ or $\phi = \sum \rho^{-n}C(n)$, in these analyses.

Primarily for reasons of space, we shall not take up the continuous optimality problem in this book, but consider only *terminal optimality*. Here the objective function, depending on quantities having the same dates, can be specified in the same general form as in static models. The remaining sections of this chapter are devoted to the study of terminally optimal growth. Much insight into the nature of growth in a multisector economy can be obtained from this study, even though the rationale for optimizing on the terminal stocks at some specified time may lack firm foundation except in terms of planning expediency.

Efficiency and optimality are closely related concepts, in dynamic as well as static models. If we wish to make a clear distinction, an *optimal path* presumes the optimality criterion to have been specified, an *efficient path* is a path that is optimal for *some* appropriate criterion. Thus an inefficient path can never be optimal, and an optimal path is always efficient. We shall reserve the term "efficient path" for a path that is *terminally* optimal for some choice of objective function. In all the cases discussed in the following sections, the objective function will be linear, corresponding to some valuation of the terminal stocks at given prices.[7]

11.2 THE PRINCIPLE OF OPTIMALITY

A very useful concept in many discussions of optimal growth paths is an intuitive idea which is generally known as the *principle of optimality*. The concept in the particular form given below was enunciated by Bellman as a basis for dynamic programming.[8]

Principle of Optimality

An optimal policy has the property that whatever the initial state and initial decision are, the remaining decisions must constitute an optimal policy with regard to the state resulting from the first decision.

[5] See Ramsey. The original Ramsey problem (continuous time) is discussed in Allen [1]. A discrete-time version is given in Dorfman, Samuelson, and Solow (Chapter 11).

[6] A good introduction to this area is Phelps. See also Hahn and Matthews and McFadden's paper in Farrell and Hahn, also Samuelson's paper in Shell.

[7] No loss of generality results from this procedure. By choosing appropriate prices, the optimal path found by this method can be fitted to any objective function which depends on terminal outputs only.

[8] The statement given is from Bellman and Dreyfus (page 15). See also Bellman [2].

The principle can, in some contexts, be used to define an optimal policy. It is primarily designed for terminally optimal problems but, in spite of its strong intuitive appeal, its applicability should be checked in each class of problem. Problems can arise in applying the principle to continuously optimal cases.[9]

In the context of growth theory, the principle implies that, if a path is terminally optimal, commences from a point $x(0)$ and passes through $x(n)$ on its way to the terminal point $x(N)$, then the segment from $x(n)$ to $x(N)$ is terminally optimal with respect to initial point $x(n)$. It is then intuitive that the segment from $x(0)$ to $x(n)$ of the whole path must be terminally optimal with respect to $x(n)$ *on some suitable criterion*, and must therefore be efficient. Thus we can state a generalized principle of optimality for efficient paths:

Principle of Optimality for Efficient Growth
Any segment of an efficient growth path is itself an efficient growth path.

In applying this criterion to terminally optimal paths, the appropriate terminal criterion for a segment of the whole path must be established in each case. For example, if we are maximizing the value of terminal stocks $x(N)$ at some prices, the prices applicable to $x(n)$, treated as the terminal point, will, in general, differ from those applicable to $x(N)$[10]

11.3 EFFICIENT GROWTH[11]

In this section we shall consider efficient growth paths for an economy without consumption (or with consumption of each good in the same proportion to production as for every other good), with neoclassical transformation, discrete time periods and subject to terminal optimization.

The transformation relationship is taken to be in the implicit form:

$$T(y, x) = 0,$$

where $y, x \in R^m$. The vector y represents outputs, the vector x inputs. T is written so that y, x are both nonnegative, using the neoclassical tradition of separating inputs and outputs and writing them both positive, rather than the convention of generalized production theory in which the inputs are negative (as in Section 8.7 and Chapter 9).

[9] See, for example, Strotz, Pollak, and Blackorby.

[10] See Section 11.4.

[11] This is a generalization of the two-commodity analysis given in Dorfman, Samuelson, and Solow, Chapter 12. For a derivation of the intertemporal efficiency conditions in a continuous time model, using calculus of variations, see Samuelson [4].

$T(y, x)$ is neoclassical in the following senses:

(a) $T(x, y) \in C^2$;

(b) For fixed inputs x^*, the output set

$$Y(x^*) = \{y \mid T(y, x^*) = 0\}$$

is assumed to form a neoclassical transformation surface, convex outwards, of the kind studied in Chapter 8 (Sections 8.4 through 8.6).

(c) For given outputs y^*, the input set

$$X(y^*) = \{x \mid T(y^*, x) = 0\}$$

is assumed to form an isoquant surface, convex to the origin, of the traditional neoclassical kind.

(d) We assume homogeneity of degree zero so that, if $T(y, x) = 0$, $T(hy, hx) = 0$ for all $h > 0$.

Considered as a relationship between x and y, the transformation function gives a *point-to-set*, not a *point-to-point* mapping. That is, given x^* we have an output set $Y(x^*)$ and not an output vector. The distinction should, therefore, be kept in mind between the relationship $T(y, x)$ used here and the vector-valued function $y = F(x)$ (a point-to-point mapping) used in discussing *balanced* growth with neoclassical production (Section 10.5). The relationship between the two will be discussed in the next section.

As is usual in neoclassical production theory, $T(y, x)$ is assumed to represent an efficient relationship between inputs and outputs, in the static sense.

The dynamic specifications of the model are that *the outputs of period t form the inputs of period $t + 1$, that is,* $x(t + 1) = y(t)$.

Given $x(0)$, the economy can produce any output in the set $Y[x(0)]$. If an arbitrary $y(0) \in Y[x(0)]$ is chosen, we have $x(1) = y(0)$, and another output set from which an arbitrary choice can be made. Any path $y(0)$, $y(1), \ldots, y(N)$ obtained in this way is a *feasible path.* Among the set of feasible paths, we seek the subset of *efficient paths.*

Our criterion of terminal optimality will be the maximization of the value of terminal outputs $y(N)$ for given prices p^*. An efficient path will be defined as a path which is optimal for some price vector p^*.

An efficient path which is optimal for the particular prices p^* (assumed normalized) will be a solution of the optimizing problem:

$$\begin{aligned} &\max p^* y(N) \\ \text{S.T.}\quad &T[y(n), x(n)] = 0, \qquad n = 0, \ldots, N \\ &x(n + 1) - y(n) = 0, \qquad n = 0, \ldots, N - 1, \end{aligned}$$

in which we assume the implicit nonnegativity constraints are never effective, and that equalities are appropriate in the dynamic constraints.

This problem can be solved by classical methods. The Lagrangean will be defined as[12]:

$$L[y(0), \ldots, y(N), x(1), \ldots, x(N), \lambda_0, \ldots, \lambda_N, \mu_0, \ldots, \mu_N]$$

$$= p^*y(N) - \sum_0^N \lambda_n T[y(n), x(n)] - \sum_0^{N-1} \sum_1^m \mu_{n,1}[x_i(n+1) - y_i(n)], \quad (11.3.1)$$

where $y(n)$, $x(n)$, μ_n are m-vectors, λ_n is a scalar.

The first order conditions with respect to the variables $y_i(n)$, $x_i(n)$ are[13]

$$\frac{\partial L}{\partial y_i(n)} = -\lambda_n \frac{\partial T(n)}{\partial y_i(n)} + \mu_{n,i} = 0, \quad (11.3.2)$$

$$\frac{\partial L}{\partial x_i(n)} = -\lambda_n \frac{\partial T(n)}{\partial x_i(n)} - \mu_{n-1,i} = 0, \quad (11.3.3)$$

with the conditions for the initial and terminal periods,

$$\frac{\partial L}{\partial y_i(N)} = p_i^* - \lambda_N \frac{\partial T(N)}{\partial y_i(N)} = 0; \quad (11.3.4)$$

$$\frac{\partial L}{\partial y_i(0)} = -\lambda_0 \frac{\partial T(0)}{\partial y_i(0)} + \mu_{0,i} = 0. \quad (11.3.5)$$

By taking $n + 1$ in (11.3.3) instead of n, we obtain

$$-\lambda_{n+1} \frac{\partial T(n+1)}{\partial x_i(n+1)} - \mu_{n,i} = 0. \quad (11.3.6)$$

We can then eliminate $\mu_{n,i}$ between (11.3.2) and (11.3.6) to obtain

$$\lambda_{n+1} \frac{\partial T(n+1)}{\partial x_i(n+1)} = -\lambda_n \frac{\partial T(n)}{\partial y_i(n)}. \quad (11.3.7)$$

We can derive a similar relationship with j in place of i, then eliminate λ_n, λ_{n+1}, to obtain

$$\frac{\partial T(n+1)}{\partial x_i(n+1)} \Big/ \frac{\partial T(n+1)}{\partial x_j(n+1)} = \frac{\partial T(n)}{\partial y_i(n)} \Big/ \frac{\partial T(n)}{\partial y_j(n)} \quad (11.3.8)$$

which implies that the marginal rate of substitution between any two goods as *inputs* in period $n + 1$ is equal to their marginal rate of substitution as *outputs* in period n.

[12] Note that the constraints in a standard optimizing problem must be *scalar-valued* functions. We cannot treat $x(n+1) - x(n) = 0$ as a single constraint, but must take each of the constraints $x_i(n+1) - x_i(n) = 0$ separately.

[13] $T(n)$ is shorthand for $T[y(n), x(n)]$.

The conditions (11.3.8) are equivalent to Dorfman, Samuelson, and Solow's *intertemporal efficiency conditions*,[14] although the notation is different.

Since $T(n + 1)$ depends on $y(n + 1)$ and $x(n + 1)$ [$= y(n)$], and since $T(n)$ depends on $y(n)$ and $x(n)$ [$= y(n - 1)$], the efficiency conditions are difference equations of the second order. They are not *partial* difference equations, in spite of the prevalence of partial derivatives, since there is only one *independent* variable, n.

11.4 PROPERTIES OF EFFICIENT PATHS

The first task is to interpret the dual variables in the optimizing problem of the previous section. From (11.3.4) we have immediately

$$\lambda_N \frac{\partial T(N)}{\partial y_i(N)} = p_i^*,$$

so that, since $\lambda_n \partial T(n)/\partial y_i(n)$ must have the same dimension as $\lambda_N \partial T(N)/\partial y_i(N)$, and since from (11.3.2) we have

$$\lambda_n \frac{\partial T(n)}{\partial y_i(n)} = \mu_{n,i},$$

it follows that $\mu_{n,i}$ represents some kind of shadow price of the ith good in the nth period. Since $\mu_{n,i}$ has the same dimension as p_i^*, it can be interpreted as a "*money*" shadow price and $\mu_{n,i}/\lambda_n = \partial T(n)/\partial y_i(n)$ as a "*real*" shadow price.

To interpret the multipliers λ_n, we use the zero degree homogeneity of $T(y, x)$ and Euler's theorem to obtain

$$\sum \frac{\partial T(n)}{\partial y_i(n)} y_i(n) + \sum \frac{\partial T(n)}{\partial x_i(n)} x_i(n) = 0. \tag{11.4.1}$$

If we denote the first sum on the left-hand side of (11.4.1) by $V(n)$ and substitute $\mu_{n,i}/\lambda_n$ for $\partial T(n)/\partial y_i(n)$, we obtain

$$V(n) = 1/\lambda_n \sum \mu_{n,i} y_i(n), \tag{11.4.2}$$

with the obvious interpretation of $V(n)$ as the *real value of output in period n*.

In the second sum on the left-hand side of (11.4.1), we can substitute $-\mu_{n-1,i}/\lambda_n = \partial T(n)/\partial x_i(n)$ from (11.3.3), and $x_i(n) = y_i(n - 1)$ from the basic dynamic relationship, to obtain

$$\begin{aligned}\sum \frac{\partial T(n)}{\partial x_i(n)} x_i(n) &= -\frac{1}{\lambda_n} \sum \mu_{n-1,i} y_i(n - 1) \\ &= -\frac{\lambda_{n-1}}{\lambda_n} \sum \frac{\mu_{n-1,i}}{\lambda_{n-1}} y_i(n - 1) \\ &= -(\lambda_{n-1}/\lambda_n)\, V(n - 1).\end{aligned}$$

[14] See Dorfman, Samuelson, and Solow, Chapter 12.

From (11.4.1) we then have

$$V(n) = (\lambda_{n-1}/\lambda_n)\, V(n - 1). \tag{11.4.4}$$

Thus the ratio λ_{n-1}/λ_n represents the ratio of real output in period n to that in period $n - 1$. If we write $\rho(n) = \lambda_n/\lambda_{n+1}$, we can interpret $\rho(n)$ as the growth factor for period n. Note that this will not, in general, be constant, but will be a function of n.

If the economy is productive, then $\rho(n) > 1$, so that $\lambda_n < \lambda_{n-1}$, and the multipliers λ_n decrease with n.

Now $\lambda_n V(n) = \sum \mu_{n,i} y_i(n)$ is the money value of output in period n. From (11.4.4) we have

$$\lambda_n V(n) = \lambda_{n-1} V(n - 1),$$

so that the money value of output remains constant over time. Thus the "money" values should be interpreted as *discounted* money values. Since the reference money prices are the prices p^* of period N, the prices are discounted *forward*, consistent with the λ's decreasing with time. Thus $\rho(n)$ can be interpreted as a forward discount factor, with $\{1 - [p(n)]^{-1}\}$ as a "rate of interest" (not constant from period to period).

Is there an *efficient balanced* growth path? From the previous chapter (Section 10.5), we saw that a neoclassical economy has a balanced growth path corresponding to every homogeneous allocation rule that gives a point-to-point mapping from the transformation relationship. It is obvious that the efficiency conditions of the previous section are homogeneous and give a unique $y(0)$ from an arbitrary $x(0)$ and therefore constitute a point-to-point mapping. From the arguments of Section 10.5 it follows that there is only one positive vector $x(0)$ that will give $y(0) = \gamma x(0)$ and hence only one balanced growth path which is also efficient.

Let γ^* be the growth factor corresponding to the efficient balanced growth path. Then $y(n) = \gamma^* x(n) = \gamma^* y(n - 1)$. Since T is homogeneous of degree zero, in this case we have $\partial T(n)/\partial y_i(n) = \partial T(n - 1)/\partial y_i(n - 1)$ all i, n. Then

$$\begin{aligned} V(n) &= \sum \frac{\partial T(n)}{\partial y_i(n)} y_i(n) \\ &= \gamma^* \sum \frac{\partial T(n - 1)}{\partial y_i(n - 1)} y_i(n - 1) \\ &= \gamma^* V(n - 1), \end{aligned}$$

so that we identify γ^* with $\rho(n)$, as expected. Thus in the balanced growth case, the growth/discount factor is constant. From the definition of $\rho(n)$ we then have $\lambda_n = \gamma^{-n}\lambda_0$.

Consider an arbitrary balanced growth path with growth factor γ. Then the vectors $y(n)$, $x(n)$ must be such as to satisfy $T[y(n), x(n)] = 0$ and $y(n) = \gamma x(n)$ for all n. Obviously γ is a function of $x(n)$. Let us investigate the maximum of γ.

Using the relationships $T[y(n), x(n)] = 0$, $y(n) = \gamma x(n)$, and $\gamma = \gamma[x(n)]$, we can use implicit differentiation to obtain the equation

$$\gamma \frac{\partial T(n)}{\partial y_i(n)} + \left(\sum_j \frac{\partial T(n)}{\partial y_j(n)} x_j(n)\right) \frac{\partial r}{\partial x_i(n)} + \frac{\partial T(n)}{\partial x_i(n)} = 0,$$

or

$$-\left(\sum_j \frac{\partial T(n')}{\partial y_j(n')} x_j(n)\right) = \gamma \frac{\partial T(n)}{\partial y_i(n)} + \frac{\partial T(n)}{\partial x_i(n)}$$

Since $\left(\sum_j \frac{\partial T(n)}{\partial y_i(n)} x_j(n)\right)$ can be taken as nonzero, we have $\partial\gamma/\partial x_i(n) = 0$, if and only if

$$\gamma \frac{\partial T(n)}{\partial y_i(n)} = -\frac{\partial T(n)}{\partial x_i(n)}. \tag{11.4.5}$$

For γ to be a maximum, we must have $\partial\gamma/\partial x_i(n) = 0$ for all i, so that

$$\gamma \frac{\partial T(n)}{\partial y_j(n)} = -\frac{\partial T(n)}{\partial x_j(n)}. \tag{11.4.6}$$

The relationships (11.4.5) and (11.4.6) together give

$$\frac{\partial T(n)}{\partial x_i(n)} \Big/ \frac{\partial T(n)}{\partial x_j(n)} = \frac{\partial T(n)}{\partial y_i(n)} \Big/ \frac{\partial T(n)}{\partial y_j(n)}. \tag{11.4.7}$$

Since there is balanced growth, $x(n + 1) = \gamma x(n)$ so that, because $T(y, x)$ is homogeneous of degree zero, we have $\partial T(n + 1)/\partial x_i(n + 1) = \partial T(n)/\partial x_i(n)$ for all i. Using this relationship, we can replace (11.4.7) by

$$\frac{\partial T(n + 1)}{\partial x_i(n + 1)} \Big/ \frac{\partial T(n + 1)}{\partial x_j(n + 1)} = \frac{\partial T(n)}{\partial y_i(n)} \Big/ \frac{\partial T(n)}{\partial y_j(n)} \tag{11.4.8}$$

But (11.4.8) is identical with the intertemporal efficiency conditions (11.3.8) so that (not unexpectedly) the efficient balanced growth path turns out to be the one with the highest growth factor. Using the uniqueness property from Section 10.3, we can therefore state:

With neoclassical production, there is only one balanced growth path that is terminally optimal. This path has the largest growth factor among all balanced growth paths, and so may be identified as the Von Neumann path. The terminal prices associated with this path, and the half-line along which the path lies, will be identified as the Von Neumann prices and the Von Neumann ray.

We now turn from balanced growth paths back to efficient paths in general, in order to investigate the principle of optimality.

It is obvious that, given the continuity of $T(x, y)$ and its first and second order derivatives, there is a unique association between the terminal prices p^* and the terminal point $y(N)$, given N and $x(0)$. Thus we can expect to find a terminally optimal path whose terminal point is βx, where x is an arbitrary vector, if we can find the appropriate terminal prices.

Suppose we consider the problem of finding the optimal path which passes through the $y(n)$ of the original problem in period n, commencing from $x(0)$. Comparison of (11.3.2) and (11.3.4) show that the prices $\mu_{n,i}$ are appropriate terminal prices. It is also obvious that, if we replace N by n, and the prices p_i^* by $\mu_{n,i}$, in the original program, the optimal conditions for all periods through n are unchanged.

Now consider the problem with the original terminal conditions, but starting from the point $x(n + 1) = y(n)$ of the original optimal path and optimizing over $N - n$ periods. Then the original initial conditions (11.3.5) become

$$-\lambda_n \frac{\partial T(n)}{\partial y_i(n)} + \mu_{n,i} = 0.$$

But, from (11.3.2), these are satisfied by the original optimal path. Thus any segment of the terminally optimal path is terminally optimal between its end points, provided the terminal prices are the relevant shadow prices from the complete path. Thus:

Terminally optimal paths satisfy the principle of optimality.

11.5 A TURNPIKE THEOREM

The class of theorems that are concerned with the relationship of optimal growth paths to the Von Neumann path, especially in terms of "closeness" to the Von Neumann path, are generally referred to as *turnpike theorems.* The name was originally given by Dorfman, Samuelson, and Solow[15] and, although vivid, is not altogether apt. The type of path most appropriately described by the term, namely movement directly to the Von Neumann path, growth along that path, then movement off to the terminal point, is not, in general optimal.

Several turnpike theorems, of rather different kinds, have been proved,

[15] See Dorfman, Samuelson, and Solow, Chapter 12.

the best-known being those of Radner-Nikaido,[16] McKenzie,[17] and Morishima.[18] The theorem presented here is based on Radner's (which is a remarkable example of ingenuity), but modified and extended in the Nikaido manner.

We shall be concerned with exactly the same growth process studied in the two previous sections, with some slight additional conditions. We assume that the transformation relationship is strictly convex around the Von Neumann ray, and that free disposal (inefficient production in the static sense) is possible. Neither of these assumptions was needed in the previous discussion.

The theorem is concerned with the *distance*, in some sense, of a terminally optimal path from the Von Neumann path. We choose a distance measure that is invariant with respect to scale, so that it gives the same distance between λx and μy as between x and y. The measure we will use is the "normalized norm,"

$$d(x, y) = \left|\frac{x}{|x|} - \frac{y}{|y|}\right|,$$

where $|v|$ denotes the Euclidean norm of v.

Since $|\lambda x| = \lambda|x|$ and $|\mu y| = \mu\,|y|$, $(\lambda, \mu > 0)$, we have

$$d(\lambda x, \mu y) = \left|\frac{\lambda x}{\lambda|x|} - \frac{\mu y}{\mu|y|}\right| = \left|\frac{x}{|x|} - \frac{y}{|y|}\right| = d(x, y),$$

so that the measure $d(x, y)$ has the desired property.

In all the discussion which follows, the distance between two vectors x, y will mean the measure $d(x, y)$.

We shall denote the Von Neumann proportions by x^* (that is, the Von Neumann ray itself is the halfline λx^*, $\lambda \geqq 0$) and the Von Neumann prices and growth factor by p^*, γ^*. The strict convexity assumption on T can be taken to imply[19]

$$p^*y < \gamma^* p^* x, \qquad \text{all } x \text{ not on} \qquad \lambda x^*.$$

[16] See Radner and Nikaido. A simplified exposition of Radner's theorem is given in Hahn and Matthews. A generalization (of a kind different from that given in this chapter) is given in Morishima [1], Chapter VI, and another in Koopmans [4].

[17] See McKenzie [4], [5], and [6].

[18] See Morishima [2] and Morishima [1]. This theorem is for the Von Neumann-Leontief model and Section 10.4 of the previous chapter is, in effect, the first part of the theorem. The turnpike part proper is not discussed here.

[19] The conditions for strict convexity with constant returns to scale have been derived in Chapter 8 (Sections 8.4 through 8.6). For lack of proper statement of these conditions, it has often been incorrectly asserted that strict convexity requires externalities.

This, in turn, implies the following *lemma*, whose proof is immediate.

For $\varepsilon > 0$, *there is some* $\delta > 0$ *such that for all* y, x *satisfying* $T(y, x) = 0$ *and such that* $d(x, x^*) \geqq \varepsilon$, *we have*

$$p^*y \leqq (\gamma^* - \delta)p^*x.$$

We are now in a position to state the main theorem.

Turnpike Theorem

For an economy having a neoclassical transformation relationship $T(y, x) = 0$ *which is strictly convex around the Von Neumann ray, then:*

(1) *For any* $\varepsilon > 0$, *a terminally optimal path over a finite number of periods N will be at a distance greater than, or equal to,* ε *from the Von Neumann path for not more than M periods, where M is a finite number independent of N but, in general, dependent on* ε.[20]

Subject to additional special assumptions, the following can also be proved:

(2) *If the terminal prices are the Von Neumann prices, the terminally optimal path is closer to the Von Neumann path in each succeeding period, and is asymptotic to the Von Neumann path as N increases indefinitely.*

(3) *On a terminally optimal path which passes through two points which are the same distance from the Von Neumann path, all intermediate points will be closer to the Von Neumann path than are the end points.*

If we use this theorem in conjunction with the principle of optimality, it is easy to visualize the shape of typical terminally optimal paths. Illustrations are given in Figure 11–1 [illustrating (1)] and Figure 11–2 [illustrating (2) and (3)].

The proof of the theorem depends on a technique of comparing two paths, a certain *comparison path* and the optimal path. Given the terminal prices and the initial point, the value of the terminal output must be higher on the optimal path, and we use this relationship to derive the properties of the path.

We define the *comparison path* in the following way:

(a) Given the initial point $x(0)$, we dispose of whatever stocks are necessary to attain some point kx^* on the Von Neumann ray, with k a maximum. Obviously $kx^* \leqq x(0)$, with equality for at least one good.

[20] This is the content of Radner's original theorem.

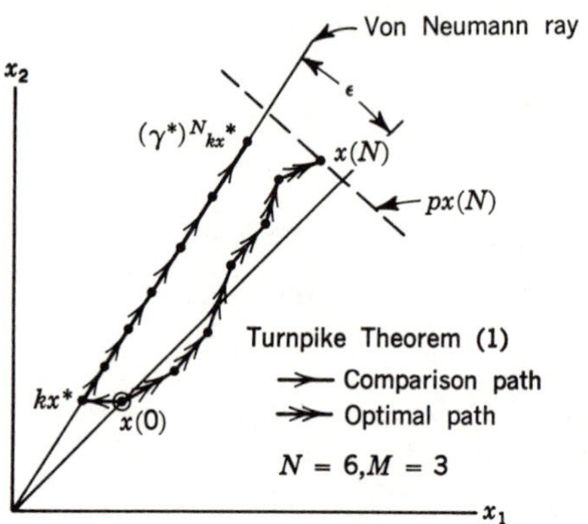

Figure 11–1 Turnpike paths.

(b) The economy then grows along the Von Neumann path until period N, reaching the terminal point $(\gamma^*)^N kx^*$.[21]

If p denotes the terminal prices, the value of the terminal output along the comparison path is given by

$$V^* = (\gamma^*)^N kpx^*. \tag{11.5.1}$$

Now consider any other feasible path commencing from $x(0)$ and continuing for N periods. Suppose that this path lies at a distance at least equal to ε [that is, $d(x, x^*) \geqq \varepsilon$] for M periods.

From the *lemma*, we have, for each of the periods in which $d(x, x^*) \geqq \varepsilon$,

$$p^*y(n) \leqq (\gamma^* - \delta)\, p^*x(n), \tag{11.5.2}$$

that is,

$$p^*x(n + 1) \leqq (\gamma^* - \delta)\, p^*x(n).$$

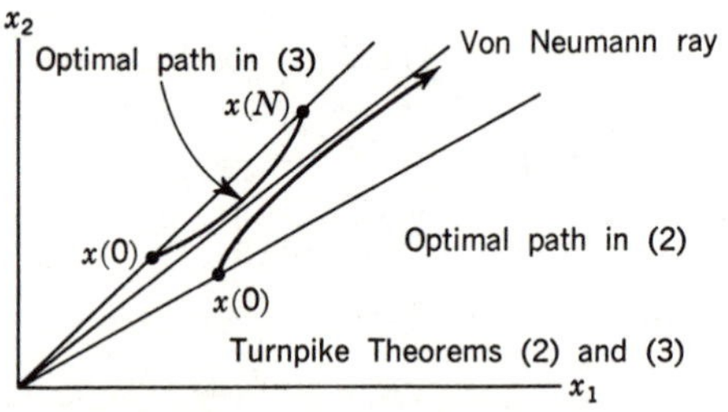

Figure 11–2 Turnpike paths.

[21] It is this comparison path which fits best the description of a "turnpike" path, rather than the path we shall presume to be optimal.

By definition of the Von Neumann path, we have, for all $x(n)$,

$$x(n+1) \leqq \gamma^* x(n),$$

so that

$$p^* x(n+1) \leqq \gamma^* p^* x(n). \tag{11.5.3}$$

Using the inequality (11.5.2) for the M periods in which $d[x(n), x^*] \geqq \varepsilon$, and the inequality (11.5.3) for the remaining $N - M$ periods, we have

$$p^* y(N) \leqq (\gamma^* - \delta)^M (\gamma^*)^{N-M} p^* x(0). \tag{11.5.4}$$

Thus we have an upper limit on the value of the terminal output, *but measured in Von Neumann prices.* We are interested in the value in terms of the prices p.

Taking both the Von Neumann prices and the terminal prices to be normalized, define

$$a = \min_i p_i^*,$$

and

$$b = \max_i p_i.$$

Then, for all x,

$$px \leqq b[1]x \qquad \text{and} \qquad a[1]x \leqq p^* x,$$ [22]

so that

$$px \leqq (b/a)\, p^* x.$$

Then, from (11.5.4) we have

$$V = py(N) \leqq (b/a)\, (\gamma^* - \delta)^M (\gamma^*)^{N-M} p^* x(0). \tag{11.5.5}$$

Now compare the terminal values of the arbitrary path and the comparison path. We have

$$\frac{V}{V^*} \leqq \frac{(\gamma^* - \delta)^M (\gamma^*)^{N-M}}{(\gamma^*)^N} \frac{bp^* x(0)}{akpx^*}. \tag{11.5.6}$$

Since $a, b, p, p^*, x^*, k, x(0)$ depend only on the properties of T, the initial conditions and the terminal prices, they are all fixed for a given problem.

Now $kx^* \leqq x(0)$, so that $kp^* x^* \leqq p^* x(0)$. By the same argument as used above, we have $kpx^* \leqq k(b/a)\, p^* x^* \leqq (b/a)\, p^* x(0)$. Thus we can write

$$K = \frac{bp^* x(0)}{akpx^*}, \tag{11.5.7}$$

where K is a constant of the problem and $K \geqq 1$.

[22] [1] is a unit vector.

If the arbitrary path is optimal, then $V \geqq V^*$, so that $V/V^* \geqq 1$. But, from (11.5.6) and (11.5.7),

$$V/V^* \leqq K\left(\frac{\gamma^* - \delta}{\gamma^*}\right)^M \tag{11.5.8}$$

so that, for optimality we must have

$$K\left(\frac{\gamma^* - \delta}{\gamma^*}\right)^M \geqq 1. \tag{11.5.9}$$

The only variable is M, the number of periods for which the optimal path lies at a distance at least equal to ε from the Von Neumann path. To satisfy the inequality (11.5.9) we must have M such that

$$r^M \leqq K, \tag{11.5.10}$$

where $r = \gamma^*/(\gamma^* - \delta)$. Since r, $K \geqq 1$, this has a solution $M \geqq 0$. Neither K nor r depend on N, so Part (1) of the theorem is proved.

Certain points should be noted. If r is close to 1, but K is much larger than 1, M may be large—much larger than N, in fact, so that the theorem is empty in this case. In general, for a smooth transformation relationship, δ will be a continuous increasing function of ε, so that r will be a decreasing function of ε. Thus the smaller is ε, the larger will be the number M. The closer that $x(0)$ is to the Von Neumann ray and the closer [in the sense of the ratio (b/a)] that the terminal prices are to the Von Neumann prices, the closer K will be to unity. Therefore, closeness of the initial or terminal points to the Von Neumann ray will reduce the number of periods for which the optimal path is "distant" from the Von Neumann ray, as expected. In particular, if $x(0) = kx^*$ and $p = p^*$, we will have $K = 1$, giving $M = 0$, so the theorem correctly states the case of the Von Neumann path, itself.

To prove special result (2) of the theorem, we first assume that $x(0)$ is "*sufficiently near*" x^* in the sense that, starting from $x(0)$, it is possible to grow to the Von Neumann ray in a single period. That is, $T[\lambda x^*, x(0)] = 0$ for some λ.

Consider the inequality

$$p^*x(1) \leqq (\gamma^* - \delta)\, p^*x(0)$$

for all $x(1)$. From the properties of T, we can assume that $p^*x(1)$ is maximized for growth *towards* the Von Neumann ray and that, in particular, we can attain a point $x^*(1) = \lambda x^*$ such that

$$p^*x^*(1) = (\gamma^* - \delta)\, p^*x(0).$$

We now take as the comparison path the path $x(0) \to x^*(1)$, followed by growth along the Von Neumann ray for $N - 1$ periods. This gives a final value,

$$V^{**} = (\gamma^* - \delta)(\gamma^*)^{N-1} p^*x(0).$$

Since the terminal prices in this case are the Von Neumann prices, the value of the arbitrary path is given from (11.5.4) as [23]

$$V = p^*y(N) \leqq (\gamma^* - \delta)^N (\gamma^*)^{N-M} p^*x(0).$$

The criterion for optimality, $V \geqq V^{**}$, now becomes

$$\frac{(\gamma^* - \delta)^N (\gamma^*)^{N-M} p^*x(0)}{(\gamma^* - \delta)(\gamma^*)^{N-1} p^*x(0)} \geqq 1,$$

which gives

$$r^{M-1} \leqq 1.$$

Since $r > 1$, this has a unique solution $M - 1 = 0$, that is, $M = 1$. Since $x(0)$ already lies at the stated distance ε from x^*, we can state:

If the path is optimal for Von Neumann prices and commences from a point sufficiently close to the Von Neumann ray, it lies closer than the initial point to the Von Neumann ray at all subsequent periods.

In particular, $d[x(1), x^*] < d[x(0), x^*]$. From the principle of optimality, we know that the original path viewed as commencing from $x(1)$ is terminally optimal at Von Neumann prices over $N - 1$ periods. We can then apply the same argument to conclude that

$$d[x(2), x^*] < d[x(1), x^*],$$

and, in general, that

$$d[x(n + 1), x^*] < d[x(n), x^*],$$

which proves part (2) of the theorem.

For part (3), we set up a new comparison path, similar to that for part (2) up to the $(N - 1)$th period, at which time its value will be

$$p^*x^*(N - 1) = (\gamma^* - \delta)(\gamma^*)^{N-2} p^*x(0).$$

The economy now grows directly to some point $\mu x(N)$, where the proportions are the same as for the terminal point of the optimal path itself.

[23] Note that, although $p^*x^*(1) = (\gamma^* - \delta) p^*x(0)$, we will expect to have $p^*x(1) < (\gamma^* - \delta) p^*x(0)$ along the optimal path (provided $N > 1$) because the segment $x(0) \to x(1)$ of the over-all path is not terminally optimal at prices p^* but at the shadow prices $\mu_{1,i}$ given in Section 11.4.

Since $d[x(N), x^*] = d[x(0), x^*]$, by assumption, this is possible for the terminal point if it was possible for the initial point, We will then assume that the point $\mu x(N)$ has the property

$$\mu p^* x(N) = (\gamma^* - \delta)\, p^* x^*(N - 1),$$

at least to an order of approximation sufficient for our purposes. This is a symmetry assumption, requiring that maximal growth from a distance ε to the Von Neumann ray is almost equal to maximal growth from the Von Neumann ray to a distance ε from it. Along the comparison path we then have

$$\mu p^* x(N) = (\gamma^* - \delta)^2 (\gamma^*)^{N-2} p^* x(0).$$

Along the presumptively optimal path we have

$$p^* y(N) \leqq (\gamma^* - \delta)^N (\gamma^*)^{N-M} p^* x(0).$$

We do not use terminal prices here, since both terminal points lie along the same ray. The optimal path is so only if $\mu \leqq 1$, that is, if

$$\frac{(\gamma^* - \delta)^2 (\gamma^*)^{N-2}}{(\gamma^* - \delta)^N (\gamma^*)^{N-M}} \leqq 1.$$

This gives

$$r^{M-2} \leqq 1.$$

Since $r > 1$, we conclude that $M = 2$. But the two terminal points lie at a distance ε from x^*, so that all other points are closer, proving (3) of the theorem.

It is obvious that the proofs of (2) and (3) can be tightened, and that it should be possible to give a global proof of these.

11.6 AN EXPLICIT TURNPIKE EXAMPLE[24]

The general shape of efficient growth paths, established by the turnpike theorem of the preceding section, can be warranted by the example developed in this section, which also serves as an example of the use of calculus of variations methods.

[24] This section requires the use of simple calculus of variations methods. The relevant material is given in Review R11, Section R11.2.

The example given here was circulated but not published by the author some years ago. The original version contained an error which was pointed out to the author by Paul Samuelson.

We assume continuous time and a transformation function of the form:

$$T(\dot{y}_1, \dot{y}_2, \dot{x}_1, \dot{x}_2) = a^2(\dot{y}_1)^2 + (\dot{y}_2)^2 - b^2\dot{x}_1^2[f(\dot{x}_2/\dot{x}_1)^2] = 0, \quad (11.6.1)$$

where $\dot{y}_1$, $\dot{y}_2$ are rates of output, $\dot{x}_1$, $\dot{x}_2$ rates of input. The only assumptions made on the function f are that $f > 0, f' > 0, f'' < 0$. It is clear that T is a neoclassical transformation function, linear homogeneous, with an elliptical production possibility curve and a neoclassical isoquant.

We make the following dynamic assumptions:

$$\begin{aligned} \dot{x}_1(t) &= k_1 y_1(t), \\ \dot{x}_2(t) &= k_2 y_2(t). \end{aligned} \quad (11.6.2)$$

That is, we identify $\dot{y}_1$, $\dot{y}_2$ as the rates of increase in the stocks of productive resources, and assume the input flows are proportional to the resource stocks.

This particular form of the transformation function enables us to make a change of variables that permits obtaining an almost explicit solution for the efficient growth paths. We change variables as follows:[25]

$$\begin{aligned} y_1 &= (1/a)\, e^u \cos v, \\ y_2 &= \qquad\ \ e^u \sin v. \end{aligned} \quad (11.6.3)$$

Together with the relationships (11.6.2), this enables us to write (11.6.1) in the form,

$$(\dot{u})^2 + (\dot{v})^2 = [(bk_1/a)(\cos v) f[(ak_2/k_1)\tan v]]^2 \quad (11.6.4)$$

Writing the expression in the bracket on the right-hand side as $[\phi(v)]^{-1}$ (the reason for choosing the reciprocal will be apparent later) and noting that v' $(= dv/du) = \dot{v}/\dot{u}$, we can reduce (11.6.4) to the form,

$$\dot{u} = [1 + (v')^2]^{-1/2}[\phi(v)]^{-1}. \quad (11.6.5)$$

Since v is a function of the ratio y_1/y_2 only, we have $v' = 0$ for balanced growth. Since u is a logarithmic quantity, the rate of balanced growth is given by $\dot{u}$. Thus, for balanced growth in proportions represented by v, the growth rate is given, from (11.6.5), as

$$g(v) = \dot{u} = [\phi(v)]^{-1}.$$

It is easy to show that there always exists some v^* for which $g'(v^*) = 0$ and $g''(v^*) < 0$, so that the rate of growth has a strong maximum at v^* which we can identify as the *Von Neumann proportions.*

[25] This is equivalent to stretching or shrinking the y_1 axis to give a circular production possibility curve, then transforming to polar coordinates, and finally using the logarithm of the radius vector.

We can solve (11.6.5) to obtain t explicitly as

$$t = \int [1 + (v')^2]^{1/2}\,\phi(v)\,du, \tag{11.6.6}$$

where the integral is taken along some feasible path between given endpoints. The efficient path will be that path that minimizes the time between the endpoints, and hence which minimizes the integral in (11.6.6). This is an example of the *First Problem of the Calculus of Variations* (see Review R11, Section R11.2). If we denote the integrand by F, we note that u does not appear explicitly in F, so that this is special case (2) of the Euler equation, which reduces to an equation of the form,[26]

$$F - v'F_{v'} = C,$$

where C is a constant of integration. In this case, we have $F_{v'} = [1 + (v')^2]^{-1/2}v'\phi(v)$ so that the Euler equation becomes, after some manipulation,

$$[1 + (v')^2]^{-1/2}\phi(v) = C.$$

We can solve for v' to obtain the fundamental efficiency condition for this problem as

$$v' = \left\{\frac{[\phi(v)]^2}{C^2} - 1\right\}^{1/2}. \tag{11.6.7}$$

This cannot be integrated, in general, but we can determine the properties of efficient paths directly from (11.6.7) since this gives the slope of the path in terms of the stock proportions v and the arbitrary constant C. Each value of C corresponds to a particular type of path, and an efficient path through any point will be uniquely determined by C and the initial value of u at that point.

First we recall that the maximum growth rate $g(v) = [\phi(v)]^{-1}$ occurred at the Von Neumann proportions, so that $\phi(v)/\phi(v^*) \geqq 1$ for all v. Thus for $C \leqq \phi(v^*)$, there is a real value of v' corresponding to every v and such a path can pass through all commodity proportions. For $C > \phi(v^*)$, however, there will be no real value of v' for $v = v^*$ and for some neighborhood of v^*. Such a path cannot cross or meet the Von Neumann ray. There will, however, be values $\bar{v}$, $\underline{v}$ for which $\phi(\bar{v}) = \phi(\underline{v}) = C$ (we will have $v' = 0$ at these values of v). For $v \geqq \bar{v}$ and $v \leqq \underline{v}$ real values of v' will exist, so that there are efficient paths corresponding to these values of C.

[26] The Euler equation is the continuous-time equivalent of the intertemporal efficiency conditions of Section 11.3. Note that the independent variable here is u, not t as in the formulation in Section R11.2.

We can calculate v'' from (11.6.7), obtaining

$$v'' = \frac{\phi(v)}{C^2}\,\phi'(v) \tag{11.6.8}$$

Since $\phi(v)$ [$= 1/g(v)$] has a minimum at the Von Neumann proportions, $\phi'(v) \gtreqless 0$ according as $v \gtreqless v^*$. From (11.6.8) it follows that $v'' \gtreqless 0$ according as $v \gtreqless v^*$. *The efficient paths are therefore convex toward the Von Neumann ray* from both above and below. Paths which cross the Von Neumann ray will be inflected at the crossing, while paths that do not will be bowed towards the ray, being closest to the ray at $\bar{v}$ (for paths that commence with $v_0 > v^*$) or $\underline{v}$ (for paths that commence with $v_0 < v^*$).

Now, $\phi(v) \neq 0$, $v'' = 0$, if and only if $\phi'(v) = 0$, which is true only for $v = v^*$. Thus the only efficient balanced growth path of finite length ($v', v'' = 0$) is the Von Neumann ray itself. All paths for which $C > \phi(v^*)$ have $v' = 0$ at $v = \bar{v}$ or $\underline{v}$, but $v'' \neq 0$ at these points, so there is merely instantaneously tangency with the ray $\underline{v}$ or $\bar{v}$ and no finite travel along the ray.

Finally, we note that (11.6.7) gives both a positive and a negative value of v' for given v, so that the bowed paths [$C > \phi(v^*)$] are symmetric in the direction of the u axis around the point corresponding to the closest approach to the Von Neumann ray.

We have now assembled the material for a complete description of the efficient paths. For a given initial point, each value of C defines a unique efficient path. Each path falls into one of the following three types:

***Type* 1.** $C > \phi(v^*)$. The path is continuously curved and convex toward the Von Neumann ray, with the bowed or catenary shape often described in discussions of turnpike paths. The path does not reach the Von Neumann ray, but is closest to it at the point $\underline{v}$ (assuming that $v_0 < v^*$) such that $\phi(\underline{v}) = C$. Such a path lies entirely on the side of the Von Neumann ray from which it started.

***Type* 2.** $C = \phi(v^*)$. In this case $\underline{v} = v^*$, so the path reaches the Von Neumann ray. However $v' = 0$ at $v = v^*$, so the path does not cross the ray. It can be shown that such a path does not reach the Von Neumann ray at a finite distance, so that the path is asymptotic to the ray, unless $v_0 = v^*$, in which case the path is the Von Neumann ray itself.

***Type* 3.** $C < \phi(v^*)$. In this case, v' has a real value for all v, so that the path crosses rays of all proportions. $v' \neq 0$ for any v, so the path is not tangent to any ray and crosses the Von Neumann ray at an angle. At the Von

Neumann ray, $v'' = 0$, so the path is inflected from convex upward to convex downward. This is the type of path for end points on opposite sides of the Von Neumann ray, or when one endpoint is on the Von Neumann ray itself.

Note that the paths described by an arbitrary point and a given value of C extend indefinitely in both directions. The path of a particular problem is the segment of the over-all path on which the endpoints lie.

The three types of path are illustrated in Figure 11–3. This has been drawn with v, u as rectangular coordinates. If we transformed back to the variables y_1, y_2, the paths would be similar to those in Figure 11–2.

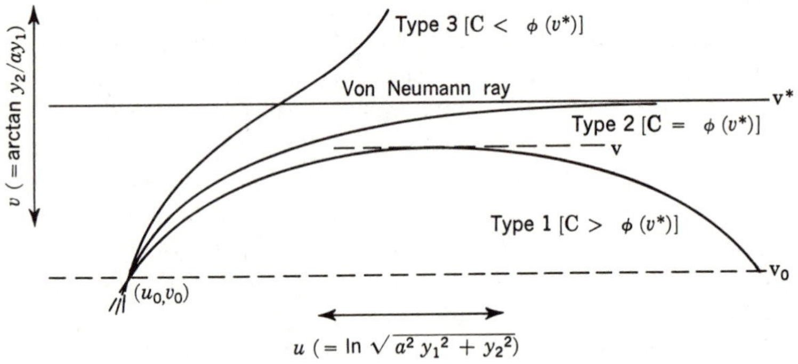

Figure 11–3 Paths in explicit turnpike example.

FURTHER READING

If the reader has not already done so, it is recommended that he read Dorfman, Samuelson, and Solow, Chapter 12, Hahn and Matthews, and Koopmans [4]. He might then study the Morishima turnpike theorem (Morishima [1], Chapter VI), which follows on from Section 10.4 of the previous chapter, McKenzie [6], and then turn to the periodical literature referred to in the above and to the papers in Farrell and Hahn and in Shell.

CHAPTER

12

Stability

This chapter requires familiarity with vector differential and difference equations (Review R10) and also with the properties of dominant diagonal matrices (Review R7, Section R7.4).

12.1 THE CONCEPT OF STABILITY[1]

"Stability" is, like "equilibrium," a term with strong intuitive connotations that must be defined suitably in each particular case. Some idea of the variety of ways in which the term may be used is given from the following illustration.

Consider a dynamic system whose behavior is represented by the path

$$y(t) = ae^{\alpha t} + be^{\beta t} + c,$$

where $y(t)$ is scalar valued and α, β are real. Among the various types of economic model that might be represented by a time path with such an equation, let us consider two.

[1] Simple models of market stability are discussed in many price theory texts, for example Henderson and Quandt, and also in Baumol [1] and [2]. The more general analysis of multiple market stability is not available at a more elementary level than given here. At a more advanced level, Karlin [1] is especially rich in discussions of variations on the basic models presented here, and further reading should commence with Karlin. Morishima [1] discusses market stability from a slightly different viewpoint.

(a) *A static disequilibrium model.* We might interpret c as an equilibrium value for $y(t)$, with a, b constants that fit the dynamic path to initial conditions $y(0)$, $Dy(0)$. Obviously $y(t) \to c$ as $t \to \infty$, if and only if $\alpha, \beta < 0$. We would have no doubts in calling this system stable if $\alpha, \beta < 0$, unstable if either α or β was positive.

(b) *A growth model.* We might interpret the model as, say, a multiplier-accelerator model. Suppose $\beta > 0$. From the standpoint of (a), the model would be unstable, but then so are all steady growth processes from this point of view.

Let us suppose that, for reasons which must be derived from the substantive interpretation of the model, we regard the path $be^{\beta t}$ as an *equilibrium* growth path. We wish to consider appropriate ideas of stability.

Define the following quantities which are obviously relevant to analysis of the system as a growth model.

$$g(t) = [1/y(t)]\, Dy(t) \qquad \text{(the rate of growth)}$$

$$= \frac{a\alpha e^{\alpha t} + b\beta e^{\beta t}}{ae^{\alpha t} + be^{\beta t} + c}$$

$$= \frac{(a/b)\, \alpha e^{(\alpha - \beta)t} + \beta}{(a/b)\, e^{(\alpha - \beta)t} + 1 + (c/b)\, e^{-\beta t}} \qquad \text{(dividing top and bottom by } be^{\beta t}\text{)}$$

$$h(t) = y(t) - be^{\beta t} \qquad \text{(deviation from equilibrium)}$$

$$ae^{\alpha t} + c.$$

Now $g(t) \to \beta$ when $t \to \infty$ if $\alpha - \beta < 0$. Since $\beta > 0$, this is possible with $\alpha > 0$. Thus we could choose to say that the system was stable for $0 < \alpha < \beta$, because the *equilibrium growth rate* is stable under these circumstances. On the other hand we might choose to say that the system was unstable because, under these circumstances, the deviation from the equilibrium growth path $[h(t)]$ increases with t if $\alpha > 0$. Then again the *proportional* deviation from the equilibrium path $[h(t)/be^{\beta t}]$ does converge to zero for $0 < \alpha < \beta$. Suppose we had $\alpha < 0$. Then $h(t) \to c$ as $t \to \infty$. Are we to regard this as stable? [2]

Thus the same mathematical model has a variety of different stability criteria, depending on the underlying substantive model of which it is a representation. Note that the definition of stability depends on the definition of equilibrium.

Let us suppose we have some particular path $y^*(t)$ which is regarded as an equilibrium path, with y now interpreted as a point in R^n. Suppose also

[2] In effect, we showed stability of the Von Neumann growth path in Section 11.5 of the preceding chapter. We do not discuss further problems of growth stability here. For a discussion of the stability properties of the single-sector growth model, see Jorgenson.

that there are other paths $y(t)$, functionally related to $y^*(t)$. Then the most common usage is to say that the system is *stable* if every path $y(t)$ eventually enters, then remains within, some bounded region containing $y^*(t)$ as t increases, and to say it is *asymptotically stable* if the region is a neighborhood of $y^*(t)$. In the growth version of the example given, we would say that $y(t)$ was stable for $\alpha < 0$, since $y(t)$ converges to $y^*(t) + c$, and asymptotically stable for $\alpha < 0$ and $c = 0$. On the other hand, we could transform the variables to $z(t) = y(t)/y^*(t)$, $[y^*(t) = be^{\beta t}]$, $z^*(t) = 1$ and conclude that $z(t) \to z^*(t)$ as $t \to \infty$, so that the transformed system is asymptotically stable for all c and $0 < \alpha < \beta$. Thus we cannot simply use stability definitions in a mechanical fashion, we must relate them to the problem in hand.

There are a variety of uses of qualifying words in discussing stability. Weak and strong stability are sometimes used in place of stability and asymptotic stability, and other terms such as regular stability may be used in particular cases. The usage must be clearly defined for each model, unless it is quite straightforward.

Two terms that are usually unambiguous are *global* and *local* stability. A system is globally stable if the stability property being used is possessed by the system for all paths within the whole region under consideration. It is locally stable if the stability property is established only for paths which lie "near" the equilibrium path in some suitable sense.

Additional problems are posed by the existence of multiple equilibria. In many economic models, equilibrium may be a set and not a point. The case that we shall be most concerned with is that of equilibrium prices in market models. In the absence of money illusion or real balance effects, we expect that, if prices p^* equilibrate the market, then λp^* are also equilibrium prices, for all $\lambda > 0$. We shall usually regard a price adjustment process as stable if it converges to λp^*, for any $\lambda > 0$.

Problems of nonconnected multiple optima, important in physical science problems, are not usually present in economics. The multiple optima with which we are concerned will typically all be members of a connected, usually convex, set.

12.2 STABILITY ANALYSIS

There are two main approaches to the analysis of stability, *explicit solution* and the *Liapunov method*. In discussing both we shall assume:

(a) That the system is expressed in the form of a differential or difference equation;

(b) That the variables have been transformed, if necessary, so that the equilibrium is $y^*(t) = 0$. That is, the system is considered in a form in which the variables represent deviations from equilibrium.

If the relevant differential or difference equations can be solved, we obtain the function $y(t)$ whose behavior as t increases can be investigated directly. In particular, if the system equations are linear with constant coefficients, we obtain solutions of the form:

$$y(t) = \sum k_j e^{\lambda_j t} v^j \qquad \text{(differential equation)}$$
$$y(t) = \sum k_j (1 + \lambda_j)^t v^j \qquad \text{(difference equation)},$$

where the λ_j are typically characteristic roots of a matrix (or of a polynomial in the scalar equation of nth order). The system is then asymptotically stable, if and only if

$$\text{Re}\,(\lambda_j) < 0, \qquad \text{all } j \qquad \text{(differential equation)}$$
$$|1 + \lambda_j| < 1, \qquad \text{all } j \qquad \text{(difference equation)}.$$

As a matter of practical application, it is not always easy to locate a suitable criterion for guaranteeing that the roots satisfy the stability conditions unless the numerical values of the parameters are known. When the numerical values are given, there are a variety of techniques which are widely used in electrical engineering for determining whether the solution is stable or not. In economic models we rarely have this information, and must depend on broad properties of the system matrix, the most useful property which such a matrix can have being a dominant diagonal. The results for dominant diagonal matrices (Review R7, Section R7.4) are important for market models, and we shall use them in the next section.

If the system is represented by linear equations without constant coefficients, or by nonlinear or unspecified equations, the direct solution technique is of little use. It may be possible to use known results for certain special equations. Usually in economics the only direct solution method that is applicable if the system is not linear with constant coefficients is to *approximate* the system by a system that does have constant coefficients. Using calculus methods, we obtain a linear approximation near the equilibrium. Any stability we find is then necessarily *local* stability. Until the 1950s, all stability theorems in economics for nonlinear systems were local theorems, obtained by this method.

A much more powerful tool is the Liapunov method.[3] Although originally published in 1892 (in Russian, but a French version was available in

[3] For a full discussion of the Liapunov method, see Lasalle and Lefschetz.

1907), this did not penetrate Anglo-American thought on stability analysis until after World War II. The beauty of the Liapunov method is that direct solution of the differential equations is not required.

The method we are interested in is called *Liapunov's second* (or *direct*) *method*, the first method involving explicit solution. The procedure is as follows. We seek to find a certain *scalar* function $V(y)$ with four properties, of which the first three are

$$\begin{aligned}
&(1)\quad V(y) \in C^1, \\
&(2)\quad V(y) \geqq 0, \qquad \text{all } y, \\
&(3)\quad V(y) = 0 \qquad \text{if and only if} \qquad y = 0.
\end{aligned}$$

Such a function is a *positive definite function of* y. A simple example of such a function would be $y'y$, the Euclidean norm of y (we shall use just such a function in the next section), or, more generally, $y'Ay$, where A is a positive definite matrix. The important thing to note is that $V(y)$ has properties similar to that of a norm, and can be regarded as a generalized norm of y.

We have given three of the desired properties. To obtain the fourth, we note that V is a function of y and y is a function of t, so that V is also a function of t, once the dynamic system is specified. Applying the chain rule for taking a derivative, we have

$$\frac{dV}{dt} = \sum_i \frac{\partial V}{\partial y_i}\frac{dy_i}{dt},$$

or, in vector notation

$$\dot{V} = \nabla V \cdot Dy.^4$$

The final property we seek of V is then

$$(4a) \qquad \dot{V} \leqq 0,$$

even better,

$$(4b) \qquad \dot{V} < 0.$$

Any function that possesses properties (1) through (4) is known as a *Liapunov Function*. Note that the properties (1) through (3) are independent of the dynamic system. The skill in using the method is to find a function satisfying (1) through (3) which also satisfies (4).

If the dynamic system is expressed, as it usually shall be in the examples in which we are interested, in the form,

$$Dy(t) = F[y(t)],$$

[4] The notation has $\dot{V}$ been used for dV/dt because DV is too easily confused with a matrix product.

then we can substitute for Dy in the expression for $\dot{V}$ to obtain

$$\dot{V} = \nabla V \cdot F(y),$$

so that $\dot{V}$ is expressed directly in terms of y, without the intervention of a differential equation.

Why do we seek a Liapunov function? The answer is provided by the following stability theorem.

Liapunov Stability Theorem

If a Liapunov function exists for a system, the system is stable. If, in addition, the function satisfies the strong condition (4b), the system is asymptotically stable.

We shall not give a rigorous proof, but a heuristic proof of the strong version is straightforward. Suppose $y(0) \neq 0$, then $V(0) > 0$. Since $\dot{V} < 0$, V decreases with t and $\rightarrow 0$ as $t \rightarrow \infty$. But, if $V = 0$, $y = 0$, so that the system is asymptotically stable.

We shall refer to a function satisfying (1) through (3) and (4a) as a *weak* Liapunov function, and to a function satisfying (1) through (3) and (4b) as a *strong* Liapunov function.

The Liapunov principle can be applied to difference equation systems, but is most applicable to first order differential equations because of the simplicity of the form of $\dot{V}$ in this case.

12.3 MARKET STABILITY[5]

A wide choice of theorems on the stability of competitive markets is available. The three presented here are chosen to illustrate both Liapunov and explicit solution techniques, and to demonstrate the stronger requirements required to show stability in a difference equation system, as compared with the differential equation system.

In all cases the dynamic behavior of the market is assumed to be expressed by relationship having one of the following forms:

$$\text{(1a)} \quad Dp(t) = Kz[p(t)]$$
$$\text{(2a)} \quad \Delta p(t) = Kz[p(t)],$$

where $z[p(t)]$ is the vector of excess demands at prices $p(t)$ and K is a diagonal matrix with positive diagonal elements.

[5] The basic contributions to the modern analysis of market stability are Arrow and Hurwicz [1] and [3]; Hahn; Arrow, Block, and Hurwicz; and Uzawa. Karlin [1], Chapter 9 gives many examples of the type of theorem developed in this section.

In the differential equation formulation (1a), the price in the ith market changes *instantaneously* in response to the existence of excess demand in that market. The rate of change of price is positive if excess demand is positive, negative if excess demand is negative, since $k_i > 0$. The actual rate of change is proportional to the level of excess demand, the factor of proportionality being k_i which is therefore to be interpreted as the *speed of response*.

In the difference equation case we have $\Delta p(t) = p(t + 1) - p(t)$, so that prices respond to excess demand only in the next period. The change in price from t to $t + 1$ is then related to excess demand at time t in the same kind of way the instantaneous rate of change is related in the differential equation case. It is more correct to regard k_i, in the lagged response case, as the *degree of response*, since the speed of response depends both on this and on the length of the unit time period.

Note that, whereas in the discussion of the existence of market equilibrium, we were prepared to contemplate the existence of a set $Z(p)$ of excess demand vectors associated with each price, here we shall assume the existence of a unique point $z(p)$. Note also that the adjustment process implied by either (1a) or (1b) represents an arbitrary assumption concerning the behavior of economic agents. Although these processes are highly plausible, we must recognize that:

There is no fundamental theory concerning the behavior of economic decision makers out of equilibrium; there are only theories of equilibrium behavior.

Finally, note that the dynamic equations are *aggregate*, with no reference to individual excess demands.

Before proceeding further, we must pause to examine certain problems concerning units of measurement. Each component of z is measured in some units of its own, and these units are arbitrary. We can simplify the problem by making a change of units which gives the ratio of excess demand to response the same in all markets, so that we shall consider the system in the form:

$$\text{(1b)}\quad Dp(t) = \mu z[p(t)]$$
$$\text{(2b)}\quad \Delta p(t) = \mu z[p(t)],$$

where μ is a positive scalar representing the general speed or degree of response, and p, z are in the appropriate new units.

The other units problem arises from the zero-degree homogeneity of excess demand as a function of prices. This implies that $z(\lambda p) = z(p)$ $(\lambda > 0)$ and, in particular that equilibrium prices are not a point p^* but a halfline

λp^*. One way to avoid this problem is to normalize prices. This can be done, as in the study of general equilibrium (Chapter 9), by imposing a purely mathematical norm. It is usual in stability analysis, however, to follow Walras and choose one commodity (say, the nth) as *numeraire*. We then replace the n vector p by the normalized $(n - 1)$ vector $\hat{p}$, where $\hat{p}_1 = p_i/p_n$. Then p, λp both correspond to the same normalized vector $\hat{p}$. Since the excess demand functions $z_i(p)$ are homogeneous of degree zero, we have $z_i(p) = z_i(\hat{p}:1)$, all i. We can write the normalized system as

$$\text{(1c)} \quad D\hat{p}(t) = \mu\hat{z}[\hat{p}(t)]$$
$$\text{(2c)} \quad \Delta\hat{p}(t) = \mu\hat{z}[\hat{p}(t)].$$ [6]

We shall refer to the system (1c) and (2c) as the *normalized* system, the system (1b) and (2b) as the *non-normalized* system.

We are now ready to examine the stability theorems themselves.

Theorem I

If Walras' Law and the aggregate revealed preference axiom hold everywhere, the non-normalized differential equation model (1*b*) *is globally stable.*[7]

Here we shall use the Liapunov method. First we derive the implications of the two main assumptions. Let p^* denote an equilibrium price vector, and p any price vector. Since the system is not normalized, p^* is also an equilibrium price vector for $\lambda > 0$, and we shall consider the system stable if it is stable with reference to λp^* for any $\lambda > 0$.

Walras' Law states that, for all p,

$$p \cdot z(p) = 0.$$

Aggregate revealed preference states that, for any p, p' so related that $p \cdot z(p') \leqq p \cdot z(p)$, then $p' \cdot z(p) \geqq p' \cdot z(p')$ with equality only if $z(p) = z(p')$ (the primes do not represent transposition here). Put $p' = p^*$. Then $z(p^*) \leqq 0$ (by definition of market equilibrium), and $p \cdot z(p) = 0$ (Walras' Law) so that we have

$$p \cdot z(p^*) \leqq 0 = p \cdot z(p),$$

so that p, p^* satisfy the first inequality. Thus we must have

$$\begin{aligned} p^* \cdot z(p) &\geqq p^* \cdot z(p^*) \\ &\geqq 0 \qquad \text{(definition of equilibrium)} \end{aligned}$$

and this must be true for all p.

[6] The μ's differ from those in (1b) and (1c). Those above are p_n times the earlier values. It is, of course, essential for the use of the numeraire method that the numeraire is never in excess supply and $p_n \neq 0$.

[7] Originally due to Arrow, Block, and Hurwicz. See also Karlin [1].

Now construct the positive definite function, identical with the Euclidean norm of the deviation of $p(t)$ from p^*, as follows:

$$V(t) = \tfrac{1}{2}[p(t) - p^*] \cdot [p(t) - p^*]'.$$

This differs slightly from the standard Liapunov type in that $V(p) = 0$ for $p = \lambda p^*$ as well as $p = p^*$.

The gradient vector of V is given by

$$\nabla V = [p(t) - p^*],$$

so that

$$\begin{aligned}\dot{V} &= \nabla V \cdot Dp(t)\\ &= \mu[p(t) - p^*] \cdot Dp(t)\\ &= \mu[p(t) - p^*] \cdot z[p(t)] \qquad \text{[from (1b)]}\\ &= \mu p(t) \cdot z[p(t)] - \mu p^* \cdot z[p(t)]\\ &= -\mu p^* \cdot z[p(t)] \qquad \text{(using Walras' Law).}\end{aligned}$$

But we have already shown that $p^* \cdot z[p(t)] \geqq 0$, so that $\dot{V} \leqq 0$ and V is a *weak* Liapunov function. Thus the system is stable (but not asymptotically stable).

If the equilibrium is unique, so that $z(p) \neq z(p')$ unless $p = \lambda p'$, $[p \cdot z(p') \leqq p \cdot z(p)$ implies $p' \cdot z(p) > p' \cdot z(p')$ unless $p = \lambda p']$ then we would obtain a strong Liapunov function and so we can state:

If the equilibrium is unique, then the system is asymptotically stable, in the sense of converging to λp^* for some $\lambda > 0$.[8]

The stability discovered above is obviously *global*, since no restrictions have been placed on the size of the norm V.

Theorem II

If Walras' Law holds, if all goods are normal goods (not inferior goods), if all goods are gross substitutes with the numeraire a strict gross substitute for every good, then the normalized differential equation model (1c) *is asymptotically locally stable.*[9]

This theorem is given primarily as an example of the use of local linear approximation, explicit solution, and the properties of dominant diagonal

[8] It follows that the normalized system is asymptotically stable under the above circumstances, converging to $\hat{p}^*$.

[9] The original theorem is in Arrow and Hurwicz and in Hahn. Proof of global stability under almost the same assumptions is given in Arrow, Block, and Hurwicz and in Uzawa.

matrices. Global stability can now be shown under the same general conditions, making the local stability proof obsolete.

Good j is said to be a *gross substitute* for good i if $\partial z_j/\partial p_i \geqq 0$, that is, if the excess demand for the jth good does not fall with a rise in the price of the ith good. If the partial derivative is actually positive, j is a *strict* gross substitute for i. The term *gross* refers to the fact that we are measuring the effects over the whole market with all influences from income effects, redistribution effects, and so on, included.

Consider the price adjustment equation

$$D\hat{p}(t) = \mu\hat{z}[\hat{p}(t)].$$

Each $\hat{z}_i$ is assumed to be a continuous function ($\in C^1$) of prices, so we can take the linear approximation around the equilibrium point given by

$$\hat{z}_i(\hat{p}) - \hat{z}_i(\hat{p}^*) = \nabla\hat{z}_i^* \cdot (\hat{p} - \hat{p}^*)$$

[where $\hat{z}_i^*$ denotes $\hat{z}_i(p^*)$.]

Denote by M the matrix whose rows are $\nabla\hat{z}_i^*$, $1 = 1, \ldots, n - 1$. Then we can write the vector approximation

$$\hat{z}(\hat{p}) - \hat{z}(\hat{p}^*) = M(\hat{p} - \hat{p}^*).$$

Now $D(\hat{p} - \hat{p}^*) = D\hat{p}$, so that, if we write $u = (\hat{p} - \hat{p}^*)$, the price adjustment equation can be written in the linear form,

$$Du(t) = \mu Mu(t),$$

for small deviations from equilibrium. This is a first order linear vector differential equation with constant coefficients. It will be asymptotically stable if all the roots of M have negative real parts, unstable otherwise. Since $\mu > 0$, it plays no part in determining stability.

We have now reduced the problem to the study of the matrix M whose typical element is $\partial\hat{z}_i/\partial\hat{p}_j$. Since every good is normal, the matrix has a negative diagonal. From the properties of dominant diagonal matrices (Review 7, Section 7.4) such a matrix will have all its roots with negative real parts if we can show it to have a dominant diagonal.

Consider Walras' Law (which must be applied over all n commodities, including the numeraire)

$$p \cdot z(p) = 0.$$

Taking derivatives with respect to p_j, we obtain

$$z_j(p) + \sum_1^n p_i \frac{\partial z_i}{\partial p_j} = 0 \qquad j = 1, \ldots, n.$$

At equilibrium we have $z_j(p^*) = 0$, all j. The assumption that the numeraire is a strict gross substitute for every good implies $\partial z_n/\partial p_j > 0$, all j, so that $p_n(\partial z_n/\partial p_j) > 0$.

Thus at equilibrium we have

$$\sum_1^{n-1} p_i^*(\partial z_i^*/\partial p_j) < 0.$$

This same inequality will be true for the normalized prices, when it is put in the form

$$\sum_1^{n-1} \hat{p}_i^*(\partial \hat{z}_i^*/\partial \hat{p}_j) < 0.$$

By the assumptions of the theorem, we have $\partial \hat{z}_j^*/\partial \hat{p}_j < 0$ and $\partial \hat{z}_i^*/\partial \hat{p}_j \geqq 0$ $(i \neq j)$, so that the above inequality implies

$$\hat{p}_j^* \mid \partial \hat{z}_j^*/\partial \hat{p}_j \mid > \sum_{i \neq j} \hat{p}_i^* \mid \partial \hat{z}_i^*/\partial \hat{p}_j \mid \qquad j = 1, \ldots, n-1.$$

This is exactly the condition that the matrix M have a dominant diagonal. Since this diagonal is negative, all the roots of M have negative real parts (Review R7, Section R7.4), proving the theorem.

Note that the signs of the off-diagonal elements of M are needed to show that the diagonal is dominant. Once this is done, they play no further part. If we could show, on some other grounds, that M had a dominant diagonal, the off-diagonal elements could have any signs provided the diagonal was negative.

Theorem III

If the conditions of Theorem II are satisfied, then there is some positive number μ_0 such that the difference equation model (2c) *is locally asymptotically stable for $\mu < \mu_0$ and unstable for $\mu > \mu_0$.*[10]

We introduce this theorem primarily to illustrate the importance of the degree of response in determining the stability of the delayed response model. In the instantaneous response models of Theorems I and II, the size of the response parameter μ was seen to play no part. Here it is important.

We can proceed exactly as in Theorem II to give a linear approximation around the equilibrium point and derive the local price adjustment equation

$$\Delta u(t) = \mu M u(t),$$

where $u(t)$, M are the same as in Theorem II.

[10] Other discrete time formulations are discussed in Karlin [1]. The rather weak theorem given here is designed primarily as an example.

From ordinary difference equation theory (Review R10, Section R10.5), the system is stable, if and only if

$$|1 + \mu\lambda_i| < 1, \qquad \text{all } i,$$

where the λ_i are the roots of M.

From the definition of $|1 + \mu\lambda_i|$ we have

$$\begin{aligned}(|1 + \mu\lambda_i|)^2 &= [\text{Re}\,(1 + \mu\lambda_i)]^2 + [\text{Im}\,(1 + \mu\lambda_i)]^2 \\ &= [1 + \mu\,\text{Re}\,(\lambda_i)]^2 + \mu^2[\text{Im}\,(\lambda_i)]^2 \\ &= 1 + 2\mu\,\text{Re}\,(\lambda_i) + \mu^2[\text{Re}\,(\lambda_i)]^2 + \mu^2[\text{Im}\,(\lambda_i)]^2 \\ &= 1 + 2\mu\,\text{Re}\,(\lambda_i) + \mu^2(|\lambda_i|)^2.\end{aligned}$$

Thus $(|1 + \mu\lambda_i|)^2 < 1$, if

$$2\mu\,\text{Re}\,(\lambda_i) + \mu^2(|\lambda_i|)^2 < 0,$$

that is, if

$$\mu < \frac{-2\,\text{Re}\,(\lambda_i)}{(|\lambda_i|)^2} \qquad (\text{since } \mu > 0).$$

Since Theorem II showed all the roots of M to have $\text{Re}\,(\lambda) < 0$, the expression on the right-hand side above is positive for all i. Denote this expression by $m(i)$ and define

$$\mu_0 = \min_i m(i);$$

it is obvious that this is the μ_0 of the theorem.

The theorem is *instructive* rather than *constructive*, since the value for μ_0 has no economic interpretation, nor can it be found without detailed knowledge of M (although with ingenuity, we can find some bounds for it). It does illustrate very well the problems associated with the stability of systems having delayed and discontinuous response.

12.4 STABILITY OF DECENTRALIZED ECONOMIC POLICY[11]

In this section we shall take up a problem that is different from the usual problems in stability in that we attempt to construct a stable matrix from

[11] The general type of policy model from which this kind of analysis is derived was originated by Tinbergen. The particular problem discussed here arises from the work of Mundell and Cooper. See also Patrick.

It seems appropriate to end the main part of the book with an interesting, incompletely solved, problem.

certain materials provided, rather than examine the stability of a given matrix.

We consider an economy with n economic variables which are the primary concern of economic policy (level of employment, rate of change of price level, balance of payments, and so on). We shall refer to these as the *target variables*. It is assumed that there is a desired level, or *target value*, for each variable. The government has at its disposal m policy *instruments* (budget, quantity of money, exchange rate, and so on).

The level of each variable is some function of the level of the various instruments. Denoting the vector of target variables by y and the vector of policy instruments by x, and assuming that both are measured from the equilibrium values (the target values for the variables, the levels giving the target values for the instruments), there is a vector-valued function

$$y = F(x).$$

We shall immediately simplify the analysis by assuming that $m = n$ and that F is linear or has been locally linearized around the target values so that the system can be written in the form

$$y = Ax,$$

where A is square of order $n \times n$.

It is assumed that there is some random input into the system in addition to the formal specification, so that the system needs to be stabilized with reference to the equilibrium $y = 0$.

In a *fully centralized* system the relation of instruments to targets is known, so that there is no problem in stabilizing.

We are interested, however, in the following kind of *decentralized policy:*

(a) There are n semi-autonomous policy *controllers* (central bank, treasury secretary, and so on), each of whom is in sole control of a particular instrument (central bank in control of quantity of money, for example).
(b) Each controller is to vary his instrument in response to changes in a single target variable which he has been assigned to watch.
(c) The central governing authority makes the initial assignments: (i) to which variable each controller is to adjust his instrument; and (ii) in what direction the controller is to vary his instrument for a given direction of deviation of the assigned variable from its target.
(d) It is assumed that each controller will vary his policy instrument immediately and at a rate proportional to the deviation of the target variable from its target value.

By suitable choice of units, the behavior of the controller of the ith instrument can then be described by the relationship:

$$Dx_i = \pm y_j,$$

where both the sign and j are to be assigned initially at the center.

Since $y_j = A_j x$, where A_j is the jth row of A, the combined behavior of the controllers can be written as

$$Dx = A^{**}x,$$

where A^{**} consists of some permutation of the rows (but not the columns) of A, with some or all of the row signs changed.

Obviously the decentralized policy model is stable if and only if the matrix A^{**} is stable. The problem here is not whether A itself is a stable matrix (that is having all roots with negative real parts), but whether a stable matrix A^{**} can be constructed out of A by permuting rows and changing signs of rows.

There are $n!$ permutations of the rows of A. Is at least one a stable matrix, possibly after changing some row signs? If so, how do we find which one? If there are several, in what sense is one the "best?" We shall not attempt to answer the last question, but shall concern ourselves with the first two.

The most fruitful attack would seem to be to see whether we can make a dominant diagonal matrix by permuting the rows of A. If this can be done, we are then free to change row signs so that we can always make the dominant diagonal *negative* and hence guarantee stability (see Review R7, Section R7.4). We shall make use of several properties of dominant diagonals from Section R7.4.

We can state an immediate *necessary* condition:

If A has no row or column containing an absolutely dominant element (that is, some element whose absolute value exceeds the sum of the absolute values of all other elements in the same row or column), then A^{**} cannot have a dominant diagonal.

From Section R7.4, a dominant diagonal matrix must have at least one *diagonal* element which is absolutely dominant in this sense. If A has no absolutely dominant element, no permutation can give A^{**} an absolutely dominant diagonal element.

If the above necessary condition is satisfied, the following stepwise procedure is then suggested.

(a) If the absolutely dominant element in A is the (i, j)th, renumber the columns to put $j = 1$, and choose the ith row of A as the first row of A^* (A^* is the chosen permutation of the rows of A; A^{**} is obtained from A^* by row sign changes only).

(b) Thus the absolutely dominant element of A becomes a^*_{11}. Now consider the submatrix of A obtained by deleting the row and column containing a^*_{21}. This will become, when the rows are suitably ordered, a principal submatrix of A^* and so it must have a dominant diagonal if A^* is to have one. Thus one of the rows or columns of the submatrix must have an absolutely dominant element (it must exceed in absolute value the sum of the absolute values of the other elements of the row or column *which are in the submatrix*. We are comparing the element now with the sum of only $n - 2$ other elements). If no such element exists, then A^* cannot have a dominant diagonal and the procedure stops.

(c) If such an element exists for the submatrix, we make it a^*_{22} and then consider the submatrix of order $n - 2$ of the remaining rows and columns. The arguments of step (b) are then applicable to this submatrix, and so we continue step by step.

If A^* does *not* have a dominant diagonal, the procedure may terminate at some step when the necessary condition on the appropriate submatrix is not satisfied. If A^* *does* have a dominant diagonal, then the above procedure certainly gives the appropriate permutation to arrive at A^*. We may obtain several permutations if, at any stage, there are more than one row or column with an absolutely dominant element.

The conditions satisfied at each step are, however, *necessary but not sufficient*. Having found a matrix A^* (or several), we must then check to see if there exists a strictly positive solution to the linear inequalities

$$|a^*_{ii}|\, x_i - \sum_{i \neq j} |a^*_{ij}|\, x_j > 0 \qquad i = 1, \ldots, n.$$

If this final condition is satisfied, then A^* has a dominant diagonal. We now make such sign changes in the rows of A^* as are necessary to make the diagonal negative, and obtain A^{**}. Then A^{**} is stable, and the decentralized policy scheme will work.

Note that the only *signs* that we ever need to know are those of the elements of A which ultimately become the diagonal elements of A^*. Thus the discussion is appropriate to systems in which "large" parameters in A are assumed known, with others being written as "$\pm\varepsilon$."

PART IV

Mathematical Reviews

REVIEW

R1

Fundamental Ideas

Familiarity with the material of this review (except Section R1.6) is essential to all later work. The material of Section R1.6 is not required until Review R5.

R1.1 SETS[1]

A *set* is a collection of objects of any kind, called *elements* or *members* of the set. In the usual notation, $a \in S$ means a is an element of S, $a \notin S$ that it is not an element. A set is completely defined by a list of all its elements. If the set is *finite*, containing n elements with n some number, it may be defined by enumeration as, for example the set of 50 states, Alabama, . . ., Wyoming, or by some property as, for example, the states of the United States. If the set is *infinite*, containing an infinite number of elements, membership in the set is

[1] The general content of Sections R1.1 through R1.5 is covered in the introductory sections of many modern mathematical texts in a variety of fields. A good elementary account of the use of set theory in economics is given in Koopmans [2]. A more advanced discussion is given in Debreu [1].

For a general introduction to mathematics for economists, covering topics that are assumed here to be familiar to the reader, the author recommends Yamane and Chiang. Both these treat, at a more elementary level, various topics covered in this book.

given by some rule defining the property which members of the set have, but nonmembers do not. We usually write this rule in the following form:

$$S = \{x \mid 0 < x < a\},$$

which we read as "S consists of all x such that x is between 0 and a". The symbol in the bracket to the left of the bar shows how we designate members of S, to the right is given the rule for membership.

At the most general level, the only property assumed of objects is that they do, or do not, belong to certain sets, so that fundamental operations on sets are concerned entirely with this property of being, or not being a member of a set.

Given two sets, the relationship between them is determined entirely by the extent of their common membership. We write

$$A \subset B,$$

to be read *A contained in B* or *A is a subset of B* if every member of A is also a member of B. If we have *both* $A \subset B$ and $B \subset A$, then the two sets contain exactly the same elements and we write $A = B$. If $A \subset B$, but $B \not\subset A$ we say that A is a *proper subset* of B, meaning that there are members of B not included in A. Sometimes the set inclusion symbols are written "$\subseteq$" for general inclusion and "$\subset$" for the proper subset property, but we shall not have occasion to make this distinction.

Given two or more sets, we can define new sets based on whether objects are in both of the original sets or in at least one of them. We define the two new sets:

The intersection of A, B, written $A \cap B$. This is the set of elements which are in both A and B.

The union of A, B, written $A \cup B$. This is the set of elements belonging to at least one of the sets A, B.

Remembering that all we are doing is noting whether given elements belong to given sets, we can state propositions in terms of the symbols $\subset, \cap, \cup$. The following propositions can immediately be seen to be true:

$$(A \cap B) \subset A, \qquad (A \cap B) \subset B$$
$$A \subset (A \cup B)$$
$$(A \cap B) \subset (A \cup B)$$
$$(A \cap B) \cap C = A \cap (B \cap C)$$
$$(A \cup B) \cup C = A \cup (B \cup C).$$

As a result of the last two propositions we can define the union or intersection of a family of sets. We write the typical member of the family as A_i and use the notations

$$\bigcap_i A_i, \qquad \bigcup_i A_i.$$

For completeness of the scheme, we define the *empty set*, formally written ϕ, as the set containing no members. We say two sets A, B such that

$$A \cap B = \phi$$

are *disjoint* (they have no elements in common) and that sets A_i such that $A_i \cap A_j = \phi$, all i, j, and $\cup_i A_i = A$ form a *partition* of A (every member of A is in one and only one of the A_i).

Although one should formally distinguish carefully between the set ϕ (the set with no elements) and the set $\{0\}$ (the set containing the single element zero), it is usually clear from the context which is appropriate, and the ordinary symbol 0 will sometimes be used for either of these.

Sometimes $A \cap B$ is called the *product* or *logical product* of A, B and written $A \cdot B$, and sometimes $A \cup B$ is called the *sum* or *logical sum* of A, B and written $A + B$. *These notations should be scrupulously avoided in economics.* Most sets in economics are sets of vectors, not of generalized objects, and we shall later use the notations $A \cdot B$, $A + B$ (also $A - B$) in quite different senses.

Consider sets A, B, where A is a proper subset of B. We may wish to discuss the set consisting of those elements of B which are not in A. This set is called the *complement of A in B*. It is usually written $A - B$. For the above reasons, this notation is undesirable, but there is no standard alternative other than the symbol " $\sim$ " in place of " $-$ " and may sometimes be used.

If the set B becomes so inclusive that it contains all the objects that are under discussion in some context, it is called the *universal set*. The complement of A in the universal set is called simply the *complement* of A and written as A' or $\bar{A}$. A' is the set of all *relevant* elements not included in A.

R1.2 ORDERED AND QUASI-ORDERED SETS

We have already drawn upon simple intuitive ideas of order. Now let us consider the formal properties of ordering. Consider a set S of elements x, y, z, etc. We introduce the idea of a binary relation between members of S, defined as follows:

R is a binary relation defined on S if, for every pair of elements $x, y \in S$, where the order ("x is the first element") is crucial, the statement xRy is either true or false.

For example if S is the set of states of the Union, the statement "x entered the Union not later than y" is either true or false, so the relation "entered the Union not later than" is a binary relation. The difference between xRy and yRx is obvious here. The most common binary relations we shall be concerned with are relations like "not less than" (ordinary numbers), "every component of ... is not less than the corresponding component of ..." (point or vector ordering) and "not less desirable than" (consumer preferences).

A binary relation leads to a *quasi-ordering* (also called a *preordering*) if the following additional properties hold:

(1) xRx for all $x \in S$ (*reflexitivity*),
(2) xRy and yRz imply xRz (*transitivity*).

The examples of binary relations already given clearly have these properties. For example, if R represents "entered the Union not later than" then xRx (a state entered the Union not later than itself) and transitivity is satisfied.

However, consider the set of all countries and the relation "exports wheat to." This is a binary relation and it may be considered reflexive, every country being considered to export wheat to itself. But it is not true that the propositions, "The U.S. exports wheat to the U.S.S.R." and "The U.S.S.R. exports wheat to China" imply that "the U.S. exports wheat to China," so the relation is not transitive.

A quasi-ordering is an *ordering*, if xRy and yRx imply $x = y$. The set of real numbers is ordered by the binary relation $\geqq$ since $a \geqq b$ and $b \geqq a$ imply $a = b$. Consumer preferences, on the other hand give a typical quasi-ordering, since two commodity bundles x, y can be so related that x is at least as desirable as y and y at least as desirable as x, without the bundles being identical. If both xRy and yRx but $x \neq y$, we say that x, y are *equivalent* for the quasi-ordering given by R.

An ordering or quasi-ordering over a set S is *complete* if for $x, y \in S$ we have xRy or yRx (or both). Otherwise the ordering or quasi-ordering is *partial.* Consider the set of pairs in R^2, and the binary relation of the form "both components of x are not less than the corresponding components of y." Denote this statement by xRy. Then R is certainly both reflexive and transitive, and xRy, yRx together imply $x = y$, so that it is an ordering. But if x, y are so related that the first component of x is less than the first component of y, and the second component of x is greater than the second component of y, then we have neither xRy nor yRx, and the ordering is partial.

Note that strong ordering relationships, like " > " are *not* useful relationships because xRx is not true and these should not be used in attempting to construct orderings or quasi-orderings. Formally, we would derive a strong relationship like > or "preferred to" from the reflexive relation $\geqq$ or "at least as desired as." If we denote the strong relationship by R^* then we can see that

$$xR^*y, \qquad \text{if and only if} \qquad xRy \qquad \text{but not} \qquad yRx.$$

An equivalence relationship ($x \sim y$, meaning both xRy and yRx) does give a proper binary relationship, but leads only to a partial quasi-ordering.

Sometimes it is convenient in discussing consumer preferences to use the strong relation P ("preferred to") and the equivalence relation I ("indifferent between"). If we use the notation $x\bar{P}y$ to mean "xPy is not true," then it is obvious that $\bar{P}$ is a binary relation. We can therefore use $\bar{P}$ to derive a quasi-ordering, and derive I as the equivalence implied by both $x\bar{P}y$ and $y\bar{P}x$.

R1.3 CARTESIAN PRODUCTS AND SPACES

The *Cartesian product of the sets A, B, written $A \times B$*, is the set of *pairs* (a, b) such that $a \in A$, $b \in B$. It is important to regard the elements of $A \times B$ not as two objects but as a single pair of objects. The individuals in the pair are ordered in the sense that the first object is always from the first set, the second from the second set. It is not assumed that the choice of A as the first set carries any implications, only that, once the sets are ordered the individuals of the pair are ordered in the same way.

It is obvious that $A \times (B \times C) = (A \times B) \times C$, so that we can form the Cartesian product of a family of sets. We use the notation $A_1 \times A_2 \times \cdots \times A_i \times \cdots \times A_n = \prod_i^n A_i$. The element of this set is called an *n-tuple* and is the analog of a pair, but with n individuals. As in the case of $n = 2$, the ith individual of the n-tuple is always a member of the ith set. An individual of an n-tuple is a *component* or *coordinate*. We typically write a_i for the ith component ($a_i \in A_i$) and $a = (a_1, a_2, \ldots, a_n)$ for the n-tuple.

If the order of the a_i were not important, we would write $a = \{a_1, a_2, \ldots, a_3\}$ and a would be a set rather than an n-tuple. We shall regard n-tuples as elements of sets, not as sets, and we shall soon be referring to them as points or vectors in most uses.

A *space* is a term that can be used in several ways. It may mean a set so inclusive that every other set in the same context is a subset of the space. Usually, however, it refers to a universal set that can be considered to possess some properties that we would recognize as geometric. If not, we would prefer the term *universal set* or *universe*.

The space with which we shall most often be concerned is defined in the following way. Let R be the set of all real numbers. Then the set given by the Cartesian product of n sets $R \times R \times R \times \cdots \times R$, written R^n, is *real n-dimensioned space*. If this also has a certain property concerning distance relationships, it is called *Euclidean n-space* and written E^n. We shall not use any space R^n which is not also Euclidean, so we can regard R^n, E^n as equivalent for our purposes.

An element of R^n is an n-tuple each of whose components is a real number. We refer to an element of R^n as a *point*. Later we shall refer to such points in R^n as *vectors*, implying some special relationships. In most economic contexts points and vectors are interchangeable terms unless a certain approach requires emphasis of the vector properties, or perhaps emphasis on nonvector properties.

In some cases, chiefly in laying mathematical foundations for economic analysis rather than in economic analysis itself, we shall also be interested in the space C^n whose elements are points with complex components. We shall not differentiate between this space and *unitary space* (the complex analog of Euclidean space), written U^n.

R1.4 FUNCTIONS, TRANSFORMATIONS, MAPPINGS, CORRESPONDENCES

The terminology here can be very confusing, as there are wide variations in usage. Fundamentally all four terms refer to the same general idea, that of *a rule associating members of one set with members of another*. "Month of birth" would be a rule associating members of the set of human beings with members of the set of months, and would be an example of a very general kind. We shall prefer to use the term *mapping* for a rule relating members of such general sets.

A mapping contains some idea of the *direction* of the mapping, that is, of taking one set first (say S) and using the rule to discover which member (or members) of the second set (S') were associated with a given member of S. We would symbolize this mapping as $S \rightarrow S'$. In the example given, the mapping is from the set of persons (P) into the set of months (M); $P \rightarrow M$. For each $p \in P$ there will be a unique $m \in M$, since one cannot be born more than once. The $m \in M$ corresponding to p is called the *image* of p under the mapping in question. We would speak of the mapping as mapping P *into* M.[2]

We could look at the above relationship from the other direction and take the rule "persons born in a given month." This would give a mapping $M \to P$, from months to persons. We would refer to the mapping $M \to P$ as *inverse* to the mapping $P \to M$. Note that, for each $m \in M$ there will, in general, be associated a large set of $p \in P$, not a unique p. Whereas the original mapping maps p into an element of M, the inverse mapping maps m into a subset of P.

A mapping that maps an element $x \in S$ into a unique element $y \in S'$ is a *point-to-point mapping*, while a mapping that maps x into a subset $Y \subset S'$ is a *point-to-set mapping*.[3]

An extremely important class of mappings are those which map from some set into the set of real numbers, R. It is for this class that we shall primarily reserve the term *functions*. That is, a function (strictly, a *real-valued function*, but we shall rarely have occasion to use complex valued functions) is a rule assigning a number to each member of a set. A utility function is a function assigning real (in this case positive) numbers to members of the set of collections of goods; the function $y = 5x_1^3 - 3x_2^2 + 2$ associates a real number y with every pair (x_1, x_2) of real numbers in R^2.

We shall usually write a function in the form $f(x)$, where it will be clear from the context whether x is a real number or an n-tuple of real numbers.

In linear algebra we shall be concerned with a class of mappings from R^n into R^m. We shall usually refer to these as *linear transformations* simply because the term is standard in that context.

Sometimes we shall be concerned with a set of functions, not necessarily linear, each of which maps from a point in R^n to a number that can be regarded as a component of a point in R^m. It is often convenient to consider this set of functions as a single mapping from a point in R^n to a point in R^m. If the mapping is of a rather general kind, we shall simply refer to it as a point-to-point mapping, but if the individual functions are of the kind for which calculus methods are appropriate, we shall refer to the mapping as a *vector-valued function*.

It is not uncommon in economic models to have relationships which represent point-to-set mappings. For example, consider the simple budget constraint $p_1x_1 + p_2x_2 = k$. If income (k) is varied over some set K, then

[2] A set S is said to be mapped *onto* a set T if: (a) every point in S has an image in T; and (b) every point in T is the image of some point in S.

[3] Most of the time in economics our sets will be in R^n (sometimes in C^n) and the elements of the sets will be points, hence the above terminology.

each value of k is associated with a set of points (x_1, x_2), so the mapping from K into X is a *point-to-set* mapping, sometimes also called a *correspondence* or a *set-valued function*.

The analysis of point-to-set mappings requires special mathematical tools, and is given in a later review (Review R9).

R1.5 CLOSEDNESS AND BOUNDEDNESS

A space S is said to have a *metric* or *distance function* if, for every two points $x, y \in S$ there exists a real nonnegative number $d(x, y)$ such that

$$(1)\quad d(x, y) = 0 \qquad \text{if and only if} \qquad x = y,$$
$$(2)\quad d(x, y) = d(y, x),$$

and if, for every three points $x, y, z \in S$, the *triangle inequality*,

$$(3)\quad d(x, y) + d(y, z) \geqq d(x, z)$$

holds.

The best-known distance function is the *Euclidean distance*

$$d(x, y) = \sqrt{\sum (x_i - y_i)^2}.$$

Although most space in economic analysis is Euclidean, for the purposes of this section we require only that the space have some metric.

Given a metric, we define a *neighborhood* of a point $x \in S$ as the set of all points y such that $d(y, x) < \varepsilon$, where ε is a real positive number, usually to be thought of as "small."

A set $T \subset S$ is said to be *open* if every point $x \in T$ has some neighborhood (with ε sufficiently small) which lies entirely in T. The set $X = \{x \mid 0 < x < 1\}$ is an open set since, for any point $x = 1 - h \in X$ and a distance measure $d(x, y) = |x - y|$, the neighborhood defined by $x \pm \varepsilon$ consists entirely of points in X for $\varepsilon < h/2$. There is a strong association, which is important in economic applications, between strong inequalities ($>$) and open sets.

A point $y \in S$ is said to be a *limit point* (or *point of accumulation*) of T, if every neighborhood of y contains some point of T other than y. In the set X above, every point in the set is a limit point, and the points 0, 1 (which are not in X) are also limit points. On the other hand, the members of the finite set $\{0, 1, 2\}$ are not limit points, and the set has no limit points.

If $\hat{T}$ denotes the set of all limit points of T, the set $\hat{T} \cup T$ is called the *closure* of T. It is the set of all points and all limit points of T. If a set contains all its limit points it is said to be *closed*, and such a set is obviously equal to its own closure. For the set X above, the closure is the set X and the points 0, 1. On the other hand, the set $\bar{X} = \{x \mid 0 \leqq x \leqq 1\}$ is closed (it is also the closure of X). We can notice the association of the closed set with the weak ($\geqq$) inequality.[4]

Typical infinite sets encountered in economics will have the property that all points in the set are also limit points (but not all limit points may be in the set). Such sets are *compact*, possessing the property that every neighborhood of every point in the set also contains other points in the set. That is, there are no "discontinuities" or "holes" in the set.

For such a set we can distinguish between points whose neighborhoods consist *entirely* of other members of the set (if the neighborhood is sufficiently small), or *interior* points, and points whose neighborhoods always contain points both in and not in the set. The latter are *boundary* points. For the set X already used, $\frac{3}{4}$ is clearly an interior point, 1 is clearly a boundary point since $1 - \varepsilon$ is always in X and $1 + \varepsilon$ is never in X. Since boundary points do not have neighborhoods consisting of points entirely in the set, an open set does not contain its boundary points. On the other hand, since boundary points always include some point of the set in their neighborhood, they are necessarily limit points of the set and so must be included in a closed set.[5]

In most economic contexts, the boundaries of a set will be defined by inequalities that must be satisfied. Such a set will be closed (which is usually the property we seek) if all the inequalities are in the weak form.

Another property of sets in metric space that is important in our later analysis is that of *boundedness*. A set is bounded if the distance between any pair of points in the set is not greater than some finite number. Usually we show boundedness by proving that the set in question can be enclosed in some simple geometric figure, such as a hypercube or hypersphere, of finite dimension. This may require ingenuity in some cases.

Note that a set may have a boundary but be unbounded (the convex cones discussed in Review R4 are examples) and a set may be infinite but bounded (typical of almost all the sets we shall be concerned with).

R1.6 COMPLEX NUMBERS

This material is required only for characteristic roots and vectors of matrices (Review R5) and differential and difference equations (Review R10), and

[4] It is possible for a set to be neither open nor closed. The set $\{x \mid 0 < x \leqq 1\}$ falls into this category.

[5] A set need not possess boundary points. The set $\{x \in R^n\}$ clearly does not.

need not be read until the appropriate time. Since most readers will have some acquaintance with the topic, only a brief treatment is given.

A complex number is defined by the use of the set of real numbers and a special number i defined by $i = \sqrt{-1}$. A complex number u can be written in the form

$$u = a + ib,$$

where a, b are real numbers. a is called the *real part* of u [Re (u)] and b the *imaginary part* [Im (u)].

Addition and scalar multiplication of complex numbers u, v are defined as follows:

$$\begin{aligned}
&(1)\quad \mathrm{Re}\,(u + v) = \mathrm{Re}\,(u) + \mathrm{Re}\,(v)\\
&\qquad\ \mathrm{Im}\,(u + v) = \mathrm{Im}\,(u) + \mathrm{Im}\,(v);\\
&(2)\quad \mathrm{Re}\,(\lambda u) = \lambda \mathrm{Re}\,(u)\ (\lambda \text{ real});\\
&\qquad\ \mathrm{Im}\,(\lambda u) = \lambda \mathrm{Im}\,(u)\\
&(3)\quad u = v \qquad \text{if and only if}\\
&\qquad\qquad\qquad \mathrm{Re}\,(u) = \mathrm{Re}\,(v)\\
&\qquad\qquad\qquad \mathrm{Im}\,(u) = \mathrm{Im}\,(v).
\end{aligned}$$

Multiplication is defined as follows:

$$\begin{aligned}
&(4)\quad \mathrm{Re}\,(uv) = \mathrm{Re}\,(u)\,\mathrm{Re}\,(v) - \mathrm{Im}\,(u)\,\mathrm{Im}\,(v)\\
&\qquad\ \mathrm{Im}\,(uv) = \mathrm{Re}\,(u)\,\mathrm{Im}\,(v) + \mathrm{Im}\,(u)\,\mathrm{Re}\,(v).
\end{aligned}$$

It is easily shown that addition and multiplication satisfy the ordinary algebraic laws. By putting u, v in the form $a_1 + ib_1$, $a_2 + ib_2$ it is easily seen that the rule for the product follows the ordinary rule for the product of the form $(a_1 + ib_1)(a_2 + ib_2)$ with $i^2 = -1$.

We can perform the operation of *forming the conjugate* on a complex number. If u is a complex number, its *complex conjugate*, written $\bar{u}$ or u^*, is defined as

$$\begin{aligned}
&(5)\quad \mathrm{Re}\,(u^*) = \mathrm{Re}\,(u)\\
&\qquad\ \mathrm{Im}\,(u^*) = -\mathrm{Im}\,(u).
\end{aligned}$$

It follows from the multiplication rule that the product uu^* is a *real number* equal to $[\mathrm{Re}\,(u)]^2 + [\mathrm{Im}\,(u)]^2$. The square root, $\sqrt{uu^*}$, is known as the *modulus* or *amplitude* of u, and written $|u|$. The sum $u + u^*$ is also a real number, equal to 2 Re (u).

If Im $(u) = 0$, u is a real number. If Re $(u) = 0$ [but Im $(u) \neq 0$], u is *pure imaginary*. For a complex number, $u = 0$ means both Re $(u) = 0$ and

Im $(u) = 0$. An important property of the relationship between complex, real and pure imaginary numbers is

$u = u^*$ *if and only if* u *is real, and* $u = -u^*$, *if and only if* u *is pure imaginary.*

The operation of taking conjugates satisfies algebraic laws. It is easy to show that $(u + v)^* = u^* + v^*$ and $(uv)^* = u^*v^*$. Both these results will prove useful later, as will the obvious reflexive property $(u^*)^* = u$.

Let us now consider the important complex number $e^{i\omega}$. We shall define this (as we shall later define the matrix exponential) by the exponential series,

$$\begin{aligned} e^{i\omega} &= 1 + i\omega + (i\omega)^2/2! + (i\omega)^3/3! + \cdots \\ &= 1 + i\omega - \omega^2/2! - i\omega^3/3! + \omega^4/4! + \cdots \\ &= (1 - \omega^2/2! + \omega^4/4! - \cdots) \\ &\quad + i(\omega - \omega^3/3! + \omega^5/5! - \cdots). \end{aligned}$$

The two series in brackets can be recognized as the ordinary series for $\cos\omega$ and $\sin\omega$, so that we have

$$e^{i\omega} = \cos\omega + i\sin\omega.$$

Since $\cos\omega$ involves only even powers, and $\sin\omega$ only odd powers, it follows that

$$e^{-i\omega} = \cos\omega - i\sin\omega. \qquad [= (e^{i\omega})^*]$$

If ρ is a real number, then $\rho e^{i\omega}$ is some complex number. We have, from above,

$$\begin{aligned} \text{Re}\,(\rho e^{i\omega}) &= \rho\cos\omega \\ \text{Im}\,(\rho e^{i\omega}) &= \rho\sin\omega \end{aligned}$$

Thus $[\text{Re}\,(\rho e^{i\omega})]^2 + [\text{Im}\,(\rho e^{i\omega})]^2 = \rho^2$, so that we can identify ρ as the modulus of $\rho e^{i\omega}$. Thus we can write a complex number u in the *polar form* $\rho e^{i\omega}$, where

$$\begin{aligned} \rho &= |u| \\ \tan\omega &= [\text{Im}\,(u)]/\text{Re}\,(u). \end{aligned}$$

The polar form is particularly useful for operations involving multiplication or powers, since, if $u = \rho_1 e^{i\omega_1}$, $v = \rho_2 e^{i\omega_2}$, then

$$\begin{aligned} uv &= \rho_1\rho_2 e^{i(\omega_1 + \omega_2)} \\ u^n &= \rho_1^n e^{in\omega_1}. \end{aligned}$$

EXERCISES

1. We define the following sets:

$$S_1 = \{x \mid 0 \leqq x < 2\}$$
$$S_2 = \{x \mid -1 < x \leqq 1\}$$
$$S_3 = \{y \mid 0 \leqq y < 1\}$$
$$S_4 = \{y \mid 0 < y \leqq 1\}.$$

a. Show that $S_1 \cap S_2$ is closed, $S_1 \cup S_2$ is open.
b. Show that $S_3 \cup S_4$ is closed, $S_3 \cap S_4$ is open.
c. Find a boundary point of $S_2 \times S_3$ which is in $S_2 \times S_3$.
d. Find a boundary point of $S_2 \times S_3$ which is not in $S_2 \times S_3$.

2. Show that, for the set of real numbers greater than unity, the binary relationship "$x \leqq y^2$" gives a quasi-ordering.
3. Show that, for complex numbers u, v, $|u + v| \leqq |u| + |v|$.

REVIEW

R2

Linear Algebra

The material of this review is basic to every part of the book, including nonlinear analysis.

R2.1 VECTORS[1]

We shall consider vectors in R^n, that is, real n-dimensional space. For our purposes, we do not differentiate between this and Euclidean space, E^n.

A vector is an n-tuple in R^n of the form $x = [x_i, \ldots, x_n]$, where x_i is the ith *coordinate* or ith *component* of x. We define the following operations on vectors:

Multiplication by a Scalar

If λ is a real number and x a vector, the vector whose ith coordinate is λx_i is the scalar multiple of x by λ, written λx. Since numbers, as well as

[1] Due to its current popularity as an undergraduate mathematics course, linear algebra is well provided with texts. The most useful to the economist is probably Hadley [1].

Among references likely to be found on the economics shelf, Gale [1] has a good brief discussion of linear algebra (Chapter 2). Allen [2] (Chapters 12 and 13) covers vectors and matrices, but not quite in terms of modern linear algebra. Texts for econometrics and linear programming courses usually contain a summary of the material covered by this review and the next two.

vectors, play a prominent role in vector algebra we refer to them as scalars to avoid confusion.

Addition

If x, y are both vectors in R^n, the sum of x and y, written $x + y$, is the vector whose ith component is $x_i + y_i$.

If vectors are visualized in 2-space as lines joining the origin to the point whose coordinates are those of the vector, vector addition corresponds to the "parallelogram rule" of high-school physics. Scalar multiplication corresponds to movements in or out along the line of the vector. This geometric interpretation is given only as an illustration. Vectors in R^n should usually be considered as algebraic entities.

The above operations are sufficient to give the following properties, if we define [0] as the vector all of whose components are zero:

$$
\begin{aligned}
(x + y) + z &= x + (y + z) \\
x + y &= y + x \\
x + (-1)x &= [0] \\
0x &= [0] \\
1x &= x \\
\lambda(x + y) &= \lambda x + \lambda y \\
(\lambda + \mu)\, x &= \lambda x + \mu x \\
\lambda(\mu x) &= \lambda \mu x,
\end{aligned}
$$

where λ, μ, 0, 1 are $\in R$ (scalars) and x, y, z, [0] are $\in R^n$ (vectors). $x + (-1)y$ is usually written $x - y$.

Vectors x, y are *equal*, if and only if $x - y = [0]$; that is, if $x_i = y_i$ for all i.

In normal usage it will usually be clear from the context which elements of the system are scalars and which are vectors. Various conventions are, or have been, used if distinctions are required. Vectors are sometimes printed in heavy type, scalars in regular type. Greek letters almost always mean scalars, but in optimizing theory we shall use the *vector* λ. In this book we shall almost always rely on context or explicit statement, but some general conventions are that a subscripted lower case letter will usually mean the component of a vector,[2] and in multiplication by a scalar, the scalar is *always* written first.

[2] For this reason, we shall, as far as possible, use superscripts to denominate different vectors. Thus x^i, x^j will typically mean two different vectors, x_i, x_j two different components of the same vector.

Because the correct interpretation will always be clear from the context, we shall drop the notation [0], and write 0 for both the scalar and the vector (and later, for the zero matrix).

A *vector space* is an algebraic system whose components satisfy the properties given above for vectors. Obviously R^n is a vector space. As an abstract system, a vector space depends on two sets, the set R^n to which the vectors belong, and the set R to which the scalars belong. Vector spaces may be defined on sets other than real numbers. The only vector space other than that based on R^n, R that will be of interest to us is that based on complex numbers, which is discussed briefly in Section R2.6.

R2.2 FUNDAMENTAL THEOREM OF VECTOR SPACES

We define a *linear subspace* or a *vector subspace* of the vector space V as a subset, L, of vectors in V that has the following properties:

$$\begin{aligned} x + y \in L \quad &\text{if} \quad x, y \in L, \\ \lambda x \in L \quad &\text{if} \quad x \in L, \quad \lambda \in R \end{aligned}$$

A linear subspace can be thought of as a "flat" subspace, unbounded, which passes through the point 0 in V. Consider a vector space in R^3; an unbounded plane passing through 0 is a vector subspace, but a plane which does not pass through 0 is not.

It is easy to see that, if we take a vector space in R^n, the set of all vectors whose nth components are zero form a linear subspace.

Let $v^1, \ldots, v^m$ be a set of vectors in V. The vectors $v^1, \ldots, v^m$ are said to be *linearly dependent* if there is a set of numbers $\lambda_1, \ldots, \lambda_m$, not all zero, such that

$$\sum_{i=1}^{m} \lambda_i v^i = 0.$$

If such numbers cannot be found, the vectors are *linearly independent*.

A vector x is a *linear combination* of the vectors $v^1, \ldots, v^m$, if

$$\sum_{i=1}^{m} \mu_i v^i = x$$

for some set of numbers $\mu_1, \ldots, \mu_m$.

Example. We can find n linearly independent vectors in R^n. Choose the following n vectors: $e^1 = [1, 0, 0, \ldots, 0]$, $e^2 = [0, 1, 0, \ldots, 0], \ldots, e^n = [0, 0, 0, \ldots, 1]$ (such vectors are called *coordinate vectors*). It is obvious that the equations $\sum \lambda_i e^i = 0$ are simply the n independent equations $\lambda_i e^i = 0$ and are thus satisfied only by $\lambda_i = 0$ for all i. Thus the vectors $e^1, \ldots, e^n$ are linearly independent.

We can now state the following theorem, from which the remaining theory can be derived:

Fundamental Theorem

Any set of $m + 1$ vectors, all of which are linear combinations of the same m vectors $v^1, \ldots, v^m$, are linearly dependent.

Let $x^1, \ldots, x^{m+1}$ be the vectors which are linear combinations of the vectors $v^1, \ldots, v^m$. Then we have

$$x^j = \sum_i \lambda_{ij} v^i \qquad j = 1, \ldots, m + 1,$$

and we wish to prove that there exist numbers μ_j not all zero such that

$$\sum_j \mu_j x^j = 0.$$

The proof is inductive. First we note that, if $m = 1$, $x^1 = \lambda_1 v$, $x^2 = \lambda_2 v$. If $\lambda_1, \lambda_2 = 0$, the linear dependence is immediate. Suppose $\lambda_1 \neq 0$, and put $\mu_1 = -\lambda_2$, $\mu_2 = \lambda_1$. Then

$$\begin{aligned} \mu_1 x^1 + \mu_2 x^2 &= -\lambda_2 \lambda_1 v + \lambda_1 \lambda_2 v \\ &= 0, \end{aligned}$$

so the theorem is true for $m = 1$.

Now assume the theorem holds for $m = k - 1$ and put $m = k$. We have

$$x^j = \sum_{i=1}^{k} \lambda_{ij} v^i \qquad j = 1, \ldots, k + 1.$$

Suppose $\lambda_{11} \neq 0$. Define new vectors as follows:

$$\begin{aligned} z^j &= x^j - \frac{\lambda_{1j}}{\lambda_{11}} x^1, \\ &= \sum_{i=2}^{k} [\lambda_{ij} - \left(\frac{\lambda_{1j}}{\lambda_{11}}\right) \lambda_{i1}] v^i, \end{aligned}$$

since the first terms of the expansion cancel each other.

Thus each of the vectors z^j is a linear combination of the $(k - 1)$ vectors $v^2, \ldots, v^k$. By the induction hypothesis, the z^j must be linearly dependent, so that

$$\begin{aligned} \sum_1^{k+1} \mu_j z^j &= \sum_1^{k+1} \mu_j x^j - \frac{1}{\lambda_{11}} \left(\sum_1^{k+1} \mu_j \lambda_{1j} \right) x^1, \\ &= 0 \end{aligned}$$

with $\mu_j \neq 0$ for some j.

Put $\mu_1' = \mu_1 - (1/\lambda_{11})(\sum \mu_j\lambda_{1j})$: $\mu_j' = \mu_j$ for $j = 2, \ldots, k$. Then we have

$$\sum_1^{k+1} \mu_j' x^j = 0 \qquad \mu_j' \neq 0, \qquad \text{some } j,$$

so that the x^j are linearly dependent for $m = k$ if they are for $m = k - 1$. Since we have shown the linear dependence for $m = 1$, the theorem is proved.

R2.3 BASIS AND RANK

If L is a linear subspace of V, the maximum number of linearly independent vectors that can be chosen from L is called the *rank* or *dimension* of L.

Example. ***R^n* has rank *n*.** We saw in the previous section that the coordinate vectors e^i form a linearly independent set of n vectors. If x is any vector in R^n, then $x = \sum x_i e^i$. Thus any vector in R^n can be expressed as a linear combination of the n vectors e^i and so, from the fundamental theorem, any $(n + 1)$ vectors in R^n must be linearly dependent. Thus n is the maximum number of linearly independent vectors in R^n, and so is its rank.

In a linear subspace of rank r (we will often write such a subspace L^r, to indicate the rank), any set of r linearly independent vectors forms a *basis* for L, with the following very important property:

Basis Theorem

The set of linearly independent vectors $v^1, \ldots, v^r$ form a basis for L, if and only if every vector in L can be expressed as a linear combination of $v^1, \ldots, v^r$.

To prove the theorem, we first note that if every vector in L can be expressed as a linear combination of $v^1 \cdots v^r$, then any set of $(r + 1)$ vectors in L is linearly dependent (fundamental theorem). Then L has rank r and the v^i form a basis.

On the other hand, if the v^i form a basis then L has rank r so that the $(r + 1)$ vectors $v^1, \ldots, v^r, x$ are linearly dependent for all $x \in L$. Thus

$$\sum^r \lambda_i v^i + \lambda_{r+1} x = 0,$$

with $\lambda_{r+1} \neq 0$, otherwise the v^i would be linearly dependent. It follows that

$$x = -(1/\lambda_{r+1}) \sum \lambda_i v^i,$$

so that x can be expressed as a linear combination of the v^i, and the theorem is proved.

Example. The vectors (1, 1), (−3, 2) form a basis for R^2. We have two vectors and R^2 has rank 2, so we need to show the vectors to be linearly independent. In this case $\sum \mu_j v^j = 0$ consists of the two equations

$$\mu_1 - 3\mu_2 = 0,$$
$$\mu_1 + 2\mu_2 = 0.$$

Direct solution gives $\mu_1 = \mu_2 = 0$, so the vectors are linearly independent and form a basis.

We can express an arbitrary vector, say $x = (2, 1)$, in terms of the basis as follows:

We seek λ_1, λ_2 such that

$$\lambda_1 v^1 + \lambda_2 v^2 = x,$$

that is, that satisfy

$$\lambda_1 - 3\lambda_2 = 2,$$
$$\lambda_1 + 2\lambda_2 = 1.$$

Solution gives $\lambda_1 = \frac{7}{5}$, $\lambda_2 = -\frac{1}{5}$, so that we have

$$x = (\tfrac{7}{5})v^1 - (\tfrac{1}{5})v^2.$$

If L has a basis $v^1, \ldots, v^r$, we sometimes say the vectors v^i *span* L, and that L is the span of the vectors $v^1 \cdots v^r$. The terms rank and basis will also appear in relation to matrices, the concepts being directly related to their use here.

R2.4 SUMS AND DIRECT SUMS[3]

Let L_1, L_2 be linear subspaces of the vector space V. We define the *sum* of L_1, L_2 to be the set consisting of all the vectors $u + v$ where $u \in L_1$, $v \in L_2$, and denote the sum by $L_1 + L_2$. It is easy to show that $L_1 + L_2$ must be a linear subspace of V.

If L_1, L_2 do not intersect (except at 0, where all linear subspaces intersect), that is, $L_1 \cap L_2 = 0$ then we refer to the sum of L_1 and L_2 as the *direct sum* and write it

$$L_1 \oplus L_2.$$

The important properties of direct sums are the following:

(*a*) *If* $w \in L_1 \oplus L_2$, *then the vectors* u, v *such that* $u \in L_1$, $v \in L_2$ *and* $w = u + v$ *are uniquely determined.*

[3] We do not make any substantial use of direct sums. The brief discussion here is primarily to familiarize the reader with some special notation he may come across.

(*b*) *If L^r is a linear subspace of V of rank r, where V has rank n, there exists a subspace L^{n-r} of V with rank n − r such that* $V = L_1 \oplus L_2$.

We shall not prove either of these properties.

R2.5 SCALAR PRODUCTS

As a bridge from the pure algebra of vector spaces to the use of matrix notation, we introduce the following terminological ideas which may seem pointless for the moment:

Consider a vector in R^n as an n-tuple. We can write out the vector in terms of its components in two ways. If we write the components side by side.

$$x = [x_1, \ldots, x_n],$$

we say that x is written as a *row vector*. If, on the other hand, we write the components vertically,

$$x = \begin{bmatrix} x_1 \\ \vdots \\ x_n \end{bmatrix},$$

we say that x is written as a *column vector*.

All vectors in the same set are considered to be written in the same way. In the context of vector algebra, it does not matter in which way the vectors are considered to be written. When matrices are involved, it matters very much, but the context usually makes it clear whether a given vector is a row or column vector.

In most economic contexts we will find that quantity vectors are written as column vectors, price vectors as row vectors.

Sometimes we may wish to rewrite a row vector as a column vector or vice versa. This rewriting is called *transposition* and usually denoted by a prime. Thus, if x is a column vector, x' is a row vector. $(x')'$ is again a column vector, x itself.

We now define an operation of great importance involving a row vector and a column vector having the same number of components (of the same order):

If y is a row vector of order n, and x a column vector of the same order, the SCALAR PRODUCT of the two vectors, written yx is defined as

$$yx = \sum_{i=1}^{n} y_i x_i.$$

That is, the scalar product is obtained by multiplying together equivalent components of the two vectors and adding the products.

It is an important convention that, in the scalar product yx, the first vector is the row vector. It is obvious that the operation above gives a scalar, hence the name.

Example. The scalar product of

$$y = [1, -2, 0] \quad \text{and} \quad x = \begin{bmatrix} 2 \\ 3 \\ 1 \end{bmatrix}$$

is

$$\begin{aligned} yx &= 1.2 + (-2).3 + 0.1 \\ &= -4. \end{aligned}$$

A clear distinction should be made between *multiplication by a scalar* (an operation giving a vector from a vector and a scalar) and the *scalar product* (the result of an operation giving a scalar from two vectors).

The scalar product is also referred to as the *inner product* or, especially in an physical sciences context, as the *dot product.* In the latter cases, the product is written $y \cdot x$, a convention we shall sometimes use if the notation yx is unclear. The *vector product* of two vectors (written $y \times x$ and also called the *cross product*) is a special operation on vectors in E^2 and E^3 that we do not use in economics.

If we wish to form the scalar product of two row vectors or two column vectors, we transpose one of them into the appropriate form. In many expositions of vector properties where matrices are not involved the distinction between row and column vectors may be dropped, but we shall adhere to it strictly in this book.

The most typical scalar product we shall meet in economics is between the typical row vector p, a price vector, and the typical column vector x, a quantity vector. The scalar product px is then identified as the *value* of the quantities x at prices p.

Among the various scalar products we can form is the scalar product of a vector with its own transpose:

$$x'x = \sum (x_i)^2.$$

The positive square root of $x'x$ is the *Euclidean norm* or *length* of the vector x. We shall also use other norms of vectors.[4]

Two vectors whose scalar product is zero are said to be *orthogonal.* A set of vectors, every pair of which form an orthogonal pair, is an *orthogonal set.* An

[4] In economics we shall often be concerned with vectors whose components are all nonnegative. If x is such a vector, we shall typically use the *linear norm* $\sum x_i$ because of its simplicity. The fundamental properties of a norm, that it be nonnegative and be zero only if all components are zero, are possessed by the linear norm, if and only if all components are nonnegative.

orthogonal set of vectors each of which has unit Euclidean norm is called an *orthonormal set.*

Example. The coordinate vectors form an orthonormal set. In R^2 we have $(e^1)'e^2 = 0$, $(e^1)'e^1 = 1$, $(e^2)'e^2 = 1$, with equivalent results for R^n.

In R^2 or R^3 orthogonal vectors are geometrically *perpendicular*. Generalizing from this, we shall associate orthogonality and perpendicularity.

R2.6 COMPLEX VECTORS

This section can be omitted until the reader is ready for Review R5.

Complex vectors, or vectors in C^n (sometimes the notation U^n is used, paired with E^n), where C is the set of complex numbers, occur in this book only in relation to the characteristic vectors of matrices and analysis (particularly of dynamic economic models) based on these.

The theory of vector spaces is simply expanded by allowing scalars and the components of vectors to be complex numbers. Such vector spaces are called *Hermitian* or *Unitary* vector spaces. Since complex numbers possess all the relevant algebraic properties of real numbers (that is, they form a *field*), all the preceding discussion concerning vectors in R^n applies to vectors in C^n, *except* the definition of the scalar product.

Let x be a vector in C^n. Then we define the *complex conjugate*, $\bar{x}$, (sometimes x^*) of x to be the vector whose components are the complex conjugates of the components of x. That is, if the jth component of x is $a_j + ib_j$, the jth component of $\bar{x}$ is $a_j - ib_j$.

Then the *scalar product* of y, x is defined to be the scalar product of $y, \bar{x}$ taken in the usual way. That is,

$$y\bar{x} = \sum y_j\bar{x}_j.$$

The product itself is often written simply yx, especially in a context in which the vectors are predominantly real. It should be noted that the products $y\bar{x}$ and $\bar{y}x$ are not identical, but are complex conjugates. This can easily be seen by writing $x_j = (a + ib)$, $y_j = (c + id)$ and comparing the terms $y_j\bar{x}_j$ and $\bar{y}_jx_j$ in the two products.

The scalar $x'\bar{x}$ is, however, a real number $[\sum (a_j)^2 + \sum(b_j)^2]$ and is the *norm* (Hermitian norm) of x.

That the direct product $y'x$ is not a suitable definition of the scalar product in the case of complex vectors can easily be seen by calculating $x'x$ for the vector of order $2r$, r of whose components are 1, and r are i. Then

$$x'x = r \cdot 1^2 + ri^2 = 0,$$

so that the vector appears to have zero length, although none of its components are zero. This contradicts the fundamental property of a distance measure (see Section R1.5).

The correctly defined norm $x'\bar{x} = 2r$.

R2.7 MATRICES

A *matrix of order* $m \times n$ is simply a rectangular array of numbers a_{ij} (the *elements* of the matrix) in which the first or *row* index runs from 1 to m and the second or *column* index runs from 1 to n. Such a matrix, usually designated by a capital letter, would be

$$A = \begin{bmatrix} a_{11} & a_{12} & a_{13} & \cdots & a_{1n} \\ a_{21} & a_{22} & a_{23} & \cdots & a_{2n} \\ \cdot & \cdot & \cdot & \cdot & \cdot \\ a_{m1} & a_{m2} & \cdot & \cdots & a_{mn} \end{bmatrix}.$$

Usually we write simply $A = [a_{ij}]$ (square brackets are normally used for matrices) to express the fact that the typical element of A is a_{ij}. When the context is clear, we simply refer to the matrix A, meaning the matrix whose typical element is a_{ij}. When matrices and vectors appear together, we will almost always write the matrix with a capital and the vector with a lower case letter.

Matrices (and vectors) are often used just to simplify the terminology, even when few algebraic properties are being used. Much of the simplification comes from adherence to the terminological conventions, of which the most important can be summarized by the mnemonic "row before column." The first index of a_{ij} is always the row index, and the first number in giving the order of the matrix (as in $m \times n$) is always the number of rows.

In all the matrices with which we shall be concerned except matrices composed of characteristic vectors (which we shall meet in Review R5), the elements of the matrix will be real numbers.

There are many ways of looking at a matrix, but the most useful, at this stage, is to regard it as an ordered set of vectors. It can be considered from this point of view in two ways. An $m \times n$ matrix can be considered as composed of m row vectors, each having n components, or as n column vectors, each having m components. These two ways are related to the idea of *duality*, that we shall come across often. We shall usually write the typical row as A_i and the typical column as A^j.[5]

[5] Since we adopt the convention that capital letters represent matrices in almost all cases, there is no confusion between the row vector A_i and the component a_i of the vector a.

Example

$$A = \begin{bmatrix} -2 & 0 & 3 \\ 1 & 2 & 1 \end{bmatrix}$$

is a matrix of order 2×3 which can be considered as composed of the three column vectors

$$\begin{bmatrix} -2 \\ 1 \end{bmatrix}, \quad \begin{bmatrix} 0 \\ 2 \end{bmatrix}, \quad \begin{bmatrix} 3 \\ 1 \end{bmatrix}$$

or of the two row vectors $[-2 \quad 0 \quad 3]$, $[1 \quad 2 \quad 1]$.

A matrix of order $1 \times n$ is simply a row vector of order n, and a matrix of order $m \times 1$ is simply a column vector of order m.

A matrix of order $n \times n$ is a *square matrix*. Square matrices have many properties not possessed by rectangular matrices and these properties are discussed in Review R5.

R2.8 MATRIX ALGEBRA

We define the following operations on matrices:

Multiplication b ʹ a Scalar

λA is the matrix whose typical element is λa_{ij}, that is, we multiply *every* element of A by the scalar.

Addition

This is defined only for two matrices of the same order. The sum $A + B$ of A, B is the matrix whose typical element is $a_{ij} + b_{ij}$. That is, we add corresponding elements of A and B.

If, in addition, we define the zero (or *null*) matrix of order $m \times n$ as the matrix of that order all of whose elements are zero, and denote this by 0, then the following algebraic relationships hold between the set of all matrices *of the same order.*

$$\begin{aligned} (A + B) + C &= A + (B + C) \\ A + B &= B + A \\ A + (-1)A &= 0 \\ 0A &= 0 \\ 1A &= A \\ \lambda(A + B) &= \lambda A + \lambda B \\ (\lambda + \mu)A &= \lambda A + \mu A \\ \lambda(\mu A) &= \lambda\mu A. \end{aligned}$$

These will be recognized as precisely analogous to the algebraic properties of vectors of the same order, set out in Section R2.1. In fact, the above properties follow directly from considering matrices as ordered sets of vectors.

As in the case of vectors we denote $A + (-1)B$ by $A - B$, and as in he case of vectors, A, B are *equal* if and only if *all* their corresponding elements are equal.

We define the operation of *transposition* as follows:

The *transpose of* A, written A', A^T, or ${}^TA$, is the $n \times m$ matrix $[a_{ji}]$. That is, we transpose A by transposing its columns into rows, preserving their order so that the first column of A becomes the first row of A^T and so on.

Example. The transpose of

$$\begin{bmatrix} -2 & 0 & 3 \\ 1 & 2 & 1 \end{bmatrix}$$

is

$$\begin{bmatrix} -2 & 1 \\ 0 & 2 \\ 3 & 1 \end{bmatrix}$$

Matrix Multiplication

We now define the operation of *matrix multiplication.*

The product of two matrices A, B (where the fact that A is the first matrix and B the second is crucial) is defined only if the number of *columns* in the first matrix is the same as the number of *rows* in the second. If A, B are so related, the product $C = AB$ is the matrix whose (i, j)th element is A_iB^j. That is, *the typical element of AB is the scalar product of the ith row of A and the jth column of B.* The mnemonic "row before column" is useful here.

It is obvious that the operation can only be carried out if the vectors A_i, B^j are of the same order, hence the requirement that the number of columns of A (the order of A_i) be the same as the number of rows of B (the order of the column vector B^j).

No restriction is placed on the number of *rows* of A or *columns* of B. The number of rows of A determines the number of row vectors A_i from which the products A_iB^j can be formed, and hence the number of rows in the product AB. Similarly the number of columns of B determines the number of columns in AB.

A simple rule for determining the possibility and outcome of matrix multiplication is to write the order of the first matrix, then the order of the second, as follows (where A is of order $m \times k$, B of order $k \times n$):

$$(m \times \underbrace{k) \times (k}_{(=)} \times n).$$

(=) — links the two inner k's; (order of product) — links the outer m and n.

$$\underbrace{(m \times k) \times (k \times n)}_{\text{(order of product)}}$$

Then the product exists only if the *inner* indices are the same, and the order of the product is obtained by deleting the inner indices.

It is obvious that the order in which the matrices are multiplied (we cannot avoid using "order" in two senses here) is crucial. For the product BA, with A, B as above, we would have

$$(k \times n) \times (m \times k).$$

The inner indices are not the same, so BA is undefined.

If A was of order $m \times n$ and B of order $n \times m$, then both products AB and BA would be defined. But AB would be of $m \times m$ while BA would be of order $n \times n$, so $AB \neq BA$. Only if A, B were square matrices of the same order would it be even *possible* to have $AB = BA$, but even then this would not generally be true.

Note that, if we consider a column vector x as a matrix of order $n \times 1$, and a row vector y as a matrix of order $1 \times n$, the matrix product yx is identical with the scalar product of y, x considered as vectors. Indeed, all the algebraic properties of x considered as a vector and as a matrix are the same.

We might note that x, y considered as *matrices* could be multiplied in the order xy, to give a matrix of order $n \times n$. We do not use this product in our analysis.

Since the order of multiplication of matrices is crucial, we may refer to *premultiplication* and *postmultiplication* if clarification is necessary. In the product AB, A is postmultiplied by B, B is premultiplied by A.

Example.

$$A = \begin{bmatrix} -2 & 0 & 3 \\ 1 & 2 & 1 \end{bmatrix} \qquad B = \begin{bmatrix} 1 & 2 \\ 2 & 3 \\ 1 & 1 \end{bmatrix}.$$

The products AB, BA are both defined. We have

$$\begin{aligned} AB &= \begin{bmatrix} (-2).1 + 0.2 + 3.1 & (-2).2 + 0.3 + 3.1 \\ 1.1 + 2.2 + 1.1 & 1.2 + 2.3 + 1.1 \end{bmatrix} \\ &= \begin{bmatrix} 1 & -1 \\ 6 & 9 \end{bmatrix}. \end{aligned}$$

We will find that

$$BA = \begin{bmatrix} 0 & 4 & 5 \\ -1 & 6 & 9 \\ -1 & 2 & 4 \end{bmatrix}.$$

It is easily verified that the following properties hold for multiplication, provided the matrices are of appropriate order:

$$\begin{aligned} (A + B)C &= AC + BC \\ A(BC) &= (AB)C \\ A(\lambda B) &= \lambda AB. \end{aligned}$$

Two products always exist for a matrix and its transpose, AA^T and A^TA. If A is of order $m \times n$, the first product is of order $m \times m$, the second of order $n \times n$.

Consider matrices A, B of order $m \times k$, $k \times n$, respectively. The product AB is of order $m \times n$. The transpose of the product $(AB)^T$ is of order $n \times m$ while the order of the transposed matrices A^T, B^T are of order $k \times m$, $n \times k$, respectively. It is obvious that the transposed matrices can only be multiplied *in reverse order*. Thus we have

$$(AB)^T = B^T A^T,$$

a result that can easily be checked by direct multiplication.

R2.9 MATRIX-VECTOR PRODUCTS AND LINEAR TRANSFORMATIONS

If A is a matrix of order $m \times n$ and x a *column* vector of order n then, considering x as a matrix of order $n \times 1$, the product Ax is defined and is of order $m \times 1$, that is, it is a column vector of order m.

Thus we can regard the matrix A as the rule for a mapping from the vector x, in R^n, to the vector Ax, in R^m. It follows from the properties of matrix algebra that

$$A(\lambda x) = \lambda Ax$$
$$A(x + y) = Ax + Ay,$$

so that the mapping $x \to Ax$ preserves sums and scalar multiples and thus maps a linear subspace into a linear subspace. We refer to such a mapping as a *linear mapping* or *linear transformation*. There are linear mappings other than those defined by a matrix, but matrices form the most important class.

Actually, the matrix A gives not one, but *two*, linear transformations. If y is a row vector of order m, then the mapping $y \to yA$ maps an m vector into an n vector. The two mappings $x \to Ax$, $y \to yA$ are *dual* mappings.

In many economic contexts, the transformation $x \to Ax$ will represent a transformation in the physical sense, from a vector of input quantities into a vector of output quantities. The dual transformation will appear in the form $p \to pA$, from input prices into output prices.

Let A be a matrix of order $m \times n$, and v be a column vector of order n. Denote by u the vector Av, a column vector of order m. If B is a matrix of order $k \times m$, we can make the transformation $w = Bu$, where w will be a column vector of order k. But $u = Av$, so that $w = B(Av) = (BA)v$. Thus we can see that the product BA, a matrix of order $k \times n$, gives a direct mapping from v to w. The definition of matrix multiplication was originally chosen to give just this result, that the result of successive linear mappings would be represented by the product of the matrices of those mappings.

Consider the relationship

$$Ax = b,$$

where b is given, x is unknown. This can be regarded as a system of linear equations with constant coefficients

$$\sum a_{ij}x_j = b_i \qquad i = 1, \ldots, m,$$

or as an inverse mapping problem of the kind, what vector x maps into b, if A is the matrix of the mapping?

There is yet another way of looking at the matrix vector product Ax that is useful. We can regard Ax as the weighted sum of the columns of A, with the x_i's considered as weights. From this point of view we can write the relationship $Ax = b$ in the form

$$\sum x_j A^j = b,$$

where the x_j's are regarded as scalars rather than as components of a vector.

In this form we can regard the problem of finding x as that of being given the information that the vector b is a linear combination of the vectors A^j, and seeking the appropriate weights for the combination. If $Ax = 0$, this approach implies that the columns of A are linearly dependent if $x \neq 0$.

The product yA can be considered, in the same way, as the weighted sum of the *rows* of A.

For the homogeneous equation system $Ax = 0$ we have an additional point of view. This system is equivalent to the m relationships $A_i x = 0$, so we can regard the solution vector x as the vector which is *orthogonal* to all the row vectors of A.

R2.10 PARTITIONED MATRICES

It is sometimes useful to consider a matrix to be *partitioned* into submatrices. The matrix A, for example, might be partitioned into four submatrices as follows:

$$A = \left[\begin{array}{c|c} A_1 & A_2 \\ \hline A_3 & A_4 \end{array}\right] \begin{array}{l} r \text{ rows} \\ (m - r) \text{ rows.} \end{array}$$

$$\quad s \text{ cols.} \quad n - s \text{ cols.}$$

Suppose we have another matrix B, of the same order as A and partitioned in the same way so that B_1 is of the same order as A_1, B_2 as A_2, and so on. The it follows directly from the definition of matrix addition that

$$A + B \quad \left[\begin{array}{c:c} A_1 + B_1 & A_2 + B_2 \\ \hdashline A_3 + B_3 & A_4 + B_4 \end{array}\right].$$

Matrices with partitioning suitable for the operation in question are *conformably partitioned*, as are A and B for addition.

What about partitioning conformable for multiplication? Suppose A is of order $m \times k$, B of order $k \times n$. Then the product AB exists. Suppose A is partitioned into two submatrices as follows[6]:

$$A = [\underset{s \text{ cols.}}{A_1} \;\vdots\; \underset{k - s \text{ cols.}}{A_2}]$$

If B is partitioned as follows:

$$B = \left[\begin{array}{c} B_1 \\ \hdashline B_2 \end{array}\right] \begin{array}{l} s \text{ rows} \\ k - s \text{ rows.} \end{array}$$

Then the four submatrices involved in the product have orders:

$$\begin{array}{ll} A_1 & m \times s \\ A_2 & m \times (k - s) \\ B_1 & s \times n \\ B_2 & (k - s) \times n. \end{array}$$

Two products of submatrices are defined, A_1B_1 and A_2B_2, both of order $m \times n$, so that these can be added. Thus the operations implied in the expression $A_1B_1 + A_2B_2$ can all be carried out.

Direct calculation, using the ordinary definition of matrix multiplication, shows that, in fact,

$$AB = A_1B_1 + A_2B_2.$$

If A was doubly partitioned into s and $k - s$ columns (as above) and also into r and $m - r$ rows, we can show that the appropriate partitioning of B is into s and $k - s$ rows (as above) and t and $n - t$ columns, where r, t can be chosen arbitrarily (only s must be the same in the two matrices).

[6] Note that a matrix can be partitioned vertically any number of times, and horizontally any number of times. The number of vertical and horizontal partitionings does not have to be the same.

With the four submatrices written, in each case, in the same order as in the addition example, we will find that

$$AB = \left[\begin{array}{c|c} A_1B_1 + A_2B_3 & A_1B_2 + A_2B_4 \\ \hline A_3B_1 + A_4B_3 & A_3B_2 + A_3B_4 \end{array}\right] \begin{array}{l} r \text{ rows} \\ m - r \text{ rows.} \end{array}$$

$$\qquad\qquad t \text{ cols.} \qquad\qquad n - t \text{ cols.}$$

Note that the rules for adding and multiplying conformably partitioned matrices are the same as if the submatrices were treated as *elements* of an ordinary matrix, except that we must be careful, in multiplication, always to order the multiplication of the submatrices so that those from the first matrix premultiply those from the second matrix.

In checking for conformability for multiplication, it is necessary only that the columns of the first matrix be partitioned the same as the rows of the second. The number of partitionings of the rows of the first matrix need not even correspond to the number of partitionings of the columns of the second.

It is often useful to partition the matrix and the vector in a matrix-vector product, following the same rules.

R2.11 VECTOR SETS

If S, T are sets of vectors in the same vector space, the ordinary properties of sets—union, intersection, and so on, hold unchanged.

We can in addition define some new relationships between the sets, as follows:

(a) λS is the set of all vectors λx, such that $x \in S$;
(b) $S + T$ is the set of all vectors $x + y$, such that $x \in S$, $y \in T$;
(c) $x' \cdot T$ is the set of all scalar products $x'y$ with $y \in T$;
(d) $S \cdot T$ is the set of all scalar products $x'y$ with $x \in S$, $y \in T$.

One of the most important properties that a set of vectors may or may not possess is *convexity*. We shall defer discussion of convex sets until the next review, however.

A vector set S is *bounded below* if there is some vector x_0 such that $x \geqq x_0$ for all $x \in S$, and *bounded above* if there is some x^0 such that $x \leqq x^0$ for all $x \in S$. If the set is bounded both above and below it is contained in a finite hypercube and therefore *bounded* in the sense of Section R1.5.

EXERCISES

1. Choose a suitable basis from among the following vectors, and express the nonbasis vector in terms of the others:

$$x^1 = \begin{bmatrix} 1 \\ 2 \\ 1 \end{bmatrix}; \quad x^2 = \begin{bmatrix} 3 \\ 0 \\ -1 \end{bmatrix}; \quad x^3 = \begin{bmatrix} -1 \\ 1 \\ 1 \end{bmatrix}; \quad x^4 = \begin{bmatrix} 1 \\ -1 \\ 1 \end{bmatrix}.$$

2. Given the vectors,

$$x^1 = \begin{bmatrix} 2 \\ -1 \\ 1 \end{bmatrix}, \quad x^2 = \begin{bmatrix} 3 \\ 1 \\ 1 \end{bmatrix}, \quad x^3 = \begin{bmatrix} 1 \\ 1 \\ -1 \end{bmatrix},$$

a. Show that they form a basis.

b. Express

$$b = \begin{bmatrix} 4 \\ 5 \\ 3 \end{bmatrix}$$

in terms of this basis.

c. Show that x^1, x^3 are orthogonal.

d. Find a vector x^4 such that x^1, x^3, x^4 form an orthogonal basis.

e. Express b in terms of this new basis.

f. Choose scalars α, β, γ so that αx^1, βx^3, γx^4 form an orthonormal set.

3. If

$$\mathrm{A} = \begin{bmatrix} 2 & 1 & 5 \\ 3 & 1 & 4 \end{bmatrix}, \quad B = \begin{bmatrix} 1 & 2 & 4 \\ 3 & 1 & 1 \end{bmatrix},$$

calculate each of the following, if they are defined:

a. AB^T; **b.** BA; **c.** AB; **d.** A^TB.

4. Given

$$b = [1 \quad 2], \qquad c = [3 \quad 1], \qquad A \quad \begin{bmatrix} 2 & 1 \\ 3 & 4 \end{bmatrix} \qquad B = \begin{bmatrix} 5 & 2 \\ 1 & 1 \end{bmatrix},$$

calculate each of the following: **a.** bc'; **b.** bA; **c.** Ab'; **d.** cA; **e.** AB; **f.** BA; **g.** $bABc'$.

REVIEW

R3

Linear Equations and Inequalities

The content of this review follows on directly from the content of Review R2 and is equally basic.

R3.1 INTRODUCTION[1]

We have already seen (Section R2.9) that the system of homogeneous equations,

$$Ax = 0$$

or

$$A_i x = 0 \qquad i = 1, \ldots, m,$$

can also be regarded as

$$\sum_{j=1}^{n} x_j A^j = 0,$$

so that the following three statements are equivalent:

(a) $Ax = 0$ has a solution with at least one nonzero component,[2]

[1] Modern linear equation theory (in terms of vector spaces rather than determinants) is discussed in linear algebra texts, including Hadley [1].

The theory of linear inequalities is more specialized. Gale [1] contains a similar treatment to that given here. For more advanced treatment see Tucker and Fan.

[2] Since $Ax = 0$ always has a solution $x = 0$, we speak of a solution $x \neq 0$ as a *nontrivial solution.*

(b) x is a vector which is orthogonal to all the rows of A.
(c) the columns of A are linearly dependent.

Since any set of n vectors in R^m must be linearly dependent if $m < n$, we can state immediately:

Any system of m homogeneous equations in n unknowns has a nontrivial solution if $m < n$.

If $m \geqq n$, the existence of a nontrivial solution depends on whether the columns are linearly dependent. Thus we are led, as fundamental to the theory of linear equations, to the study of linear dependence among the columns of a matrix.

R3.2 THE RANK OF A MATRIX

We define the *column rank* of A as the maximum number of linearly independent columns and the *row rank* as the maximum number of linearly independent rows. We now state the following fundamental theorem:

The row rank and column rank of a matrix are equal, so that we can simply refer to the rank of a matrix.

Suppose A is a matrix of order $m \times n$, with row rank r and column rank s. Suppose also that, contrary to the theorem, r, s are not equal, in particular that $r < s$. (We certainly have $r \leqq m, s \leqq n$.)

Choose r linearly independent rows of A, which we can take to be the first r, $A_1, \ldots, A_r$, and s linearly independent columns which we can suppose to be the first s, $A^1, \ldots, A^s$.

Now consider the *truncated* row vectors $\hat{A}_1, \ldots, \hat{A}_m$ which consist only of the first s components of A_i, that is, of the components which appear also in $A^1, \ldots, A^s$ These vectors are of order s. Take the first r of these vectors, that is, the truncated versions of the linearly independent set. These form a matrix $\hat{A}$ of order $r \times s$ whose relation to the matrix A is illustrated here.

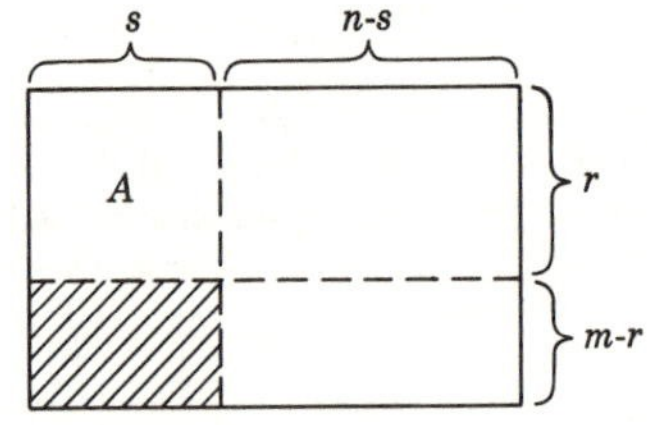

The equation system,

$$\hat{A}x = 0,$$

is a homogeneous system of r equations in s unknowns, with $s > r$, so that it has a nontrivial solution $\hat{x}$.

Now consider a new matrix $\bar{A}$ consisting of all m truncated row vectors $\hat{A}_1, \ldots, \hat{A}_m$. This matrix is $\hat{A}$ with the addition of the elements in the shaded region of the illustration.

Since the rows $A_1, \ldots, A_r$ form a basis in R^n, any other row of A, say A_k, can be uniquely expressed as a linear combination of $A_1, \ldots, A_r$.

$$A_k = \sum_1^r \mu_{ik} A_i \qquad \text{all } k.$$

This relationship clearly holds also for the truncated versions of the same vectors, so that

$$\hat{A}_k = \sum_1^r \mu_{ik} \hat{A}_i \qquad \text{all } k.$$

Forming the scalar product of these row vectors with the vector $\hat{x}$ from above, we have

$$\begin{aligned} \hat{A}_k \hat{x} &= \sum_1^r \mu_{ik} \hat{A}_i \hat{x}. \\ &= 0 \end{aligned}$$

Since $\hat{A}\hat{x} = 0$ means $\hat{A}_i\hat{x} = 0$, $i = 1, \ldots, r$.

Now consider the product $\bar{A}\hat{x}$. For the first r rows of $\bar{A}$ we have $\hat{A}_i\hat{x} = 0$ from the definition of $\hat{x}$. For the remaining rows we have $\hat{A}_k\hat{x} = 0$ from above, so that

$$\bar{A}\hat{x} = 0.$$

But the columns of $\bar{A}$ are simply the first s columns of A, so that the above equations are equivalent to

$$\sum_1^s \hat{x}_j A^j = 0.$$

This implies that the first s columns of A are linearly dependent, contradicting the hypothesis that the column rank is s. Thus we cannot have $s > r$.

By arguing symmetrically with the roles of rows and columns interchanged we can show that we cannot have $r > s$.

Thus we must have $r = s$, proving the theorem.

The rank of a matrix is sometimes defined by means of determinants (see Review R5, Section R5.2). It is more appropriate to view the determinantal conditions as the indication of underlying linear dependence, rather than as the definition of the rank.[3]

R3.3 HOMOGENEOUS EQUATIONS

All the essential properties of the nontrivial solutions to systems of homogeneous linear equations are expressed in the following theorem:

If A is a matrix of order $m \times n$ and rank r, the set X of all solutions of the equation system $Ax = 0$ is a linear subspace of rank $n - r$.

It is obvious that λx is a solution if x is a solution, and that $x + y$ is a solution if x, y are solutions, so that X has the properties of a linear subspace (Section R2.2). We need to show that X has a basis of $n - r$ linearly independent vectors.

Since A has rank r, we can choose some set of r linearly independent columns of A (which we shall take to be the first r) and express remaining columns as linear combinations of these. Thus we have:

$$A^k = \sum_{j=1}^{r} \mu_{jk} A^j \qquad k = r + 1, \ldots, n. \tag{R3.3.1}$$

Now define column vectors v^k of order n as

$$v^k = \begin{bmatrix} -\mu_{1k} \\ -\mu_{2k} \\ \vdots \\ -\mu_{rk} \\ 0 \\ 0 \\ \vdots \\ 1 \\ \vdots \\ 0 \end{bmatrix} \begin{matrix} \\ \\ \\ \\ \\ \\ \\ \leftarrow k\text{th coordinate,} \\ \\ \\ \end{matrix}$$

for $k = r + 1, \ldots, n$.

[3] Determinants are useful as a computational tool (when we wish to determine the actual rank) but are less useful as an analytical tool.

We have

$$\begin{aligned} Av^k &= -\sum_1^r \mu_{jk}A^j + A^k \\ &= 0 \qquad \text{from (R3.3.1),} \end{aligned}$$

so that each v^k is a solution of the equation system and so in X. Also $\sum_{r+1}^n \lambda_k v^k$ has λ_k as its kth coordinate ($k = r + 1, \ldots, n$) so that

$$\sum_{r+1}^n \lambda_k v^k = 0,$$

only if $\lambda_k = 0$ for $k = r + 1, \ldots, n$. Thus the $n - k$ vectors v^k are linearly independent and in X. They thus represent a possible basis.

To complete the proof, we must show that an arbitrary $x \in X$ can be expressed as a linear combination of the v^k. Since x, v^k are solutions of the equation system, the linear combination,

$$x^* = x - \sum_{r+1}^n x_k v^k, \tag{R3.3.2}$$

is also a solution.

For $k > r$, the kth coordinate of x^* is

$$x_k^* = x_k - x_k = 0.$$

Thus

$$\begin{aligned} Ax^* &= \sum_1^r x_j^* A^j + \sum_{r+1}^n x_k^* A^k \\ &= \sum_1^r x_j^* A^j. \end{aligned}$$

But $A^1, \ldots, A^r$ are linearly independent, so $Ax^* = 0$ implies $x_j^* = 0$, $j = 1, \ldots, r$. We have already shown $x_j^* = 0$, $j = r + 1, \ldots, n$, so that $x^* = 0$.

From (R3.3.2), $x^* = 0$ implies

$$x = \sum_{r+1}^n x_k v^k,$$

so that the v^k form a basis and the theorem is proved.

We can interpret some special cases. If $r = n - 1$, the solutions form a linear subspace of rank 1, that is, a line through the origin. This expresses the fact that, if x is a solution, so is λx. It is common to refer to a solution subspace of rank 1 as a *unique solution*, meaning *unique except for a scalar multiple*. No problems usually arise from this rather loose usage.

If A consists of a single row, it has rank 1. The solution space is then of rank $n - 1$. A linear subspace of rank $n - 1$ in a vector space of rank n is a *hyperplane*, so that the equation $ax = 0$, where a is a row vector, is the equation of a hyperplane through the origin.

R3.4 NONHOMOGENEOUS EQUATIONS

Now we turn to the equation system

$$Ax = b,$$

where A is of order $m \times n$ and rank r.

If we consider this is the form

$$\sum x_j A^j = b,$$

then the statements "$Ax = b$ has a solution" and "b is a linear combination of the columns of A" are equivalent.

It is obvious that we can always form a linear combination of any arbitrary number of vectors, so it is always *possible* for $Ax = b$ to have a solution, whatever the relationships between m, n, r provided the vector b is appropriate. For an *arbitrary* vector b, on the other, there is no reason to suppose that a solution always exists. The necessary condition for the existence of a solution is that the $n + 1$ vectors $A^1, \ldots, A^n, b$ are linearly dependent.

Our interest here in the equation system $Ax = b$ is in the following three main problems:

(a) The condition that a solution exists for every possible b vector;
(b) The condition that the solution is unique;
(c) The nature of the solution set X^* if the solution is not unique.

To deal first with (a) we note that an arbitrary vector b can be expressed as a linear combination of the columns of A, if and only if the columns include a basis for vector space of rank m. Thus we can state immediately:

$Ax = b$ has a solution for every b vector, if and only if $r = m$. This necessarily requires that $m \leqq n$.

We shall confine the rest of the analysis to cases in which $r = m \leqq n$, so that a solution exists for all b. Let x^* be a solution.

Now consider the homogeneous equation system $Ax = 0$. If x is a solution of this, then it is obvious that $x + x^*$ is a solution of $Ax = b$. On the other hand, if x^*, x^{**} are two different solutions of $Ax = b$, then

$$A(x^{**} - x^*) = b - b = 0,$$

so that $x^{**} - x^*$ is a solution of the homogeneous system.

Thus if x^* is *any* solution of $Ax = b$, and X is the solution set of the homogeneous system $Ax = 0$, the solution set X^* of $Ax = b$ is given by

$$X^* = x^* + X.$$

Since we already know the nature of the solution set of the homogeneous system, given r and n, we can state the following

If $r = m = n$, the equation system $Ax = b$ has a unique solution for every b.

This result follows from the above discussion because $Ax = b$ certainly has a solution ($r = m \leqq n$), and the solution space of the homogeneous system has rank $n - n = 0$.

If $r = m = n$, the matrix A is said to be *nonsingular* or *invertible*, for reasons that will be more apparent in the discussion of square matrices in a later review. We can give a sense to the term "invertible" however by considering the mapping $x \to Ax\ (=b)$. Since x is unique, this means there is a unique inverse mapping which we can write $b \to A^{-1}b$. We shall see later that A^{-1} also plays the role of an algebraic inverse.

If the uniqueness condition is not satisfied, we have

If $r = m > n$, the system $Ax = b$ has a solution space of rank $n - r$.

Since any m vector can be expressed as a linear combination of the r columns of A which form a basis, it follows that, if $r < n$, there always exists a solution of $Ax = b$ such that $x_j = 0$ if A^j is not a column in the basis. Such a solution is called a *basic solution.*

If we select the $r(=m)$ columns of A which form the basis, we can form a square matrix from these of order $m \times m$ and rank m. Such a matrix is often called a *basis* in A and written A_B. The equation system

$$A_B\hat{x} = b,$$

where $\hat{x}$ has only m components, has a unique solution. The vector $\hat{x}$ is a truncated version of a basic solution vector x of $Ax = b$, with the zero components corresponding to columns of A not in A_B being omitted.

R3.5 NONNEGATIVE VECTORS AND VECTOR INEQUALITIES

For many economic purposes it is not sufficient to know simply that a linear equation system has a solution; we wish to know whether it has a solution with no negative components, since many economic variables may be undefined for negative values. Related to this is the appearance in many economic models of relationships of an inequality kind. The use of a resource cannot, for example, exceed its availability, but part of the available quantity need not be used, so that an equality is inappropriate. Indeed, it is not unreasonable to say that the introduction of the inequality into the specification of economic models has been the greatest advance of recent years.

Thus the idea of ordering two vectors, one of which may be the zero vector, becomes essential.

If we confine ourselves to vectors in R^n, then the components of vectors, being real numbers, are completely ordered (complex numbers are not), so that the ordering relationships $x_i > 0$, $x_i \geqq 0$ are always defined. For the vector whose components are x_i we make the following definitions:

(1) $x \geqq 0$ (*x nonnegative*), if $x_i \geqq 0$, all i.
(2) $x \geqslant 0$ (*x semipositive*), if $x_i \geqq 0$, all i, $x_i > 0$ for some i.
(3) $x \gg 0$ (*x strictly positive*) if $x_i > 0$, all i.

The inequality relationships are in decreasing order of inclusiveness. Nonnegative vectors includes semipositive and positive vectors, semipositive vectors includes positive vectors.

The notation varies somewhat. One common tradition is to write $\geqq$, $\geqslant$, $>$, for (1), (2), and (3); another is to write $\geqslant$, $>$, $\gg$. Our notation is a bastard version, justified on the grounds that strict positivity is usually a very important property and ought to be given emphasis. The notation "$>$" is rejected because it means different things in the two regular traditions, and the possible confusion between semipositivity and nonnegativity is not important. The only nonnegative vector which is not semipositive is 0, and it is usually quite clear whether 0 is acceptable or not.

Inequality relationships between two vectors are derived from the nonnegativity properties of their difference. Thus $x \geqslant y$ means $x - y \geqslant 0$, and so on. Inequalities can be reversed in the usual way, $x \leqq 0$ meaning $0 \geqq x$, or $(-x) \geqq 0$.

It is obvious that vectors are only partially ordered by "$\geqq$," the most inclusive of the inequalities, since, if $x_i > 0$, $x_j < 0$, then we have neither $x \geqq 0$ nor $0 \geqq x$.

For *addition* of nonnegative vectors, it is obvious that the strongest inequality dominates so that $x + y \gg 0$ if $x \geqq 0, y \gg 0$.

For the *scalar product* we have only two possibilities since semipositivity has no meaning for a scalar. If both vectors are strictly positive, the product is positive, and if one vector is only semipositive but the other is positive, the product is positive. In all other cases the product can only be taken as nonnegative, in the absence of other information.

The product $x'x$ is, however, positive even if x is semipositive because the nonzero components of x are matched with themselves in the product.

An important set of nonnegative relationships that underlies the structure of equilibrium in economics is of the form,

$$px = 0, \qquad p \geqq 0, \qquad x \geqq 0.$$

This clearly implies $p_i = 0$, if $x_i > 0$ and $x_i = 0$, if $p_i > 0$, an implication we shall often use.

By analogy with linear equations, relationships of the form,

$$Ax \geqq 0,$$

are *homogeneous linear inequalities*, while those of the form,

$$Ax \geqq b,$$

are *nonhomogeneous linear inequalities*.[4]

Since a nonnegative solution of the equation system $Ax = 0$ can be regarded as a solution of the inequalities $Ax \geqq 0$, $Ax \leqq 0$, $x \geqq 0$, nonnegative solutions and the solutions of inequalities are closely related, as we shall see in the next sections. In a context of inequalities, we shall very often prove that $x = y$ by showing that $x \geqq y$ and $x \leqq y$ are both true.

R3.6 FUNDAMENTAL THEOREM ON LINEAR INEQUALITIES[5]

The Theorem which is stated and proved in this section is basic to the theory of linear inequalities, and the results for these inequalities that we make use of at various points in our analysis can be easily derived once the theorem is established. There are various ways of proving the theorem; the one given here is a little tedious but is based only on simple linear algebra.

[4] Sometimes *inequations* is used instead of *inequalities*.

[5] See also Gale [1], Chapter 2, for this and the following section. See also Review R4. Section R4.7 for an alternative approach.

First we shall prove the following as a lemma:

(*a*) *Either* (*i*) $Ax = b$ *has a solution, or* (*ii*) *the equations* $yA = 0$, $yb = c$ *have a solution for any number* c.

Since most of the theorems on inequalities and nonnegative solutions are in this "either, or" form, it should be stressed that we use these terms in the mutually exclusive sense. That is, *exactly one or the other but not both are true.*

To show the alternatives of the lemma are mutually exclusive, we multiply (i) by y to give $yAx = yb$. If we multiply (ii) by x we obtain $yAx = 0$. The two results are inconsistent with $yb = c$ unless $c = 0$.

We now show that if $Ax = b$ has no solution, then $yA = 0$, $yb = c$ has a solution. Suppose that A is of rank r and let $A^1, \ldots, A^r$ be a column basis for A. Then the $r + 1$ vectors $A^1, \ldots, A^r, b$ must be linearly independent otherwise $Ax = b$ would have a solution.

Form the matrix $\hat{A}$ with the columns $A^1, \ldots, A^r, b$. This is a matrix of order $m \times (r + 1)$ and rank $r + 1$, with $r + 1 \leqq m$ (otherwise the vectors $A^1, \ldots, A^r, b$ could not be independent).

Thus the equation system

$$y\hat{A} = \hat{c}$$

is a system of $r + 1$ equations in m unknowns, with A having rank $r + 1 \leqq m$. The condition that a solution exists for every $\hat{c}$ is satisfied. Choose $\hat{c}$ (a row vector) so that $\hat{c}_j = 0$, $j = 1, \ldots, r$; $\hat{c}_{r+1} = c$. The equations $y\hat{A} = \hat{c}$ are now equivalent to

$$\begin{aligned} yA^j &= 0 \qquad j = 1, \ldots, r \\ yb &= c \end{aligned}$$

and have a solution.

But the columns A^k of A are linear combinations of the columns $A^1, \ldots, A^r$ (since these are the basis) so that

$$\begin{aligned} yA^k &= \sum_1^r \mu_{kj} A^j \\ &= 0 \qquad \text{all } k. \end{aligned}$$

Thus y is a solution for

$$\begin{aligned} yA &= 0, \\ yb &= c, \end{aligned}$$

proving the lemma.

The theorem proper is:

Fundamental Theorem

Either (*i*) $Ax = b$ *has a solution* $x \geqq 0$ *or* (*ii*) *the inequalities* $yA \geqq 0$, $yb < 0$ *have a solution.*[6]

First we check the mutually exclusive property. Multiplying (i) by y we obtain $yAx = yb$. Multiplying (ii) by x ($\geqq 0$) we obtain $yAx \geqq 0$. These are inconsistent with $yb < 0$. (This method of proof of mutual exclusiveness is standard for all theorems of this kind.)

If (i) has *no* solution, then there is a solution for $yA = 0$, $yb = c$, from the lemma. Putting $c < 0$ then gives a solution for (ii).

The crux of the proof is to show that, if (i) does have a solution, but not a nonnegative solution, then (ii) has a solution. The proof is inductive, proving the theorem for an $m \times 1$ matrix, then for an $m \times k$ matrix, if it is true for an $m \times (k - 1)$ matrix.

For an $m \times 1$ matrix (i) becomes $xA^1 = b$, with x here a scalar. Put $y = -b'$ (we must transpose since b is a column vector), then

$$\begin{aligned} yA^1 &= -b'A^1 \\ &= -x(A^1)'\,A^1 \\ &> 0 \qquad \text{(since the hypothesis is, } x < 0\text{);} \\ yb &= -b'b \\ &< 0, \end{aligned}$$

so that $y = -b'$ is a solution of (ii), proving the theorem for this case.

Now assume the theorem true for an $m \times (k - 1)$ matrix, and examine the case when the number of columns is increased to k.

By hypothesis,

$$Ax = \sum_1^k x_j A^j = b \tag{R3.6.1}$$

has no nonnegative solution. Then the system

$$\sum_1^{k-1} x_j A^j = b$$

necessarily has no nonnegative solution, otherwise this solution with the additional component $x_k = 0$ would be a nonnegative solution of (R3.6.1).

Since the theorem has been supposed to hold for $k - 1$ columns, it follows that the inequalities

$$\begin{aligned} yA^j &\geqq 0 \qquad j = 1, \ldots, k - 1 \\ yb &< 0 \end{aligned} \tag{R3.6.2}$$

[6] This theorem is a variant of *Farkas' Theorem.*

do have a solution, which we shall write y^*. If $y^*A^k \geqq 0$, the theorem is proved, so we need only consider the case $y^*A^k < 0$.

To simplify the notation, write

$$\begin{aligned} \alpha_j &= y^*A^j \qquad j = 1, \ldots, k-1. \\ -\alpha_k &= y^*A^k \\ -\beta &= y^*b \end{aligned} \tag{R3.6.3}$$

Since y^* is a solution of (R3.6.2) we have $\alpha_j \geqq 0$, $\beta > 0$, and by the hypothesis on y^*A^k, $\alpha_k > 0$.

Now form the following column vectors:

$$\begin{aligned} \hat{A}^j &= \alpha_j A^k + \alpha_k A^j \\ \hat{b} &= -\beta A^k + \alpha_k b. \end{aligned} \tag{R3.6.4}$$

Consider the new equation system

$$\sum_1^{k-1} x_j \hat{A}^j = \hat{b}. \tag{R3.6.5}$$

Suppose this has a nonnegative solution $\hat{x}$. Substituting for $\hat{A}^j$, $\hat{b}$ from (R3.6.4) in (R3.6.5), we have

$$\sum \hat{x}_j(\alpha_j A^k + \alpha_k A^j) = -\beta A^k + \alpha_k b,$$

giving

$$\left(\sum \hat{x}_j \alpha_j + \beta\right) A^k + \alpha_k \sum \hat{x}_j A^j = \alpha_k b,$$

and,

$$\frac{1}{a_k}\left(\sum \hat{x}_j \alpha_j + \beta\right) A^k + \sum \hat{x}_j A^j = b. \tag{R3.6.6}$$

Now the numbers $\hat{x}_j$, α_j are nonnegative and the numbers α_k, β are positive. Hence

$$\gamma = \frac{1}{\alpha_k}\left(\sum \hat{x}_j \alpha_j + \beta\right)$$

is a positive number.

Denote by x^* the vector whose components are:

$$\begin{aligned} x_j^* &= \hat{x}_j \qquad j = 1, \ldots, k-1. \\ x_k^* &= \gamma \end{aligned}$$

Then x^* is a nonnegative vector. Equations (R3.6.6) are equivalent to

$$\sum_1^k x_j^* A^j = b,$$

or,

$$Ax^* = b. \tag{R3.6.7}$$

Thus we have proved that Equations (R3.6.7) have a nonnegative solution if Equations (R3.6.5) have a nonnegative solution. But the hypothesis is that (R3.6.7) do not have a nonnegative solution, hence (R3.6.5) cannot have a nonnegative solution. Since Equations (R3.6.5) have only $k - 1$ rows, the theorem is assumed to hold for these, so there must exist a vector $\hat{y}$ such that

$$\begin{aligned} \hat{y}\hat{A}^j &\geqq 0 \qquad j = 1, \ldots, k-1. \\ \hat{y}\hat{b} &< 0 \end{aligned} \tag{R3.6.8}$$

As the final step we shall show that we can construct a solution for (ii) out of the materials now assembled. Using the vector $\hat{y}$ from (R3.6.8) and the vector y^* from (R3.6.2), construct the vector

$$y = (\hat{y}A^k)y^* - (y^*A^k)\hat{y}. \tag{R3.6.9}$$

Using this new vector, we have

$$\begin{aligned} yA^j &= (\hat{y}A^k)(y^*A^j) - (y^*A^k)(\hat{y}A^j) \\ &= \alpha_j\hat{y}A^k = \alpha_k\hat{y}A^j \qquad \text{[from (R3.6.3)]} \\ &= \hat{y}(\alpha_jA^k + \alpha_kA^j) \\ &= \hat{y}\hat{A}^j \qquad \text{[from (R3.6.4)]}. \end{aligned}$$

In a similar way we can show that

$$\begin{aligned} yA^k &= 0 \\ yb &= \hat{y}\hat{b}. \end{aligned}$$

Using the above together with (R3.6.8) we have

$$\begin{aligned} yA^j &= \hat{y}\hat{A}^j \geqq 0 \qquad j = 1, \ldots, k-1 \\ yA^k &= 0 \\ yb &= -\hat{b} < 0, \end{aligned}$$

so that y is a solution of the inequalities (ii).

We have thus shown the theorem to be true for an $m \times k$ matrix, if true for an $m \times (k - 1)$ matrix, and we have shown it to be true for an $m \times 1$ matrix, so the theorem is proved.

R3.7 RESULTS ON LINEAR EQUATIONS AND INEQUALITIES

Given the fundamental theorem of the last section, we can derive a variety of results concerning the existence of solutions to linear equations and inequalities. The most useful of these are listed below, with the two from

Section R3.6 repeated for completeness. Proofs are given three of the results, the proofs which are not supplied being along similar lines.

(a) Either (i) $Ax = b$ has a solution, or (ii) $yA = 0$, $yb = c$ have a solution for any number c (lemma of Section R3.6).
(b) Either (i) $Ax = b$ has a solution $x \geqq 0$, or (*ii*) $yA \geqq 0$, $yb < 0$ have a solution (fundamental theorem of Section R3.6).
(c) Either (i) $Ax \leqq b$ has a solution $x \geqq 0$, or (ii) $yA \geqq 0$, $yb < 0$ have a solution $y \geqq 0$.
(d) Either (i) $Ax \geqq b$ has a solution, or (ii) $yA = 0$, $yb = 1$ have a solution $y \geqq 0$.
(e) Either (i) $Ax = 0$ has a solution $x \geqslant 0$, or (ii) $yA \gg 0$ has a solution.
(f) Either (i) $Ax = 0$ has a solution $x \gg 0$, or (ii) $yA \geqslant 0$ has a solution.
(g) Either (i) $Ax \leqq 0$ has a solution $x \geqslant 0$, or (ii) $yA \gg 0$ has a solution $y \geqslant 0$.

Many related results can be derived directly from the main results above by transposition and by appropriate scalar multiplication by (-1). We can transpose (c), for example, into the form: either $A'x \geqq 0, b'x < 0$ have a solution $x \geqq 0$ or $yA' \leqq b'$ has a solution $y \geqq 0$. Homogeneous inequalities can always be reversed, since if y is a solution of $yA > 0$, $(-y)$ is a solution of $yA < 0$. Care must be taken with sign changes for nonhomogeneous inequalities.

To prove (c), we first easily show incompatibility of (i) and (ii). Now consider the inequalities $Ax \leqq b$. These can be turned into equations by adding a vector z to the left-hand side (such a vector is often referred to as a vector of *slack variables* in linear programming theory). Obviously $z \geqq 0$ if the inequalities are satisfied. Construct the $m \times (n + m)$ matrix

$$\hat{A} = [A : I],$$

where I is an $m \times m$ matrix containing 1's on the diagonal and 0's elsewhere. Denote by $\hat{x}$ the composite $(n + m)$ vector $[x : z]'$.

Then

$$\hat{A}\hat{x} = [A : I] \begin{bmatrix} x \\ \vdots \\ z \end{bmatrix} = b$$

is equivalent to $Ax + z = b$. If the inequalities $Ax \leqq b$ have a nonnegative solution, then the equations $\hat{A}\hat{x} = b$ have a nonnegative solution. Suppose (a) is not true, that $Ax \leqq b$ has no nonnegative solution. Then $\hat{A}\hat{x} = b$ has no nonnegative solution, so, from (b), the inequalites $y\hat{A} \geqq 0$, $yb < 0$ have a solution. But from the construction of $\hat{A}$, the inequalities $y\hat{A} \geqq 0$ consist of two sets, $yA \geqq 0$ and $yI \geqq 0$. Now $yI \geqq 0$ implies $y \geqq 0$, so the inequalities $yA \geqq 0$, $yb < 0$ have a solution $y \geqq 0$, proving the result.

To prove (e), after first showing the incompatibility of (i) and (ii) we note that, if $Ax = 0$ has a *semipositive* solution (as contrasted with mere nonnegativity), we can choose x so that $\sum x_j = 1$. Define[7]

$$\hat{A} = \begin{bmatrix} A \\ \cdots\cdot \\ 11\cdot 1 \end{bmatrix} \qquad \hat{b} = \begin{bmatrix} 0 \\ 0 \\ \cdot \\ 1 \end{bmatrix}.$$

If $Ax = 0$ has a semipositive solution, then $\hat{A}x = \hat{b}$ has a nonnegative solution. Suppose $Ax = 0$ has no semipositive solution so that $\hat{A}x = \hat{b}$ has no nonnegative solution. Then, from (b), the inequalities $\hat{y}\hat{A} \geqq 0$, $\hat{y}\hat{b} < 0$ have a solution, where y is an $(m + 1)$ vector.

Since the only nonzero component of $\hat{b}$ is $\hat{b}_{m+1}$ $(=1)$, we must have $\hat{y}_{m+1} < 0$. But $\hat{y}A^j = yA^j + \hat{y}_{m+1} \geqq 0$, all j, so that $yA^j \geqq -\hat{y}_{m+1} > 0$, all j, where y is the vector of the first m components of y. Thus $yA \gg 0$ has a solution, proving the result.

To illustrate a slightly different type of approach, let us prove (f). If $x \gg 0$, we can put $x = z + b$, where $z \geqq 0$ and $b \gg 0$, for some b. Thus, if $Ax = 0$ has a solution $x \gg 0$. $A(z + b) = 0$ has a solution $z \geqq 0$ for some $b \gg 0$. That is, $Az = -Ab$ has a solution $z \geqq 0$ for some $b \gg 0$. Suppose, as before, that $Ax = 0$ does not have a solution $x \gg 0$. Then $Az = -Ab$ does not have a solution $z \geqq 0$ if $b \gg 0$. Then, again from (b), $yA \geqq 0$, $-yAb < 0$ have a solution. But $b \gg 0$, so that the vector yA must have at least one nonzero component. Hence $yA \geqslant 0$, and the result is proved.

Other proofs follow generally similar lines. (g) can be proved from (c) and (d) from (a).

EXERCISES

1. Find the solution sets of $Ax = 0$ and $Ax = b$, where

$$A = \begin{bmatrix} 3 & 1 & 1 \\ 1 & 1 & 2 \end{bmatrix} \qquad b = \begin{bmatrix} 1 \\ 3 \end{bmatrix}.$$

[7] $\hat{A}$ is the matrix A with an $(m + 1)$th row of 1's added, b the $(m + 1)$ vector containing all zeros except for 1 in the last place.

2. For the matrix

$$\begin{bmatrix} 1 & 2 & 0 \\ 3 & -1 & 7 \\ 2 & 1 & 3 \end{bmatrix}$$

show, independently, that the rows are linearly dependent and the columns are linearly dependent.

3. Prove those results of Section R3.7 which are not proved in the text.

4. Noting that "$=$" is equivalent to the simultaneous satisfaction of "$\geqq$" and "$\leqq$," relate the following pairs of results in Section R3.7:
a. (a) and (d); **b.** (b) and (c); **c.** (e) and (g). Why is (f) unpaired?

REVIEW

R4

Convex Sets and Cones

This review follows directly on from the two preceding. All the material except that of Section R4.7 should be regarded as basic.

R4.1 GEOMETRIC IDEAS[1]

In this review we shall take a more geometric view of vectors and vector sets than we have so far. As a preliminary we shall define three geometric ideas, the line joining two points, the hyperplane, and the hypersphere.

We have already seen that, given a point x, the points λx represent points on the infinite line that passes through 0 and x. If we treat x^* as fixed and λ as a variable, then $x = \lambda x^*$ is the equation of a line passing through 0 and x^*. For $0 \leqq \lambda \leqq 1$, $x = \lambda x^*$ is the equation of the segment of that line bounded by 0 and x^*, and $x = \lambda(x^* - x^{**})$ is the equation of the line through 0, $x^* - x^{**}$. If we add x^{**} to the last equation, then $y = x + x^{**} = x^{**} + \lambda(x^* - x^{**})$ is the equation of the line passing through x^* and $(x^* - x^{**}) + x^{**} = x^{**}$. Thus we have:

The points $y + \lambda(x - y)$ [or $\lambda x + (1 - \lambda)y$] for $0 \leqq \lambda \leqq 1$ are points on the line segment joining x, y.

[1] Convex sets are discussed in some, but not all, linear algebra texts. Among those which include such a discussion are Hadley [1] and Lang. An advanced treatment of the subject is given in Valentine.

We have also defined (Section R3.3) the linear subspace of rank $n - 1$ in a space of rank n, given by the equation:

$$px = 0 \qquad (p \text{ a row vector of order } n)$$

as a *hyperplane through* 0 *orthogonal to the vector p.*

It is obvious that $(\lambda p)x = 0$ gives the same hyperplane as $px = 0$, so that we shall consider p to be *normalized* to give the Euclidean norm $pp' = 1$.

If y^* is any point in R^n, then it is obvious that

$$p(x - y^*) = 0$$

is the equation of a *hyperplane, orthogonal to p, passing through* y^*.

Since $p(x - y^*) = 0$ is equivalent to $px = py^*$, or $px = c$, where $c = py^*$, any equation of the form $px = c$ is the equation of a hyperplane orthogonal to p.

Consider the point $x = \lambda p'$. Substituting in the equation $ax = c$ we have $\lambda pp' = c$, or $\lambda = c$ since p is normalized. Such a point can always be found in the hyperplane $ax = c$ and it is obviously the intersection of the line λp, the line through the origin orthogonal to the hyperplane (called the *normal* to the hyperplane) and the hyperplane itself. Now the distance d from this point to the origin is the length, or Euclidean norm, of the vector p, given by $d^2 = \lambda^2 pp' = \lambda^2 = c^2$. Thus we can state:

The equation $px = c$, with p normalized, is the equation of a hyperplane with normal p which intersects that normal at a distance $|c|$ from the origin.

Consider arbitrary $x \in R^n$. Form the scalar product px, and denote the value of px by γ. Then x can be considered as a point lying in the hyperplane $px = \gamma$, and every point in R^n can be considered as lying in some such hyperplane for some value of γ. (These hyperplanes, all sharing the same normal λp, are said to be *parallel.*) We can divide these hyperplanes into two sets with respect to the hyperplane H whose equation is $px = c$, those for which $\gamma \geqq c$, and those for which $\gamma \leqq c$.

Thus every point x in R^n belongs to at least one of the two sets

$$H^+ = \{x \mid px \geqq c\}, \qquad H^- = \{x \mid px \leqq c\}.$$

These two sets are called the *halfspaces* defined by H. H itself belongs to both halfspaces so that $H^+ \cap H^- = H$. Sometimes we may wish to exclude those points of the halfspace actually in H, in which case we refer to the *open halfspace*. The term halfspace will be taken to refer to the closed halfspace unless otherwise specified.

Finally we have, for a point y^* and a variable point x,

$$(x - y^*)'(x - y^*) = \sum (x_i - y_i^*)^2.$$

For R^2, R^3, equations of the form

$$(x - y^*)'(x - y^*) = c^2$$

represent a circle or sphere of radius c with center at y^*. By analogy we define the above equation in R^n as a *hypersphere* of radius c and center y^*.

R4.2 CONVEX SETS

We define the following concept, of the highest importance for almost all economic analysis:

A point or vector set S is said to be CONVEX if $\lambda x + (1 - \lambda)y \in S$ whenever $x, y \in S$ and $0 \leqq \lambda \leqq 1$.

Since the points $\lambda x + (1 - \lambda)y$, $0 \leqq \lambda \leqq 1$, are the points on the line segment joining x, y, a set is convex if it contains every point on the line joining any two points in the set. This leads to the geometric interpretation of convex sets as being connected sets having no re-entrant parts of the boundary. Figure R4–1 illustrates various sets, some convex, some not. Note that a set is convex everywhere or it is not a convex set, as illustrated by (c) and (d) in Figure R4–1.

A convex set whose boundaries are all linear (lines or hyperplanes, like (a) or (e) in Figure R4–1) is a *convex polyhedral set*, *convex polytope* or sometimes, loosely, a *convex polyhedron*. A more formal definition of this case will be given later. A convex set need not be bounded [as (e) in Figure R4–1). R^n, which has no boundaries, is clearly a convex set. Usually we are interested in convex sets with some boundaries.

It is immediate that the following are convex sets: R^n, any linear subspace, a line, a halfspace, a hyperplane.

A hypersphere, or any curved surface, is not a convex set, but the set of points inside a hypersphere (called an *open ball*) or the interior and the hypersphere together (a *closed ball*) are convex sets.

If S is a convex set and $x, y \in S$ then, by definition, $w = \lambda x + (1 - \lambda) y \in S$, for $0 \leqq \lambda \leqq 1$. Now consider another point $z \in S$. Then $\mu w + (1 - \mu) z \in S$ for $0 \leqq \mu \leqq 1$. Substituting for w, we have

$$\mu\lambda x + \mu(1 - \lambda) y + (1 - \mu) z \in S \qquad \text{for} \qquad 0 \leqq \mu, \lambda \leqq 1.$$

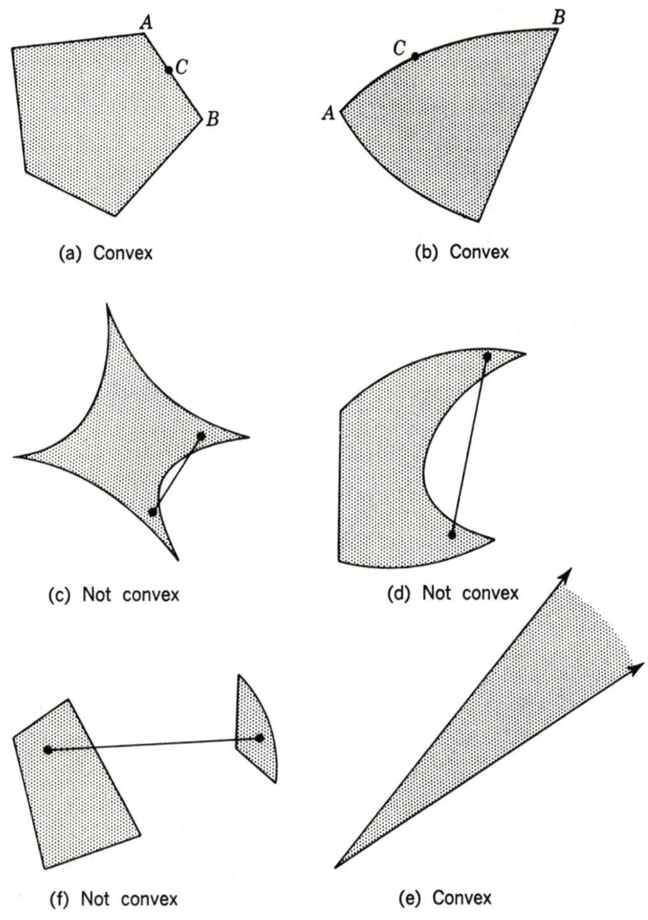

(a) Convex (b) Convex

(c) Not convex (d) Not convex

(f) Not convex (e) Convex

Figure R4–1 Convex and nonconvex sets.

Write $\gamma_1 = \mu\lambda$, $\gamma_2 = \mu(1 - \lambda)$, $\gamma_3 = (1 - \mu)$. Then we have $0 \leqq \gamma_1, \gamma_2, \gamma_3 \leqq 1$, $\gamma_1 + \gamma_2 + \gamma_3 = \mu\lambda + \mu(1 - \lambda) + (1 - \mu) = 1$, and

$$\gamma_1 x + \gamma_2 y + \gamma_3 z \in S.$$

Clearly we can extend the argument to give the following:

(a) If $x^1, \ldots, x^m$ *are m points in the convex set S, the point* $x = \sum_1^m \lambda_i x^i$, *where* $0 \leqq \lambda_i \leqq 1$, $\sum_1^m \lambda_i = 1$, *is also in S.*[2]

[2] The weighted sum $\sum \lambda_i x^i$ with nonnegative weights that add to unity is called a *convex combination* of the points x^i. It is a special case of a linear combination.

If is immediate, from (a), that:

(*b*) *A convex combination of any number of points in R^n is a convex set.*

If S, S' are convex sets and x, y are in both S, S', then the line joining x, y must also lie in both S and S', so we have the important result:

(*c*) *The intersection of convex sets is a convex set.*[3]

Let $x^1, \ldots, x^m$ be points in R^n. Then the convex combination of $x^1, \ldots, x^m$ is a convex set which we shall denote by $\text{Co}(x^1, \ldots, x^m)$. If S is any convex set containing $x^1, \ldots, x^m$, then S contains $\text{Co}(x^1, \ldots, x^m)$, from (a). It follows that:

(*d*) $\text{Co}(x^1, \ldots, x^m)$ *is the intersection of all convex sets containing* $x^1, \ldots, x^m$. *It is referred to as the smallest convex set containing* $x^1, \ldots, x^m$, *the convex hull of* $x^1, \ldots, x^m$, *or the convex closure of* $x^1, \ldots, x^m$.

We sometimes use this concept to derive a convex set from an arbitrary set:

(*e*) *If S is any set of points in R^n (not necessarily convex) the convex hull or convex closure of S, written $\langle S \rangle$ or $\text{Co}(S)$ is the smallest convex set containing S, or the convex combination of the points of S. If S is a closed convex set, then $\text{Co}(S) = S$.*

Now consider the solutions of the system of linear inequalities

$$Ax \leqq b.$$

If A is of order $m \times n$, this consists of the m inequalities $A_i x \leqq b_i$. Now the set $X^i = \{x \mid A_i x \leqq b_i\}$ is a halfspace, hence a convex set. Since the solution of the system must be the intersection of the sets X^i, each of which is a convex set, the solution set is convex, thus:

(*f*) *The set of solutions to a system of linear inequalities is a convex set.*

The solution set of a system of linear equations is a linear subspace and its convexity is trivial, but consider the set of *nonnegative* solutions of $Ax = b$. The set of all nonnegative vectors $x \geqq 0$ is the solution of a set of linear inequalities and so a convex set. The set of nonnegative solutions to $Ax = b$

[3] The union of convex sets will not, in general, be a convex set.

is the intersection of the set of all solutions and the set of all nonnegative vectors, hence:

(*g*) *The set of nonnegative solutions to a system of linear equations is a convex set.*

Finally we note that convexity depends on linear relationships and is preserved by operations that preserve linearity. It is easy to show that:

(*h*) *If S, S' are convex sets, $S + S'$, $S - S'$, λS are convex sets. The linear mapping $x \to Tx$ always maps a convex set into a convex set.*

R4.3 SEPARATING AND SUPPORTING HYPERPLANES

We have seen that a hyperplane H defines two halfspaces. If, given two convex sets S, S', we can find some hyperplane H such that S lies in one of the halfspaces defined by H and S' lies in the other, H is said to be a *separating hyperplane* of S, S'. If, for a single convex set, we can find a hyperplane such that S is contained in one of the halfspaces and at least one point of S lies in the hyperplane itself, it is said to be a *supporting hyperplane.*

Before discussing the problem of the existence of such hyperplane we shall prove the following variant of the Minkowski theorem:

Minkowski Theorem

If S is a closed convex set and z^* is a point exterior to S, there exists a vector p and a point $x^* \in S$ such that $px \geqq px^* > pz^*$ for all $x \in S$.

Choose a boundary point x^* of S such that x minimizes the expression $(z^* - x)'(z^* - x)$ over all $x \in S$ (x^* is the "closest" point of S to z^*). This minimum exists because $(z^* - x)'(z^* - x) \geqq 0$ and can only be zero if $z^* = x$. But $x \in S$ and we have specified that $z^* \notin S$, so the product is nonzero but bounded below and has a minimum.

If x is any point in S, then $x^* + \lambda(x - x^*) \in S$ for $0 \leqq \lambda \leqq 1$, by the definition of a convex set. Now

$$[x^* + \lambda(x - x^*) - z^*]'[x^* + \lambda(x - x^*) - z^*] \geqq (x^* - z^*)'(x^* - z^*)$$

by the minimum property of x^*. If we expand, use the transposition identities $(x + y)' = x' + y'$ and $x'y = y'x$, and cancel, we obtain

$$\lambda^2(x - x^*)'(x - x^*) + 2\lambda(x^* - z^*)'(x - x^*) \geqq 0.$$

Divide by λ, to obtain

$$\lambda(x - x^*)'(x - x^*) + 2(x^* - z^*)'(x - x^*) \geqq 0.$$

Now this must be true for all λ. Let $\lambda \to 0$, then we must have

$$(x^* - z^*)'(x - x^*) \geqq 0. \tag{R4.3.1}$$

Put $p = (x^* - z^*)'$, and write

$$\begin{aligned} x - x^* &= (x - z^*) - (x^* - z^*) \\ &= (x - z^*) - p'. \end{aligned}$$

Then we have, using (R4.3.1),

$$p(x - z^*) - pp' \geqq 0.$$

But $pp' > 0$, so that $p(x - z^*) > 0$. Also, direct from (R4.3.1), we have $p'(x - x^*) \geqq 0$. Bringing the inequalities together we have

$$px \geqq px^* > pz^*,$$

thus proving the theorem.

The Minkowski theorem thus establishes the existence of a hyperplane $px = pz^*$ which contains z^* and is such that S is contained in one of its half-spaces. It follows immediately that any hyperplane $px = c$, where p is as above and $px^* > c > pz^*$ has z^* in one of its halfspaces and S in the other. We note that the normal to the Minkowski hyperplane, $p' = \lambda(x^* - z^*)$, is the line passing through z^* and the point in S closest to it.

The content of the theorem can be visualized in R^2 by referring back to Figure R4-1, the hyperplane in R^2 being, of course, a line. The necessity of convexity is apparent by taking z^* to be a point in the gulf on the right-hand side of set (d) in the diagram.

From the above theorem, we can prove four important Corollaries:

(*a*) *If x^* is a boundary point of the closed convex set S, there is a supporting hyperplane through x^*.*

(*b*) *A closed convex set is the intersection of its supporting halfspaces.*

(*c*) *Two closed bounded convex sets with no points in common have a separating hyperplane.*

(*d*) *Two closed convex sets which intersect only at a single point possess a separating hyperplane which is also a supporting hyperplane of both sets.*

If we choose an arbitrary point z^* external to S and construct the Minkowski hyperplane $px = c$ with $px^* > c > pz^*$, then let $c \to pz^*$, we derive the supporting hyperplane through x^* since x^* lies in the hyperplane and $px \geqq px^*$ for all other $x \in S$.

It is intuitively to be expected, and can be formally proved, that every boundary point has at least one exterior point to which it is closer than any other point in S, and so a supporting hyperplane can be found through every boundary point of S, proving (a).

To prove (b) we note that, by definition of a supporting hyperplane, the intersection of all supporting halfspaces must contain S. If the intersection also contained some point not in S then we could find a Minkowski hyperplane lying inside a supporting hyperplane, contrary to definition. Thus the result follows.

To prove (c), let S, S' be the two sets. Consider the set $S^* = S - S'$. This is a closed convex set if S, S' are. Since S, S' have no common point, 0 is exterior to S^*. Then, by the Minkowski theorem, there is a hyperplane $p(x - y) = c$ through 0 such that $p(x - y) > p \cdot 0 = 0$. For any $x \in S$, $y \in S'$, we have $px > py$ so that some c can be found such that $px \geqq c \geqq py$ and S, S' lie in different halfspaces.

Finally we prove (d) by noting that, if S, S' have a single point in common it is necessarily a boundary point, and that 0 is a boundary point of $S^* = S - S'$. The supporting hyperplane of S^* at 0 is such that $p(x - y) \geqq 0$, with $p(x - y) = 0$ only for $x = y$. This gives the hyperplane which is separating for S, S' and supporting for both.

If the boundary of S is continuously curved at the point x^*, we will expect that the supporting hyperplane at x^* can be identified with the tangent hyperplane at x^*. In other cases there may well be a set of supporting hyperplanes at a boundary point, as there will be at a "corner" point.

R4.4 EXTREME POINTS

A point x^k is an *extreme point* of the convex set S if x^k *cannot* be expressed in the form

$$x^k = \lambda x^i + (1 - \lambda)x^j, \qquad 0 < \lambda < 1, \qquad x^i, x^j \in S.$$

Note that in this definition λ is confined to the *open* interval 0, 1. In geometric terms, an extreme point cannot lie in the *interior* of a line segment joining two points in the set, it can only be one of the end points.

It is obvious that an extreme point must be a boundary point, but not all boundary points are extreme points. In Figure R4–1, A and B are extreme points in diagram (a), but C is not. On the other hand, in diagram (b), C is an extreme point as well as A and B, since the boundary AB is curved.

Although points like A, B in this diagram may look "more" extreme than C, we do not make the distinction. In a convex set with a continuously curved boundary like a disc or a ball *every* boundary point is an extreme point.

We can now give the formal definition of a convex polyhedral set or polytope:

A convex polyhedral set is defined to be a convex set with a finite number of extreme points.

There are two theorems of considerable importance concerning extreme points. These theorems are fundamental to the theory of linear programming:

(*a*) *If S is a closed convex set bounded either from above or below* (*or both*) *then every supporting hyperplane of S contains an extreme point.*

(*b*) ***Krein-Milman Theorem***
A closed bounded convex set S is the convex closure of its extreme points. If S is a polyhedral set, every point in S is a convex combination of the finite number of extreme points.

To prove (a), consider a supporting hyperplane H of S, defined by $px = px^*$, where x^* is a boundary point. Then $px \geqq px^*$ for all $x \in S$. Let T be the intersection of H, S. T is clearly a closed convex subset of S and is bounded in the same way S is bounded.

Let $\hat{x}$ be an extreme point of T. We seek to show that it is also an extreme point of S. Suppose it is not, then

$$\hat{x} = \lambda x^1 + (1 - \lambda)x^2, \qquad 0 < \lambda < 1, \qquad x^1, x^2 \in S. \tag{R4.4.1}$$

Take the scalar product with p. We obtain

$$p\hat{x} = \lambda px^1 + (1 - \lambda)px^2. \tag{R4.4.2}$$

But $px = px^*$, since $T \subset H$. Also, since H is supporting, $px^1 \geqq px^*$, $px^2 \geqq px^*$. Since λ, $(1 - \lambda)$ are both > 0, $\lambda px^1 + (1 - \lambda)px^2 > px^*$ unless both $px^1, px^2 = px^*$. Thus (R4.4.2) can only be true if these equalities hold. If they do, then $x^1, x^2 \in H$, hence $\in T$, and x cannot be an extreme point of T as stated. Thus an extreme point of T must also be an extreme point of S.

We now seek an extreme point of T in the following way. S, hence T, is bounded either above or below. Suppose it is bounded below (the argument is symmetrical for bounded above), then there must be a point in T which has the smallest first component, x_1, among all points of T. Take the subset T_1 of all points in T whose first component has this smallest value. There must be a point in T_1 with smallest second component. Form the subset T_2 of T_1

containing points whose second component has this smallest value. Continue the process for all n components until we have found a point $\bar{x}$ with the smallest first, second, . . ., components. Such a point must always exist. We shall now show it must be an extreme point of T.

Suppose $\bar{x}$ is not an extreme point of T. Then we can write

$$\bar{x} = \lambda x^1 + (1 - \lambda) x^2, \qquad 0 < \lambda < 1, \qquad x^1, x^2 \in T. \tag{R4.4.3}$$

By definition of $\bar{x}$, $\bar{x}_1 \leqq x_1^1, x_1^2$. These inequalities are consistent with (R4.4.3) only if $x_1^1 = x_1^2 = \bar{x}_1$. We can apply the same arguments to the remaining components until we finally conclude that (R4.4.3) implies $\bar{x} = x^1 = x^2$.

Thus $\bar{x}$, which can always be found, is an extreme point of T, hence of H, and (a) is proved.

To prove (b), write Co for the convex closure of extreme points. Suppose $x^* \in S$, but $x^* \notin$ Co. Then x^* is exterior to Co and, from the Minkowski theorem, there is a hyperplane H defined by $px = px^* = c$ such that $px > c$ for all $x \in Co$.

Now consider the linear mapping $x \rightarrow px$, $x \in S$. Since S is a closed bounded convex set, the set $\{px \mid x \in S\}$ is also a closed bounded convex set. But px is a scalar, so this set is some closed interval $[\alpha, \beta]$. Since $x^* \in S$ and $px^* = c$, $\alpha \leqq c \leqq \beta$.

Consider the hyperplane $H' = \{x \mid px = \alpha\}$. H' contains the points in S for which $px = \alpha$, and by definition of α, $px \geqq \alpha$ for all $x \in S$. Thus H' is a supporting hyperplane for S. But, by theorem (a), this must contain an extreme point of S, and this extreme point must be in Co by definition. But we previously showed that $px > c$ for all $x \in Co$ and now we have shown the existence of a point in Co for which $px = \alpha \leqq c$.

Thus there cannot be points in S not in Co, and there certainly cannot be points in Co not in S, so the theorem is proved.

The importance of this theorem is that a convex set is completely defined by specifying its extreme points.

R4.5 CONVEX CONES[4]

Convex cones are the class of convex sets possessing the property that $\lambda x \in S$ for all $\lambda \geqq 0$ if $x \in S$. Combining this property with the defining

[4] Convex cones are important in production theory, and many activity analysis results which are proved in Chapter 7 by linear programming can be proved by the use of finite cone properties.

Hadley [1] discusses cones, but most linear algebra texts do not.

Advanced discussions are given in Gerstenhaber, Gale [2], and Goldman and Tucker [2].

property of convex sets, we can define convex cones independently as sets such that $\lambda x + \mu y \in S$ for all $\lambda, \mu \geqq 0$, if $x, y \in S$. We shall designate a convex cone by the notation C rather than S.

It is immediate that R^n and all linear subspaces are convex cones. If H is a hyperplane through the origin (hence a linear subspace), H is a convex cone. Both halfspaces defined by H are also convex cones. Halfspaces defined by hyperplanes which do not pass through the origin are convex sets, but are *not* convex cones. For any vector b, the set $\{x \mid x = \lambda b, \lambda \geqq 0\}$ (called a *halfline*) is a convex cone. We shall usually designate the halfline associated with b as (b). Note that the *line* $\{x \mid x = \lambda b, \text{ all } \lambda\}$ is a linear subspace, but the halfline is not.

It follows from their properties as convex sets that, if C_1, C_2 are convex cones, $C_1 + C_2$ and $C_1 \cap C_2$ are also convex cones. All convex cones intersect at the origin, so that we shall regard the origin as the convex cone (0) in some cases, as well as regarding it as the point 0.

The type of convex cone from which the name is derived can be visualized in three dimensions in the following way. Imagine an opaque mask with a cutout in the shape of an arbitrary convex set in two dimensions. If this mask was inserted in a slide projector, the shape formed by the light rays in a dusty atmosphere would form a convex cone. Note that, even in three dimensions, the shape is not necessarily that of a cone in the Euclidean sense, but may be a pyramid or a variety of other shapes. Every cross section would, however, be a convex set. Convex cones of the above "projector" kind are pointed cones, a term we shall define formally below. Clearly, halfspaces and linear subspaces are not pointed in this sense.

Since a convex cone is a convex set, let us consider its extreme points. If $x \in C$, then $\lambda x \in C$ for all $\lambda \geqq 0$. Thus any point $x \neq 0$ can be expressed as

$$x = \lambda(1 + \lambda)\, x + (1 - \lambda)\frac{1 - \lambda - \lambda^2}{1 - \lambda}\, x,$$

and we can always choose λ so that $0 < \lambda < 1$ and $1 - \lambda - \lambda^2 > 0$. Thus any $x \neq 0$ can be found to lie on the interior of the line segment joining the distinct points $(1 + \lambda)\, x$, $[(1 - \lambda - \lambda^2)/(1 - \lambda)]\, x$, both in C, and thus cannot be an extreme point. Only the origin can be an extreme point, so we can state the following:

A convex cone either has no extreme points, or it has the origin as its only extreme point.

A convex cone which is also a linear subspace clearly has no extreme points. We can now give a formal definition to the term "pointed."

A convex cone having the origin as its extreme point is said to be pointed.

The following results are then almost immediate:

A convex cone is pointed if and only if $C \cap (-C) = 0$.[5]
If C *is a pointed convex cone there is a hyperplane* H *through* 0 *which is separating for* C, $-C$ *and supporting for both.*[6]

General convex cones of the kind discussed above occur in production theory. The production set of an economy with constant returns to scale everywhere, but with general neoclassical production functions (and the possibility of inefficient production) forms just such a cone.

The most important class of convex cones for analytical purposes is, however, the class called *finite cones*, to be discussed in the next section.

R4.6 FINITE CONES AND HOMOGENEOUS INEQUALITIES

No convex cone is finite in the sense of containing a finite number of points, nor are the vectors of the cone all finite. A *finite cone* is finite in the sense that every point in the cone can be expressed as a linear combination of a finite number of vectors. Such a cone has analogies to a convex polyhedral set (which has a finite number of extreme points, in terms of which all points in in the set can be expressed) and is often called a *convex polyhedral cone.*

Formally, we make either of the following equivalent definitions:

C *is a finite cone if there exist a finite number of vectors* v^i *such that* $x = \sum \lambda_i v^i$, $\lambda_i \geqq 0$, *for all* $x \in C$.

C *is a finite cone if there exist a finite number of halflines* (v^i) *such that* $C = \sum (v^i)$.

The first definition is in "vector language," the second in "cone language." In the second definition the halflines (v^i) are called the *generators* of C. The first definition can be put in matrix vector form. Let A be the matrix whose columns are v^i and u the column vector whose components are λ_i. Then $C = \{x \mid x = Au, u \geqq 0\}$. Every finite cone can be put in this form for a suitable matrix A.

[5] If $C \cap (-C) \neq 0$, there exists some $x \in C$ such that $-x \in C$. Then $0 = \frac{1}{2}x + \frac{1}{2}(-x)$ and 0 is not an extreme point.

[6] This follows directly from the previous result and (d) of Section R4.3.

We can state immediately that, if C_1, C_2 are finite cones then $C_1 + C_2$, $C_1 \cap C_2$ are finite cones.

If L^r is a linear subspace of rank r, we can express every point in L^r as a linear combination of r basis vectors $v^1, \ldots, v^r$, but the weights of the combination need not be nonnegative. But if we take the $2r$ vectors $v^1, \ldots, v^r$, $-v^1, \ldots, -v^r$ we can express any point in L^r as a linear combination of these with nonnegative weights, so that:

A linear subspace is a finite cone.

Consider a hyperplane H through the origin, with normal p. Then H is a linear subspace, hence a finite cone, and the two halfspaces associated with H are $H + (p)$ and $H + (-p)$. Since each halfspace is the sum of finite cones, it is a finite cone.

A halfspace defined by a hyperplane through the origin is a finite cone.

The system of homogeneous inequalities $Ax \leqq 0$ consists of the m inequalities $A_i x \leqq 0$. But $H_i^- = \{x \mid A_i x \leqq 0\}$ is a halfspace, hence a finite cone. The solution set of the system of inequalities is the intersection of the halfspaces H_i^-, so that,

The solution set of the homogeneous linear inequalities $Ax \leqq 0$ is a finite cone.[7]

It follows that $\{x \mid x \geqq 0\}$ is a finite cone, the *nonnegative orthant*, which we shall write Ω (P is also used) and the *nonpositive orthant* $\{x \mid x \leqq 0\}$ is also a finite cone, which we shall write Ω^- (N is also used), or simply $-\Omega$.

Since the solution set X of $Ax = 0$ is a linear subspace, hence a finite cone, $X \cap \Omega$ is a finite cone. Thus:

The set of nonnegative solutions to the homogeneous linear equation system $Ax = 0$ is a finite cone.

R4.7 THE DUAL CONE[8]

If C is a convex cone, the set $C^* = \{y \mid yx \leqq 0, \text{ all } x \in C\}$ is the *dual cone* of C. Any cone has a dual, but we are chiefly interested in the dual of a finite

[7] Most of the interest in finite cones is derived from this property.

[8] The material of this section is not used elsewhere in the book, since the theory of inequalities was treated by other methods. Dual cones provide a very elegant proof of some propositions.

cone. Not all the results which will be given for the dual of a finite cone will hold for the dual of a nonfinite cone. The dual cone is sometimes called the *negative polar cone.*[9]

In geometric terms, the dual of a convex cone is the set of vectors which make a nonacute angle with the vectors of the original cone and is easily depicted for pointed cones in two-dimensional space. Figure R4–2 gives an illustration.

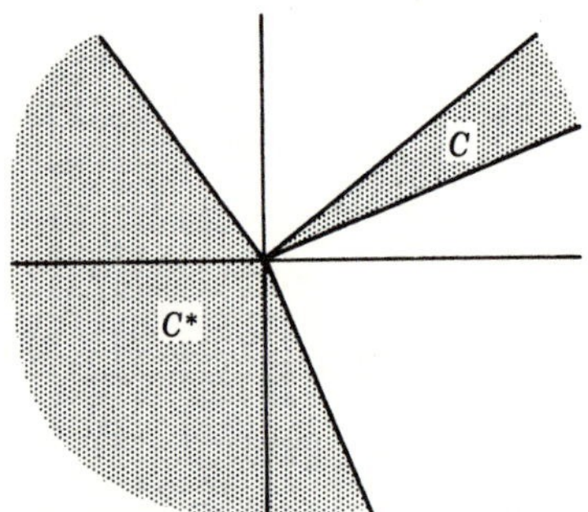

Figure R4–2 The dual cone.

If (b) is a halfline, then $(b)^* = \{y \mid yb \leqq 0\}$, so the dual of a halfline is a halfspace bounded by the hyperplane orthogonal to the halfline. It is obvious the dual of a halfspace is a halfline. Since $y0 \leqq 0$ for all $y \in R^n$, we have $(0)^* = R^n$, that is, the whole space is the dual of the origin considered as a finite cone. Since 0 is the only vector such that $0x \leqq 0$ for every $x \in R^n$, we have $(R^n)^* = (0)$.

The following summarize the relationships between C, C^* when c^* is a *finite* cone:

(*a*) *C^* is a finite cone.*

(*b*) (***Duality Theorem for Finite Cones***)
$(C^*)^* = C$.

If C is a finite cone, $C = \sum (v^i)$. Then C^* is the solution set of the inequalities $yv^i \leqq 0$ and is a finite cone, proving (a). If C is expressed in the form $C = \{x \mid x = Au, u \geqq 0\}$, then $C^* = \{y \mid yA \leqq 0\}$.

[9] The *positive polar cone* is defined as $C^+ = \{y \mid yx \geqq 0, \text{ all } x \in C\}$. The corresponding notation C^- may be used for the dual or negative polar cone instead of C^*.

To prove (b) we note that, for all $z \in (C^*)^*$ (which we shall hence write C^{**}) we have $yz \leqq 0$, if $y \in C^*$, but if $y \in C^*$ we have $yx \leqq 0$ for all $x \in C$. Thus we certainly have $C \subset C^{**}$.

Now C^* is a finite cone, so its dual C^{**} is a finite cone, both from (a). We have shown $C \subset C^{**}$. If $C^{**} \subset C$, the result is proved. Suppose that $C^{**} \not\subset C$, then, since C, C^{**} are both finite cones and $C \subset C^{**}$, we must have $C^{**} = C + \sum (b^j)$ where (b^j) are halflines not in C.

If we now take duals again, then again, we obtain C^{****} which is related to C^{**} in the same way that C^{**} is related to C. That is, on our hypothesis, $C^{****} \not\subset C^{**}$ so that $C^{****} = C^{**} + \sum (c^j) = C + \sum (b^j) + \sum (c^j)$, where the (c^j) are halflines not in C^{**}. If we continue taking duals in this way we add new halflines to C^{2n} (where n is the number of times double duals have been taken) at each round, so that C^{2n} as $n \to \infty$ is not a finite cone. But C^{2n} has been obtained by continually taking duals of finite cones, so itself must be a finite cone. Thus the hypothesis that $C^{**} \neq C$ cannot be true, proving (b).[10]

We can now state the relationships between the duals of different finite cones:

(*c*) *If* $C_1 \subset C_2$, $C_2^* \subset C_1^*$.
(*d*) $(C_1 + C_2)^* = C_1^* \cap C_2^*$.
(*e*) $(C_1 \cap C_2)^* = C_1^* + C_2^*$.
(*f*) $C \oplus C^* = R^n$.

Result (c) follows directly from definition.

To prove (d) we have, if $x \in C_1^* \cap C_2^*$ then $xy_1 \leqq 0$ for all $y_1 \in C_1$ and $xy_2 \leqq 0$ for all $y_2 \in C_2$. Then if $y \in (C_1 + C_2)$, $y = y_1 + y_2$ and $xy \leqq 0$ so that $x \in (C_1 + C_2)^*$. Thus $(C_1^* \cap C_2^*) \subset (C_1 + C_2)^*$.

Conversely, if $x \in (C_1 + C_2)^*$ then $xy \leqq 0$ for all $y = y_1 + y_2$, $y_1 \in C_1$, $y_2 \in C_2$. Putting $y_1 = 0$ we have $xy_2 \leqq 0$ so that $x \in C_2^*$ and putting $y_2 = 0$ we find $x \in C_1^*$. Thus $x \in C_1^* \cap C_2^*$ so that $(C_1 + C_2)^* \subset (C_1^* \cap C_2^*)$.

The two arguments imply $(C_1 + C_2)^* = C_1^* \cap C_2^*$, proving the result.

Result (e) is an immediate consequence of (d) and the duality theorem (b).

To prove Result (f) we note that $C \cap C^* = (0)$ since, if we had some $x \in (C \cap C^*)$, then $x'x \leqq 0$, which is only possible for $x = 0$. Taking duals of this, using (e) and the relationship $(0)^* = R^n$, gives the result. [We used the direct sum sign $\oplus$ because $C \cap C^* = (0)$. It is not essential.]

[10] This method of proof shows clearly the necessity that C be a finite cone for the duality theorem to hold.

The results on linear inequalities set out in Section R3.7 can be proved using methods based on finite cones and duality.[11]

As an example, consider the set of nonnegative solutions of the homogeneous linear inequalities $Ax \leqq 0$. The set of all solutions is the finite cone $C = \{x \mid Ax \leqq 0\}$, whose dual is $C^* = \{z \mid z = yA,\ y \geqq 0\}$. The set of nonnegative solutions to the inequalities is $C \cap \Omega$.

Suppose there are no semipositive solutions ($x = 0$ must be a solution), then we have

$$C \cap \Omega = (0),$$

Taking the dual of this equation, we obtain

$$C^* + \Omega^* = R^n$$

or,

$$C^* + \Omega^- = R^n,$$

since the dual of the nonnegative orthant is the nonpositive orthant.

In particular, $0 \in R^n$, so the above equation implies the existence of a vector $z \in C^*$ such that

$$z - \beta = 0, \qquad \beta \gg 0$$

or,

$$z \gg 0.$$

But since $z \in C^*$, $z = yA$, $y \geqq 0$. Thus there exists some y such that

$$yA = z \gg 0, \qquad y \geqslant 0$$

($y = 0$ is obviously ruled out).

Thus we have proved that, if $Ax \leqq 0$ has no semipositive solution, $yA \gg 0$ has a semipositive solution. This is Result (g) of Section R3.7.

Other results can be proved along similar lines.

EXERCISES

1. Given the hyperplane $4x_1 + 3x_2 + x_3 + 5x_4 + x_5 = 21$, determine which of the following points lie in the same halfspace:

$$(1\ \ 2\ \ 1\ \ 2\ \ 1); \qquad (2\ \ 1\ \ 2\ \ 1\ \ 1); \qquad (1\ \ 1\ \ 2\ \ 2\ \ 1).$$

2. Find examples, graphical or otherwise, to show that:

a. The union of convex sets is not necessarily convex;

[11] See Gale [2], also the brief discussion in Gale [1].

b. the intersection of nonconvex sets may be convex;

c. a set S is not necessarily convex if

$$\lambda x^1 + (1 - \lambda)x^2 \in S \qquad \text{for all} \qquad x^1, x^2 \in S \qquad \text{and } \textit{some } \lambda \quad (0 \leqq \lambda \leqq 1).$$

3. Determine which of the following are convex sets:

a. $\{x_1, x_2 \mid x_1x_2 \leqq 1, x_1 \geqq 0, x_2 \geqq 0\}$;

b. $\{x_1, x_2 \mid x_1x_2 \leqq 1, x_1 \leqq 0, x_2 \leqq 0\}$;

c. $\{x_1, x_2 \mid x_1x_2 \leqq 1, x_1 \leqq 0, x_2 \geqq 0\}$;

d. $\{x_1, x_2 \mid x_1x_2 \leqq -1, x_1 \geqq 0, x_2 \leqq 0\}$.

4. The transformation $y = Ax$ will transform a convex set S into a convex set T. Show that, if A is square and nonsingular, the extreme points of S transform into extreme points of T. What happens if (a) A is square but singular?; (b) A is not square?

5. Each of the following sets of points are points in a finite cone and each set includes points on all generators of the cone. Determine in each case which points lie on the generators and whether the cone is pointed or not:

a. $(-1, 1), (1, 2), (2, -1), (-1, 2)$;

b. $(-1, 1), (5, 5), (3, -6), (0, 2)$.

6. (Optional) Prove as many of the results of Section R3.7 as you can by using the properties of dual cones.

REVIEW

R5

Square Matrices and Characteristic Roots

This review follows on from Review R3. Review R4 is not a necessary preliminary. The reader who omitted complex numbers (Section R1.6) and complex vectors (Section R2.6) should now study these. The reader anxious to proceed to a first reading of optimizing theory (Chapters 2 through 5) should omit this and the next two reviews and pass to Review R8.

R5.1 INTRODUCTION[1]

If we take the set of rectangular matrices of some given order, $m \times n$, we can add and subtract them, multiply them by a scalar and form linear combinations. We have a zero matrix and a definition of equality. We do not have a complete algebra, however, since we cannot multiply them, in general. We can, of course, multiply one matrix by the *transpose* of another, but the transpose is not of the same order as the original, and the product is of a different order from both original and transpose.

The set of *square* matrices of the same order forms a more complete algebra. Since they have the same number of rows and columns any two

[1] Square matrices and characteristic roots are discussed in all texts on matrices. Elementary discussions are given in Yamane and Allen [2]. Among the mathematics texts, the author considers Bellman [3] the most suitable for the purposes of this book.

matrices can be multiplied together, in any order (although the products AB and BA will not, in general be equal) and their product will be a member of the set.

We can multiply a square matrix by itself to obtain $AA = A^2$ and by continuing the process to define A^n for a positive integer. By putting $B = A^n$, we can define $B^{1/n} = A$ and so have A^n for any positive rational number. Many algebraic identities hold for square matrices as well as scalars, provided the equality $AB = BA$ is not required. We have, for example,

$$(A + B)^2 = A^2 + (AB + BA) + B^2$$

but

$$\begin{aligned}(A + B)(A - B) &= A^2 - B^2 + (BA - AB) \\ &\neq A^2 - B^2 \qquad \text{unless } AB = BA.\end{aligned}$$

A *diagonal matrix* is a square matrix with zeros everywhere except down the leading diagonal. We usually identify an element of a diagonal matrix by a single index, so that the typical element of D will be written d_i and the typical element of Λ will be λ_i.

By following through the rules of matrix multiplication, it is easily seen that pre-multiplying by a diagonal matrix, as in DA, multiplies the kth *row* of A by d_k, and postmultiplication AD multiplies the kth *column* by d_k.

It is immediate from above that $DD = D^2$ is a diagonal matrix with element d_k^2 and by continuing the multiplication that D^n is a diagonal matrix with typical element d_k^n. This is a very important property.

The *identity matrix* I is a diagonal matrix with $d_k = 1$ for all k. It is immediate that $IA = AI = A$, so that I plays the same role in matrix algebra as unity in scalar algebra.

Square matrices satisfy typical *algebraic series* relationships, including infinite series when suitable convergence properties can be established (see Section R5.6). I must be used in place of 1 to preserve the matrix character of the series. For example:

$$(I + A)^n = I + nA + \frac{n(n - 1)}{2!} A^2 + \cdots + A^n.$$

We can even define nonalgebraic functions of A, provided they can be defined by algebraic series. The matrix exponential e^{tA} is defined by the series,

$$e^{tA} = I + tA + (t^2/2!)A^2 + \cdots.$$

The matrix exponential is very important in differential equation theory. It can be shown quite simply that [2]

$$e^{sA}e^{tA} = e^{(s+t)A}.$$

However, the *matrix* index property $e^A e^B = e^{A+B}$ holds if and only if $AB = BA$.

It should be emphasized that all functions defined in this way are themselves matrices of the same order as A. The idea of e^A as a matrix will be easy to accept after Section R5.5.

R5.2 DETERMINANTS AND CRAMER'S RULE[3]

It is assumed that the reader has some familiarity with simple determinants. They are rather tedious, and many statements made in determinantal form can be expressed otherwise. Nevertheless they are useful and have played a prominent role in neoclassical mathematical economics. We are usually concerned with the sign of a determinant, or with whether it is zero or not, rather than with its actual value.

To add some interest to the topic, we shall give a rather different definition of a determinant from that usually given, state the standard properties of determinants, and prove those that follow easily from the definition given.

A determinant is a scalar defined on a square matrix. If A is a matrix of order $n \times n$, the determinant of A, which we shall typically write as "det A" (the notation $|A|$ used once to be common) is a scalar which

(a) is a linear function of the columns of A;
(b) has the value zero if two adjacent columns of A are identical;
(c) has the value unity for $A = I$.

For any square matrix A, we also define some special determinants. Denote by (A_{ij}) the $(n-1) \times (n-1)$ submatrix obtained by deleting the ith row and jth column of A. Then det (A_{ij}) is the *minor* of a_{ij} in det A (it is called a *principal minor* if $j = i$); $(-1)^{i+j}$ det (A_{ij}) is the *cofactor* of a_{ij} in det A. This is usually written A_{ij}, with the minor written as $|A_{ij}|$.

The determinant det A has the following properties:

(*a*) *Multiplying any column by* λ *multiplies* det A *by* λ.
(*b*) *Interchanging adjacent columns changes the sign of* det A.

[2] Proofs of these properties are given in Bellman [3], Chapter 10.

[3] It is assumed that the reader is already familiar with the arithmetic of calculating determinants. This is covered in Yamane and Allen [1].

(c) *If ANY two columns are identical*, det $A = 0$.

(d) *Addition of a multiple of any column to a DIFFERENT column leaves* det A *unchanged.*

(e) *If the columns of A are linearly dependent*, det $A = 0$.

(f) Det $A = \sum_{j=1}^{n} (-1)^{i+j} a_{ij} \det (A_{ij}) = \sum_{j=1}^{n} a_{ij} A_{ij}$.

(g) Det $A^T = \det A$.

(h) Det $AB = (\det A)(\det B)$.

Since we are considering det A as a linear function of its columns, write det $A = D(A^1, \ldots, A^n)$, where A^j is the jth column. Property (a) is immediate from (a) of the definition.

Now consider a determinant in which the jth and $(j + 1)$th columns have *both* been replaced by $A^j + A^{j+1}$. We have

$$D^* = D(\cdots A^{j-1}, A^j + A^{j+1}, A^j + A^{j+1}, A^{j+2}, \ldots) = 0,$$

since it has two identical adjacent columns.

Using the linearity property, we can expand D^* as follows

$$D^* = D(\cdots A^j, A^j \cdots) + D(\cdots A^{j+1}, A^{j+1} \cdots) + D(\cdots A^j, A^{j+1} \cdots) + D(\cdots A^{j+1}, A^j \cdots).$$

The first two determinants in the expansion vanish, since both have two adjacent identical columns. Since $D^* = 0$, we then have

$$D(\cdots A^{j+1}, A^j \cdots) = -D(\cdots A^j, A^{j+1} \cdots)$$

which proves (b).

To prove (c), we note that if A^j, A^k are identical, we can make them adjacent by a process of interchanging pairs of columns. Each interchange alters the sign of det A but does not alter any zero or nonzero property. When the two columns are adjacent, det $A = 0$, so this must be true when they are not adjacent, proving (c).

If we replace A^j by $A^j + \lambda A^k$, we have

$$D(\cdots A^j + \lambda A^k \cdots A^k \cdots) = D(\cdots A^j \cdots A^k \cdots) + \lambda D(\cdots A^k \cdots A^k \cdots).$$

The second determinant must be zero, so we have (d).

Now suppose the rows of A are linearly dependent. This means we can find numbers λ_j, not all zero, such that

$$\sum \lambda_j A^j = 0.$$

Choose one of the λ's which is not zero, say λ_k. Then

$$\lambda_k A^k = -\sum_{j \neq k} \lambda_j A^j.$$

Put $\mu_j = -\lambda_j/\lambda_k$, then we can replace A^k in $D(\cdots A^k \cdots)$ by $\sum \mu_j A^j$. Using the linearity property, we obtain

$$D(\cdots A^k \cdots) = \sum_{j \neq k} \mu_j D(\cdots A^j \cdots).$$

The determinants on the right-hand side have, successively, $A^1, A^2 \cdots A^{k-1}$, $A^{k+1} \cdots A^n$ in place of A^k. But each of these then has two identical columns and so is zero. Thus det A is zero if the columns of A are linearly independent, proving (e).

We can prove (f) by induction, then continue the expansion process for det A_{ij}, finally proving the fully expanded form of the determinant. From this, we can prove (g) and (h).

Note that (g) implies we can replace *columns* in our definition and in properties (a) through (f) by *rows*.

Using the above approach, we can easily prove the well-known rule for obtaining the solutions of inhomogeneous linear equations in terms of the ratios of determinants.

Express the system of linear equations $Ax = b$ in the form

$$\sum x_j A^j = b.$$

Now consider the determinant $D^{(k)} = D(\cdots A^{k-1}, b, A^{k+1} \cdots)$ obtained by replacing the kth column of det A by the column vector b. Substituting for b and expanding, we obtain

$$\begin{aligned} D^{(k)} &= D(\cdots A^{k-1}, \sum x_j A^j, A^{k+1} \cdots, \\ &= \sum_{j \neq 1}^{n} x_j D(\cdots A^j \cdots). \end{aligned}$$

Every one of the determinants in the expansion, except that for which $j = k$, will vanish because it contains A^j in both the jth and kth places. Thus

$$D^{(k)} = x_k(\det A).$$

Thus we have

Cramer's Rule

For the equation system $Ax = b$, x_k for $k = 1, \ldots, n$ is given by

$$x_k = \frac{D^{(k)}}{\det A},$$

where $D^{(k)}$ is the determinant obtained by replacing A^k in det A *by the column vector b.*

R5.3 THE INVERSE OF A SQUARE MATRIX

Consider the matrix equation $AB = I$, with A given, B unknown. Since equating two matrices means equating their respective elements there are n^2 linear equations in n^2 unknowns b_{ij}. Equating corresponding elements on both sides of the matrix equation, the equations are

$$\sum_k a_{ik}b_{kj} = 0, \quad i, j = 1, \ldots, n, \qquad i \neq j$$
$$\sum_k a_{ik}b_{ki} = 1 \qquad i = 1, \ldots, n.$$

These can be considered as n separate systems of equations, one system for each column of B. Denoting the jth column of B by B^j, each such system is of the form

$$AB^j = c^j,$$

where c^j is a nonzero vector having its single nonzero entry, 1, in the jth place.

The system can be solved uniquely for B^j provided $\det A \neq 0$. The same is true for each B^j, the coefficient matrix A being the same for all n systems, but the location of the entry 1 in the constant vector changing for each system.

Thus, provided A is nonsingular (that is, $\det A \neq 0$) we can solve for B in $AB = I$. We call B the *inverse* of A and write it A^{-1}.

We need to check whether, if we had sought a matrix B that satisfied $BA = I$, we would have reached the same result, since we cannot, in general, expect that $AB = BA$. If we have $BA = I$, postmultiplying each side by A^{-1} gives $(BA)A^{-1} = A^{-1}$. But $(BA)A^{-1} = B(AA^{-1}) = B$, so that $B = A^{-1}$.

Thus, if A is nonsingular, there is a *unique* inverse A^{-1} such that $A^{-1}A = AA^{-1} = I$.

Since the product of matrices is a matrix, it has an inverse which we would expect to be simply related to the inverses of the factors. The inverse of a product is, in fact, given by the rule

$$(AB)^{-1} = B^{-1}A^{-1}.$$

The inverse of a product is the product of the inverses, in reverse order. We see that $(B^{-1}A^{-1})(AB) = B^{-1}(A^{-1}A)B = B^{-1}B = I$, whereas $(A^{-1}B^{-1})(AB) = A^{-1}B^{-1}AB$ and can be reduced to nothing else.

Since the inverse is another square matrix, we define negative powers of A by $A^{-n} = (A^{-1})^n$. It is easy to show that, if D is a diagonal matrix with elements d_k, D^{-1} is a diagonal matrix with elements d_k^{-1}.

We can derive an expression for A^{-1} in terms of determinants involving elements of A only by the use of Cramer's rule.[4]

As we saw above, each column of $B = A^{-1}$ is given from the linear equation system,

$$AB^j = c^j,$$

where c^j contains 1 in the jth place and 0 elsewhere. ($c^j = \delta_{ij}$, where δ_{ij} is the Kronecker delta.)[5]

From Cramer's rule we then have

$$b_{ij} = D^{(i)}/\det A,$$

where $D^{(i)}$ is obtained by inserting c^j in place of the ith column of A. Then the ith column of $D^{(i)}$ contains a single nonzero element, the entry 1 in the place occupied by a_{ji} in A. Expanding by the ith column, we have, therefore

$$D^{(i)} = A_{ji} \qquad \text{(the cofactor of } a_{ji} \text{ in } \det A)$$

so that

$$b_{ij} = A_{ji}/(\det A).$$

Thus

$$B = A^{-1} = \frac{1}{\det A}[A_{ji}],$$

or

$$A^{-1} = \frac{1}{\det A}[A_{ij}]'.$$

Example. Although finding the inverse of a 2×2 matrix is almost trivial, the main points of the general procedure can be illustrated without the usual weight of arithmetic.

Consider

$$A = \begin{bmatrix} -1 & 2 \\ -3 & 4 \end{bmatrix}.$$

We have

$$\det A = (-1)(4) - (2)(-3) = 2.$$

[4] For large matrices there are more efficient methods for computing inverses, notably the Gauss elimination method. Such arithmetic does not arise in analytical problems, and for empirical work the use of the Computer Center is even more efficient!

[5] The Kronecker delta δ_{ij} is an object defined to be zero if $i \neq j$ and unity if $i = j$. It is often a very useful shorthand device.

The cofactors in this case are very simple, since they are the diagonally opposite elements, with the appropriate signs $(-1)^{i+j}$ attached.

Thus

$$[A_{ij}] = \begin{bmatrix} 4 & 3 \\ -2 & -1 \end{bmatrix}.$$

To obtain A^{-1}, we transpose $[A_{ij}]$ and divide by det A.

$$A^{-1} = \tfrac{1}{2}\begin{bmatrix} 4 & -2 \\ 3 & -1 \end{bmatrix}$$
$$= \begin{bmatrix} 2 & -1 \\ \tfrac{3}{2} & -\tfrac{1}{2} \end{bmatrix}.$$

A quick check will show that $AA^{-1} = I$.

R5.4 CHARACTERISTIC ROOTS AND VECTORS

Consider the equation system $Ax = \lambda x$, where λ is a scalar. This represents the transformation of a vector x into another vector λx which is a homogeneous lengthening or shortening of the original vector.

We can write the system in the homogeneous form

$$(A - \lambda I)\,x = 0.$$

This is a system of n equations in n unknowns and, being homogeneous, has a nontrivial solution (that is, other than $x = 0$), if and only if the columns of $(A - \lambda I)$ are linearly dependent, that is if $\det(A - \lambda I) = 0$.

Now $\det(A - \lambda I)$ expands into a polynomial of order n in λ, since λ appears down the diagonal and nowhere else, so that $\det(A - \lambda I) = 0$ is an algebraic equation of the nth degree. It will have exactly n roots, which need not be distinct. The roots will not usually all be real, of course, but since A is real (in typical economic contexts), complex roots will come in conjugate pairs.

The equation

$$\det(A - \lambda I) = 0$$

is the *characteristic equation* of the matrix A. The n roots of the equation are the *characteristic roots*, *latent roots* or *eigenvalues* of the matrix. The term "eigenvalues" is widely used in engineering and physics, but in mathematics and economics the term "characteristic roots" is more usual.

If we take any one of the roots of the characteristic equation and insert it in the original equation system we obtain

$$(A - \lambda_k I)\,x = 0$$

which has a nontrivial solution since det $(A - \lambda_k I) = 0$. Let this solution be x^k. We refer to it as the *characteristic vector*, *latent vector*, or *eigenvector* corresponding to λ_k.

Since the equation system is homogeneous, x^k is determined only up to a scalar multiple. The *direction* of the characteristic vector is determined, not its length. Also, since λ_k need not be a real number, x^k may not have real components. However, since complex roots appear in conjugate pairs, we will find that complex characteristic vectors also appear in conjugate pairs and that in all applications we shall meet the over-all result will be real.

If there are n distinct characteristic roots, as we shall usually assume to be the case, there will be n distinct characteristic vectors. Repeated roots present problems that can usually be assumed away in most economic models.

We can compress all the information concerning the characteristic roots and vectors into one package in a way that is useful in many contexts.

Let V be an $n \times n$ matrix made up of the characteristic vectors of A in such a way that the kth column of V is x^k. Let Λ be a diagonal matrix with λ_k occupying the kth position on the diagonal.

Consider $V\Lambda$. This involves postmultiplication by a diagonal matrix, so that it is equivalent to multiplying the kth column, x^k, of V by λ_k. Now consider the product AV. From the multiplication rules, the kth column of AV is Ax^k.

Thus the statement

$$AV = V\Lambda$$

is equivalent to the n statements,

$$Ax^k = \lambda_k x^k \qquad k = 1, \ldots, n.$$

The characteristic roots and vectors of a square matrix are extremely important properties, that play an important role in some static and most dynamic models of multisector economies.

The characteristic vectors, as already pointed out, are determined only to a scalar multiple. For many purposes it is inconvenient to have indefinite scalars in the system and we consider the characteristic vectors to have been *normalized* in the following way:

Let x^k be a characteristic vector based on an arbitrary choice of scalar multiple. Then

$$\begin{aligned}(x^k)'x^k &= \sum_{i=1}^{n} (x_i^k)^2 \\ &= \mu^2 \qquad (\text{take } \mu > 0).\end{aligned}$$

Now choose the vector v^k such that

$$v^k = \frac{1}{\mu} x^k.$$

Then v^k is also a characteristic vector, but with the special property that $(v^k)'v^k = 1$.

v^k is a *normalized characteristic vector.*

If v^k is a *complex* characteristic vector, we take the vector $\bar{v}^k$ whose elements are the complex conjugates of those of v^k and normalize so that $(\bar{v}^k)'v^k = 1$.

We now wish to prove the following:

If the characteristic roots of A are distinct, its characteristic vectors are linearly independent.

Suppose, to the contrary, that the vectors are linearly dependent, so that V has rank less than n, say r. Then we can find some $z \neq 0$ which is a solution of

$$Vz = 0.$$

At most, $n - r$ of the components of z can be arbitrarily assigned.

Since $z \neq 0$, the equality $AVz = V\Lambda z$ is nontrivial. But $Vz = 0$, so that

$$V\Lambda z = 0.$$

Thus $z' = \Lambda z$ is a solution of $Vz' = 0$.

Then z, z' can differ in at most $n - r$ components. But the ith component of z' is $\lambda_i z_i$, so we require r of the roots to have the same value, contrary to the assumption of distinct characteristic roots.

For matrices with complex roots, the following property is fundamental:

If a real matrix has a complex root, the associated characteristic vector is necessarily complex. Then the conjugates of the complex root and of its associated vector are also a characteristic root and vector of the matrix.

Let λ, x be a complex root and associated characteristic vector of the real matrix A. These satisfy the equation:

$$Ax = \lambda x.$$

Take complex conjugates

$$A^*x^* = \lambda^*x^*.$$

But A is a real matrix, so that $A^* = A$. Hence

$$Ax^* = \lambda^* x^*$$

which proves the result.

We can regard the roots of a matrix as an important n-parameter summary of the properties of an n^2-dimensional entity, the matrix. It is clear that different matrices may have the same characteristic roots, but we shall see below that these matrices will all be related in a definite way and the sharing of characteristic roots is such an important property that we refer to matrices with the same roots as *similar matrices.*

R5.5 DIAGONALIZATION

Suppose we carry out some acceptable operation $f(A)$ on the square matrix A that gives another square matrix. How are the roots of $f(A)$ related to the roots of A?

This is an extremely important question, which is very easy to answer if we can *diagonalize* a matrix. To understand what is meant by this, we first of all establish the following fundamental proposition:

If T is any nonsingular square matrix, the matrix $B = T^{-1}AT$ has the same characteristic roots as A.[6]

Since B and A are similar matrices, the transformation $B = T^{-1}AT$ is often called a *similarity transformation.*

We prove the result by showing that B satisfies the same characteristic equation as A. We proceed as follows:

$$\begin{aligned} T^{-1}(A - \lambda I)\,T &= T^{-1}AT - \lambda T^{-1}IT \\ &= T^{-1}AT - \lambda I \\ &= B - \lambda I. \end{aligned}$$

Hence

$$\begin{aligned} \det(B - \lambda I) &= \det[T^{-1}(A - \lambda I)T] \\ &= \det T^{-1} \cdot \det(A - \lambda I) \cdot \det T. \end{aligned}$$

The last result follows from elementary determinant theory, that $\det AB = (\det A) \cdot (\det B)$.

Since T is nonsingular, $\det(B - \lambda I) = 0$ implies that $\det(A - \lambda I) = 0$, proving the theorem.

[6] But the characteristic vectors will not, in general, be the same.

Using this result as a springboard, we can now give the diagonalization theorem:

If A has distinct characteristic roots, we can always find some T such that $T^{-1}AT = \Lambda$, where Λ is a diagonal matrix consisting of the characteristic roots of A.

We use the compressed summary of the characteristics which was given in the previous section:

$$AV = V\Lambda,$$

where V is the matrix made up from the characteristic vectors of A. If A has distinct characteristic roots and hence distinct characteristic vectors, V will be nonsingular and V^{-1} exists. Premultiplying both sides of the above by V^{-1}, we obtain

$$\begin{aligned} V^{-1}AV &= V^{-1}V\Lambda \\ &= \Lambda. \end{aligned}$$

This not only proves the diagonalization theorem, but also identifies the appropriate diagonalizing matrix as the matrix of characteristic vectors.[7]

Given the characteristic roots and vectors, we can reconstruct a matrix to which they belong by *undoing* the diagonalization, using the relationship

$$A = V\Lambda V^{-1}.$$

Diagonalization is so useful that the problems raised by repeated roots, which may prevent it, are serious. In most economic contexts it is permissible to remove them by using an approximation theorem by Bellman:[8]

If A is a matrix with repeated roots, there is another matrix B with distinct roots whose elements do not differ from those of A by a total of more than ε, where ε is a preassigned quantity as small as we please.

The proof of this is not especially difficult, but we shall not give it.

The theorem is important in economics, since repetition of roots is due to special coincidences among the elements of the matrix. There are few, if any contexts in economics where a repetition of roots would be regarded as more than an accident of numbers, themselves subject to error. The theorem says that we can remove the repeated roots by very small variations in the numbers.

[7] The diagonalizing matrix is sometimes known as the *polar* or *modal* matrix.

[8] See Bellman [3], theorem 7 of Chapter 11.

In some fields of study, especially physics, the actual values of the characteristic roots may be required, and diagonalization may be the key to solution. Here, we are more interested in the simplicity with which we can demonstrate certain important theoretical results, especially the following.

If A has characteristic roots λ_k, then A^n has characteristic roots λ_k^n, where n is positive or negative and rational.

For positive integral n, we use diagonalization and proceed by iteration.

$$\begin{aligned} T^{-1}AT &= \Lambda \\ AT &= T\Lambda \\ A^2T &= AT\Lambda \\ T^{-1}A^2T &= T^{-1}AT\Lambda \\ &= \Lambda^2. \end{aligned}$$

Then

$$\begin{aligned} A^2T &= T\Lambda^2 \\ A^3T &= AT\Lambda^2 \\ T^{-1}A^3T &= \Lambda^3 \end{aligned}$$

and so on.

If A^n has roots λ^n, then putting $B = A^n$ with roots $\mu = \lambda^n$ we find that $B^{1/n}$ has roots $\mu^{1/n}$.

For negative powers, we have

$$\begin{aligned} AT &= T\Lambda \qquad \text{as above} \\ \cdot\ T &= A^{-1}T\Lambda \\ T\Lambda^{-1} &= A^{-1}T \\ \Lambda^{-1} &= T^{-1}A^{-1}T. \end{aligned}$$

It will be noted that diagonalization of A^n is carried out by the *same* matrix T as the diagonalization of A. Since this matrix has been identified as the matrix of characteristic vectors, it follows that *A^n has the same characteristic vectors as A.*

The above result is important in dynamic linear models.

Since all powers of A are diagonalized by the same matrix, and since the sum of diagonal matrices is a diagonal matrix whose nonzero elements are the sums of the corresponding elements of the component matrices, we can state the following important proposition:

If $f(A)$ is an algebraic polynomial in A, a matrix with characteristic roots λ_i, then the characteristic roots of $f(A)$ are $f(\lambda_i)$ and the characteristic vectors of $f(A)$ are the same as those of A.[9]

The matrix e^A can now be seen to be simply the matrix which can be diagonalized to give e^{λ_i} down the diagonal, and that the series for e^A can be diagonalized to give the n separate series

$$e^{\lambda_i} = 1 + \lambda_i + \lambda_i^2/2! + \cdots.$$

which removes much of the mystery from the idea of transcendental functions of matrices.

R5.6 CONVERGENCE OF MATRIX SERIES

Although we have given examples of matrix series, we have not established conditions under which infinite matrix series actually converge.

Assuming, as usual, that the matrix can be diagonalized, it is not difficult to see from the analysis so far that convergence depends on the characteristic roots. If every term of the series involves only powers of A, we can diagonalize all terms in the series by the same matrix (the matrix of characteristic vectors of A). The matrix series then becomes n separate series, each one corresponding to a place on the diagonal and hence to a particular characteristic root. Since A^r has characteristic roots λ_i^r, each of these series is a power series in one of the characteristic roots and ordinary convergence tests can be applied.

One series of considerable importance in input-output and matrix multiplier theory is the ordinary geometric series

$$\Sigma = I + A + A^2 + A^3 + \cdots.$$

If we diagonalize this, we obtain

$$T^{-1}\Sigma T = I + \Lambda + \Lambda^2 + \Lambda^3 + \cdots.$$

Each of the component series is of the form

$$1 + \lambda_i + \lambda_i^2 + \lambda_i^3 + \cdots.$$

which converges to the sum $(1 - \lambda_i)^{-1}$ if $|\lambda_i| < 1$. Thus the matrix series converges to $(I - \Lambda)^{-1}$ if *every* characteristic root has modulus less than unity.[10]

[9] It should be made quite clear that two different matrices, even of the same order, cannot be combined in this way.

[10] There are weaker convergence conditions available. The condition on the moduli of the individual characteristic roots is satisfactory for our purposes.

In this case we have

$$T^{-1}\Sigma T = (I - \Lambda)^{-1}.$$

Taking inverses,

$$T^{-1}\Sigma^{-1}T = I - \Lambda,$$

so that

$$\begin{aligned}\Sigma^{-1} &= I - T\Lambda T^{-1}\\ &= I - A.\end{aligned}$$

Thus, if all the characteristic roots of A have moduli less than unity, the series

$$I + A + A^2 + A^3 + \cdots$$

converges to $(I - A)^{-1}$ in a manner analogous to the convergence of the series $1 + a + a^2 + a^3 + \cdots$ for $|a| < 1$.

R5.7 CHARACTERISTIC ROW VECTORS

We chose above to consider *characteristic column vectors*, that is, to consider the equation system $Ax = \lambda x$. We could just as well have chosen to have investigated the system

$$yA = \lambda y.$$

It is obvious that the characteristic roots and the characteristic equation are unchanged. Inserting the characteristic roots in the equation system, we obtain *characteristic row vectors* y^i. How are the characteristic row and column vectors related to each other?

Assemble all the normalized characteristic row vectors into the matrix U in the same way the column vectors were assembled into V. Then the equations for all the characteristic roots can be written, as before, in the form

$$UA = \Lambda U.$$

We can diagonalize in the same way as before, to obtain

$$UAU^{-1} = \Lambda.$$

If we compare this with diagonalization by the matrix V of normalized column vectors, we see that U, V are related, for some choice of scalar multiples,[11] by

$$U = V^{-1}.$$

[11] Using the matrix of column vectors, we have $V^{-1}A = \Lambda V^{-1}$. Obviously the rows of V^{-1} are row characteristic vectors. We can always choose the arbitrary scalar multiples to give $U = V^{-1}$.

The two sets of vectors U, V thus contain the same information and we can use either set in a particular case. Now consider the single row vector u^k and the column vector v^k, both associated with the characteristic root λ_k. Then $u^k v^k$ is the kth diagonal element in the matrix UV. But $UV = I$, so that $u^k v^k = 1$. By the same argument $u^j v^k = 0$, so that each characteristic row vector is orthogonal to all the characteristic column vectors except that corresponding to its own characteristic root.

The above analysis has to be modified, by taking complex conjugates, for *complex characteristic roots*. Note that U and V and other matrices of characteristic vectors are the only matrices with complex elements that arise in most economic applications.

Characteristic row vectors are sometimes called *left-hand eigenvectors*, and characteristic column vectors *right-hand eigenvectors*.

R5.8 NUMERICAL EXAMPLES

Example 1. Consider the 2×2 matrix

$$\begin{bmatrix} 1 & 4 \\ 1 & 1 \end{bmatrix}.$$

The characteristic equation $\det(A - \lambda I) = 0$ has the form

$$\begin{vmatrix} 1 - \lambda & 4 \\ 1 & 1 - \lambda \end{vmatrix} = 0.$$

This gives the simple quadratic in λ,

$$\lambda^2 - 2\lambda - 3 = 0,$$

with roots -1, 3.

For each root, we can now solve for the characteristic vector. For $\lambda_1 = -1$, the equation system is

$$\begin{aligned} x_1 + 4x_2 &= (-1)\,x_1 \\ x_1 + x_2 &= (-1)\,x_2. \end{aligned}$$

These two equations are identical. They ought to be, since the characteristic root was chosen specifically to make one of the equations linearly dependent on the others (the other, in this case). Since the system is homogeneous, the solution has an arbitrary scalar multiple, which we shall suppose is chosen to make $x_2 = 1$.

Thus the first characteristic vector is

$$x^1 = (-2, 1)'.$$

For $\lambda_2 = 3$, using only the first equation and putting $x_2 = 1$ as before, we obtain

$$x^2 = (2, 1)'.$$

Remembering that these are column vectors, we can write the matrix of characteristic vectors

$$V = \begin{bmatrix} -2 & 2 \\ 1 & 1 \end{bmatrix}.$$

The complete system properties can be expressed in the form $AV = V\Lambda$, which, in this case, is

$$\begin{bmatrix} 1 & 4 \\ 1 & 1 \end{bmatrix}\begin{bmatrix} -2 & 2 \\ 1 & 1 \end{bmatrix} = \begin{bmatrix} -2 & 2 \\ 1 & 1 \end{bmatrix}\begin{bmatrix} -1 & 0 \\ 0 & 3 \end{bmatrix}.$$

Direct calculation will reveal the correctness of the equation system.

Let us now calculate V^{-1}. This is

$$V^{-1} = \begin{bmatrix} -\frac{1}{4} & \frac{1}{2} \\ \frac{1}{4} & \frac{1}{2} \end{bmatrix}.$$

Now examine the characteristic row vectors. Taking only the first equation in each system and putting $y_2 = 1$, we obtain

for $\lambda_1 = -1$,

$$y_1 + y_2 = (-1)\,y_1, \qquad y^1 = (-\tfrac{1}{2}, 1);$$

for $\lambda_2 = 3$,

$$y_1 + y_2 = 3y_1, \qquad y^2 = (\tfrac{1}{2}, 1).$$

The matrix of row vectors is

$$\begin{bmatrix} -\frac{1}{2} & 1 \\ \frac{1}{2} & 1 \end{bmatrix}.$$

The choice of $y_2 = 1$ was arbitrary. If we choose instead $y_2 = \frac{1}{2}$, we obtain

$$Y = \begin{bmatrix} -\frac{1}{4} & \frac{1}{2} \\ \frac{1}{4} & \frac{1}{2} \end{bmatrix}.$$

This equals V^{-1}.

Note that the matrix A is a *positive* matrix, that is, $a_{ij} > 0$ for all i, j. It has a *positive characteristic root*, which is *the largest in absolute value* of the roots, and the associated characteristic vectors $x^2 = (2, 1)$, $y^2 = (\frac{1}{2}, 1)$ are *both positive.* These properties are common to all positive matrices, as is shown in Review R7.

Example 2 (Complex roots). Consider the matrix

$$\begin{bmatrix} 1 & -4 \\ 1 & 1 \end{bmatrix}$$

The characteristic equation is

$$\lambda^2 - 2\lambda + 5 = 0,$$

which has complex roots $1 + 2i$, $1 - 2i$, these being complex conjugates.

Solving for the characteristic vectors, we have, for $\lambda = 1 + 2i$,

$$x_1 - 4x_2 = (1 + 2i)x_1.$$

Putting $x_2 = 1$, we obtain $x^1 = (2i, 1)$.

For the other characteristic column vector we obtain $x^2 = (-2i, 1)$, so that x^1, x^2 are complex conjugates. We have

$$V = \begin{bmatrix} 2i & -2i \\ 1 & 1 \end{bmatrix}.$$

Direct calculation gives

$$V^{-1} = \begin{bmatrix} (-\frac{1}{4})i & \frac{1}{2} \\ (\frac{1}{4})i & \frac{1}{2} \end{bmatrix},$$

so that the characteristic row vectors are also complex conjugates.

In the complex case, the arbitrary scalar multiple of the characteristic vector can be a complex number just as well as a real number. If we had chosen to make $x_2 = i$, we would have the x_1 coordinate real and the x_2 coordinate imaginary, instead of the other way around.

Rather than choose one element of the vector equal to 1 or some other arbitrary number, we may choose to *normalize* the vector, by choosing the arbitrary scalar so that the sum of the squares of the elements adds to unity. In this case, we choose x_2 so that $(x_1)^2 + (x_2)^2 = 1$. This would give x^1, x^2 as $[(2/\sqrt{5})i, 1/\sqrt{5}]$, $[(-2/\sqrt{5})i, 1/\sqrt{5}]$.

EXERCISES

1. For the matrix

$$A = \begin{bmatrix} 3 & 4 & 5 \\ 2 & 3 & 1 \\ 1 & 2 & 3 \end{bmatrix} \text{ and the vector } b = \begin{bmatrix} 1 \\ 2 \\ 1 \end{bmatrix}:$$

a. Calculate A^{-1} and check that $AA^{-1} = A^{-1} = I$;

b. Solve $Ax = b$ by direct calculation of $A^{-1}b$;

c. Solve $Ax = b$ by using Cramer's rule.

2. a. Calculate the characteristic roots and vectors of

$$A = \begin{bmatrix} -1 & 2 \\ -3 & 4 \end{bmatrix}.$$

b. show by direct calculation that the roots of A^2, A^{-1} are the square and inverse, respectively, of the roots of A, and that the characteristic vectors of A, A^2, A^{-1} are the same.

3. Prove by direct calculation, that the roots of $(I + A)^3$ are $(1 + \lambda)^3$, where λ are the roots of A and that A, $(I + A)^3$ have the same characteristic vectors. Perform this calculation for the matrices given in Example 1, Section R5.8, and in **2** above.

4. By diagonalizing, then undoing the diagonalization, write out the matrix e^A, where A is the matrix of **2** above.

5. Determine the characteristic roots and vectors of

$$A = \begin{bmatrix} 0 & -2 \\ 2 & 0 \end{bmatrix}.$$

Show that A has complex roots but A^2 has real roots.

How is it that the complex characteristic vectors of A can be characteristic vectors of the matrix A^2 which has real roots?

What is special about this case?

REVIEW

R6

Symmetric Matrices and Quadratic Forms

The content of this review follows directly from that of Review R5. It is possible to omit Section R6.3 until the reader wishes to make a detailed study of the second order conditions in the classical optimizing problem (Chapter 4, Section 4.5).

R6.1 SYMMETRIC MATRICES[1]

A matrix of real elements is *symmetric* if $a_{ij} = a_{ji}$, for all i, j. That is, a symmetric matrix is equal to its transpose. A matrix for which $a_{ij} = -a_{ji}$, that is a matrix which equals the negative of its transpose, is *skew symmetric.* Its diagonal necessarily contains only zeros.

Both the above types of matrices have important special properties, as do related matrices with complex elements. In an economic context, our chief interest is in *real symmetric matrices*, two of the most important properties of which are the following:

(*a*) *All the characteristic roots of a real symmetric matrix are real.*

(*b*) *The characteristic vectors of a real symmetric matrix are orthogonal.*[2]

[1] The contents of this and the following section are standard and are available, in some form, in all sources recommended for the previous review.

For Section R6.3, see the footnote at the beginning of that section.

[2] Vectors are *orthogonal* if their scalar product is zero, which is geometrically equivalent to their being perpendicular to each other.

Both these results can be derived from a simple *lemma*:

If λ_1, λ_2 are two roots of a real symmetric matrix and x^1, x^2 are the associated characteristic vectors, then

$$(\lambda_1 - \lambda_2)\,(x^1)'x^2 = 0,$$

where $(x^1)'$ denotes the transposition of x^1 into a row vector.

First we note that, for a symmetric matrix, $x'Ay = y'Ax$, where x, y are any vectors of order n. The result follows from transposing $y'Ax$ and using the property that $A = A'$.

To prove the lemma, we have, since λ_1, λ_2 are the characteristic roots, and x^1, x^2 the associated characteristic vectors

$$Ax^1 = \lambda_1 x^1, \qquad Ax^2 = \lambda_2 x^2.$$

Premultiplying the first by $(x^2)'$ and the second by $(x^1)'$, we obtain

$$(x^2)'Ax^1 = \lambda_1(x^2)'\,x^1, \qquad (x^1)'Ax^2 = \lambda_2(x^1)'\,x^2.$$

But $(x^2)'\,x^1 = (x^1)'\,x^2$, since both are scalars, and we have just shown that $(x^2)'\,Ax^1 = (x^1)'\,Ax^2$, so that

$$\lambda_1(x^1)'\,x^2 = \lambda_2(x^1)'\,x^2,$$

which gives the result of the lemma.

To prove Result (a), suppose to the contrary that λ was a complex root. Then, since A is a real matrix, $\bar{\lambda}$, the complex conjugate of λ must also be a root. Again, if x is the vector associated with λ, the vector associated with $\bar{\lambda}$ is $\bar{x}$ the complex conjugate of x.

Using the lemma, we have

$$(\lambda - \bar{\lambda})\,x'\bar{x} = 0.$$

Now $x'\bar{x}$ is the sum of squares of the moduli of the components of x (the components of $\bar{x}$ having the same moduli), and is not zero, so that we must have

$$\lambda - \bar{\lambda} = 0.$$

But, from the definition of complex conjugates, λ has the form $a + bi$, while $\bar{\lambda}$ has the form $a - bi$, and the two can be equal only if $b = 0$, that is, if the roots are real, thus proving the result.[3]

[3] The roots of a skew symmetric matrix can be shown to have zero real parts, that is, to be zero or pure imaginary.

To prove result (b), take any two distinct roots. Then $\lambda_1 - \lambda_2 \neq 0$ so, from the lemma,

$$(x^1)'x^2 = 0,$$

which is precisely the condition that x^1, x^2 be orthogonal.

We can now prove a third result of great importance:

(c) *A real symmetric matrix is diagonalized by an orthogonal transformation, that is by a transformation $T^{-1}AT$ in which T has the property $T' = T^{-1}$.*

This follows easily from Result (b). If we diagonalize A (assume distinct roots, as always), we obtain

$$V^{-1}AV = \Lambda.$$

Now V is a matrix of characteristic vectors. If we take the product $V'V$, the (i, j)th element $(i \neq j)$ is $(x^i)'x^j$ which is zero from Result (b). The diagonal elements are $(x^i)'(x^i) = \sum_k (x_k^i)^2$. Since the characteristic vectors are determined only to a scalar multiple, consider them to be normalized. Then the diagonal elements of $V'V$ are unity and $V'V = I$. It follows that $V' = V^{-1}$ and the result is proved.

We usually write the diagonalization of a symmetric matrix in terms of the transpose, $V'AV = \Lambda$.[4]

R6.2 QUADRATIC FORMS

An expression of the kind $Q = \sum_i \sum_j a_{ij}x_ix_j$ is a *quadratic form*. If $a_{ij} \neq a_{ji}$, we can always set

$$\bar{a}_{ij} = \bar{a}_{ji} = \tfrac{1}{2}(a_{ij} + a_{ji}),$$

so that the coefficients of x_ix_j and x_jx_i are the same.

We can therefore write a quadratic form in matrix vector notation as

$$Q(x) = x'Ax,$$

where A is a *real symmetric matrix*.

Given A, $Q(x)$ is a scalar valued function of the vector x. We are specially interested in certain properties which may hold for Q over all possible choices for x. The following terminology is used:

If $Q(x) > 0$ (<0) for all $x \neq 0$, $Q(x)$ is said to be positive (negative) definite.

[4] An *orthogonal matrix* is one for which $A' = A^{-1}$. Its rows or columns, considered as vectors, are orthogonal to each other.

If $Q(x) \geqq 0$ ($\leqq 0$) for all $x \neq 0$, $Q(x)$ is said to be nonnegative (nonpositive) definite. If $Q(x)$ is nonnegative (nonpositive) definite with $Q(x) > 0$ (<0) for some x and $Q(x) = 0$ for some $x \neq 0$, $Q(x)$ is said to be positive (negative) semidefinite.

Although the term "positive definite" and related terms are applied to the quadratic form $Q(x)$ in the first instance, they are also applied to the matrix A by transference. Any real symmetric matrix can be the matrix of a quadratic form, so we can inquire of any such matrix whether it is positive definite or not.

If $Q(x)$ is positive definite, $-Q(x)$ is negative definite, so we will primarily discuss positive definiteness.

Consider a transformation of the vector x to another vector y of the same order, with the relationship defined by $x = Ty$, where T is nonsingular. Then

$$Q(x) = x'Ax = y'T'ATy = Q(y),$$

where the matrix of the quadratic form $Q(y)$ is $T'AT$.

It is clear that, as x takes on all values other than 0, y also takes on all values other than 0, so that A is positive definite if and only if $T'AT$ is positive definite.

We showed in the previous section, however, that, for some suitable orthogonal matrix T

$$T'AT = \Lambda.$$

Since A is positive definite if and only if Λ is positive definite, and since

$$x'\Lambda x = \sum \lambda_i x_i^2,$$

it is obvious that A is positive definite if all its roots (which are real, since A is real and symmetric) are positive. Positivity of all the roots is also necessary for positive definiteness, since, if we had $\lambda_k < 0$, putting $x_k \neq 0$ and $x_i = 0$, $i \neq k$, would give $x'Ax < 0$ for $x \neq 0$. Thus we can state:

A real symmetric matrix is positive (negative) definite if and only if all its characteristic roots are positive (negative).

Now $\det \Lambda = \det A$, since $\det T = 1$. [T is orthogonal, so that $\det T' = \det T^{-1} = (\det T)^{-1} = \det T = 1$.] But $\det \Lambda$ is simply the product of the n characteristic roots, so we can give another important result:

If A is positive definite, then det *A is positive. If A is negative definite, then* det *A has the sign of* $(-1)^n$.

The sign of the determinant is necessary, but not sufficient for definiteness, as is obvious from the case of a matrix of even order when det A must be positive in both cases.

Let us write out the expansion of $x'Ax$ in such a way that the terms containing x_n are clearly separated. We have

$$x'Ax = \sum_{i=1}^{n-1} \sum_{j=1}^{n-1} a_{ij}x_i x_j + 2 \sum_{i=1}^{n-1} a_{in}x_i x_n + a_{nn}x_n^2$$

(using the symmetry property that $a_{in} = a_{ni}$).

Now put $x_n = 0$. $x'Ax$ becomes

$$\sum_{i=1}^{n-1} \sum_{j=1}^{n-1} a_{ij}x_i x_j.$$

But this is a quadratic form based on a matrix formed from A by deleting the nth row and column from A. Any such submatrix of A formed by deleting a set of rows and the corresponding set of columns is square and symmetric and its diagonal lies along the diagonal of A. We refer to it as a *principal submatrix of order k*, where $n - k$ rows and columns of A have been deleted, and write it A_k.

In the present case, putting $x_n = 0$ gives a quadratic form $x'A_{n-1}x$ (where x now has only $n - 1$ components). But the condition that $x'Ax$ be positive definite is that it be positive for all $x \neq 0$, which includes a condition that it be positive for $x_n = 0$ and free choice for the remaining components of x provided they are not all zero. This is precisely the condition that $x'A_{n-1}x$ be positive definite. By arguing the same way for $x'A_{n-1}x$, we show that $x'A_{n-2}x$ must be positive definite, and so on. But we might just as well have chosen to single out any x_k, instead of x_n, and formed another sequence of principal submatrices to which the above arguments would apply. Thus we can state:

Every principal submatrix of a positive (negative) definite matrix is also positive (negative) definite.

Now the determinant of a principal submatrix is a *principal minor* of the determinant of A. Using the earlier result concerning the relationship between positive (negative) definiteness and the sign of the determinant, together with the above result, we have the following:

A matrix is positive definite if and only if all its principal minors are positive. A matrix is negative definite if and only if all its principal minors of order k have the sign of $(-1)^k$.

Although we have only proved necessity here, this condition can also be shown to be sufficient, as stated. The condition is widely used in maximization theory, usually in the alternative form:

A matrix is positive (negative) definite if and only if principal minors of successively higher order have the same (alternate in) sign. The condition must hold for all possible sequences of principal minors.

Before leaving the present subject, we should emphasize the difference between a *positive definite matrix* and a *positive matrix*. The latter class of matrices, which are studied in the next review, have all their elements positive, but do not necessarily have all their roots positive. Positive definite matrices have all their roots positive, but do not necessarily have all their elements positive. Positive definiteness is a property of *square symmetric* matrices only, whereas positivity can be a property of any type of matrix, usually square but not symmetric.

R6.3 CONSTRAINED QUADRATIC FORMS[5]

Instead of permitting all possible variations in the variables of a quadratic form, we now wish to consider quadratic forms in which the variables are subject to homogeneous linear constraints. This problem is fundamental to the analysis of the second-order conditions for a classical constrained maximum.

We shall consider the quadratic form in n variables

$$Q(x) = x'Ax,$$

in which the variables are constrained by the m linear equations

$$Bx = 0.$$[6]

Before proceeding, we shall state, without proof, the following[7]:

[5] The analysis of constrained quadratic forms is treated quite inadequately in most books on matrices.

The properties of these forms are crucial in the second order conditions for the classical optimizing problem, and are therefore of considerable importance in mathematical economics.

The analysis given here, which takes off from Bellman's treatment of the subject (Bellman [3], Chapter 5), is believed to provide the best available foundation for the classical optimizing problem.

See also the comments in Chapter 4, Section 4.5.

[6] B is a rectangular matrix of order $m \times n$, with $m < n$.

[7] For proof of the Finsler theorem, see Bellman [3], Chapter 5.

Finsler's Theorem

If $x'Ax$ is positive whenever $x'Cx = 0$, C being nonnegative definite, then there is some positive scalar μ such that $(A + \mu C)$ is positive definite.

To see the relevance of this theorem, let $Bx = b$ for arbitrary choice of x. Now

$$x'B'Bx = b'b = \sum (b_j)^2 \geqq 0$$

and is equal to zero for x satisfying the constraints. Then $B'B$ is a matrix which plays the part of the matrix C in the Finsler theorem.

Thus the condition that $x'Ax$ be positive for all x satisfying $Bx = 0$ is equivalent to the condition that $x'(A - \mu B'B)\,x$ be positive definite for sufficiently large $\mu > 0$.

Now we shall do something that appears quite arbitrary. Consider the matrix

$$M_1 = \begin{bmatrix} -I & B \\ \mu B' & A \end{bmatrix}.$$

This is a matrix of order $(n + m) \times (n + m)$, with the northwest matrix $-I$ of order $m \times m$. If we now multiply this by the matrix

$$M_2 = \begin{bmatrix} I & B \\ 0 & I \end{bmatrix},$$

where the lower I is of order $n \times n$ and the upper I is of order $m \times m$, we obtain, from the ordinary rules for multiplying conformably partitioned matrices,

$$M_3 = \begin{bmatrix} -I & 0 \\ \mu B' & A + \mu B'B \end{bmatrix}.$$

Thus we have

$$(\det M_1)\,(\det M_2) = \det M_3.$$

Now M_2 has only zeros below the diagonal and 1's along the diagonal, so that $\det M_2 = 1$.

Consider $\det M_3$. If we expand this along the top row we obtain, since there is a single entry (-1) in that row,

$$\det M_3 = (-1) \det \begin{bmatrix} -I_{m-1} & 0 \\ \mu B'_{m-1} & A + \mu B'B \end{bmatrix},$$

where the subscripts $(m - 1)$ on I and B' mean that the first column has been deleted from B', and one row and column from I. If we continue to expand along the remaining $m - 1$ top rows, we see that, ultimately,

$$\det M_3 = (-1)^m \det (A + \mu B'B).$$

Thus we have

$$\det M_1 = (-1)^m \det (A + \mu B'B).$$

Now the condition that $A + \mu B'B$ be positive definite implies that $\det (A + \mu B'B) > 0$, hence that $\det M_1$ has the sign of $(-1)^m$. Consider

$$\det M_1 = \det \begin{bmatrix} -I & B \\ \mu B' & A \end{bmatrix} = \mu^m \det \begin{bmatrix} (-1/\mu)I & B \\ B' & A \end{bmatrix}.$$

By making μ large enough, we can always make the sign of the last determinant the same as the sign of

$$\det \begin{bmatrix} 0 & B \\ B' & A \end{bmatrix}.$$

Thus a first *sufficient*, but not necessary, condition for the problem is that $\det \hat{A}$ have the sign of $(-1)^m$, where $\hat{A}$ is the bordered matrix above.

Now return to the original quadratic form. If we put x_n equal to zero, then the reduced form $x'A_{n-1}x$ must be positive for the $(n - 1)$ vector x subject to $B_{n-1}x = 0$, where A_{n-1} is derived from A by deleting the nth row and column and B_{n-1} from B by deleting the nth column. We can use the same analysis as before to show that a sufficient condition that the quadratic form preserve its constrained positivity when $x_n = 0$ is that

$$\det \begin{bmatrix} 0 & B_{n-1} \\ B'_{n-1} & A_{n-1} \end{bmatrix} \text{ have the sign of } (-1)^m.$$

But this determinant is the principal minor of $\det \hat{A}$ obtained by deleting the nth row and column.

We can then put x_n and x_{n-1} equal to zero and show that the same condition holds for the principal minor of $\det \hat{A}$ obtained by deleting the last two rows and columns. We can continue up to the point at which we have put $n - m + 1$ of the variables equal to zero, beyond which the constraints $Bx = 0$ do not give us any further freedom in arbitrarily choosing zero components of x.

Now let us seek a sufficient condition that $x'Ax$ be negative for x satisfying the constraints. Then $-x'Ax$ will be positive, and we can proceed as before with $-A$ in place of A. In particular, we require that the principal minor of order $m + n - r$ of

$$\det \begin{bmatrix} 0 & B \\ B' & -A \end{bmatrix}$$

have the sign of $(-1)^m$.

But

$$\det \begin{bmatrix} 0 & B_{n-r} \\ B'_{n-r} & -A_{n-r} \end{bmatrix} = (-1)^{n-r} \det \begin{bmatrix} 0 & B_{n-r} \\ -B'_{n-r} & A_{n-r} \end{bmatrix}$$

$$= (-1)^{m+n-r} \det \begin{bmatrix} 0 & B_{n-r} \\ B'_{n-r} & A_{n-r} \end{bmatrix},$$

so that, if we require the principal minor of order $m + n + r$,

$$\det \begin{bmatrix} 0 & B_{n-r} \\ B'_{n-r} & -A_{n-r} \end{bmatrix}$$

to have the sign of $(-1)^m$. Then the corresponding principal minor of det $\hat{A}$

$$\det \begin{bmatrix} 0 & B_{n-r} \\ B'_{n-r} & A_{n-r} \end{bmatrix}$$

must have the sign of $(-1)^{n-r}$.

We can now sum up our findings as follows:

A sufficient condition that the quadratic form $x'Ax$ be POSITIVE for all x satisfying the constraints $Bx = 0$ is that every principal minor of det $\hat{A}$, where $\hat{A}$ is the bordered matrix

$$\begin{bmatrix} 0 & B \\ B' & A \end{bmatrix}$$

have the sign of $(-1)^m$ where m is the number of constraints, and that det A *have this same sign.*

A sufficient condition that the quadratic form $x'Ax$ be NEGATIVE under the above conditions is that det $\hat{A}$ have the sign of $(-1)^n$ and that the principal minor of order $(m + n - r)$ have the sign of $(-1)^{n-r}$.

The principal minors referred to above are obtained by deleting the LAST $1, 2, \ldots, r \ldots$ rows and columns of det $\hat{A}$ only up to the last $n - m + 1$ rows and columns.

The simplest properties implied above are that the sequence consisting of det $\hat{A}$ and its principal minors of successively lower order have the same sign for a positive quadratic form, and alternate in sign for a negative quadratic form.

The fact that the determinantal conditions are sufficient but not necessary follows from noting that it is possible for linear dependence to exist in A. For example, if we had $b_{r1} = 0$, all r, and $a_{11} = 0$, it would be possible for the first and $(m + 1)$th columns of $\hat{A}$ to be linearly dependent. In this case $\det \hat{A} = 0$, and it becomes necessary, for the positive case, that $\det A$ have the sign of $(-1)^m$. This type of case requires the constraint to lie along the contour surface of the objective function when the above analysis is applied to problems of constrained optimization and can usually be ruled out.

EXERCISES

1. Determine by calculating the characteristic roots the signs and determinateness of the quadratic forms associated with the matrices:

$$\textbf{a. } \begin{bmatrix} 2 & 1 \\ 1 & 3 \end{bmatrix}; \quad \textbf{b. } \begin{bmatrix} -2 & \sqrt{6} \\ \sqrt{6} & 3 \end{bmatrix}; \quad \textbf{c. } \begin{bmatrix} -2 & 1 \\ 1 & -3 \end{bmatrix}; \quad \textbf{d. } \begin{bmatrix} 1 & 3 \\ 3 & 2 \end{bmatrix}.$$

2. Using the determinantal conditions, find the sign and determinateness of the quadratic forms associated with the following matrices:

$$\textbf{a. } \begin{bmatrix} -4 & -1 & 2 \\ -1 & -5 & 3 \\ 2 & 3 & -6 \end{bmatrix} \qquad \textbf{b. } \begin{bmatrix} 6 & -1 & 2 \\ -1 & 3 & 2 \\ 2 & 2 & 3 \end{bmatrix}.$$

3. Find, using the determinantal conditions, the sign of the quadratic form $x'Ax$, subject to $bx = 0$, where

$$A = \begin{bmatrix} 1 & 2 \\ 2 & -12 \end{bmatrix}; \qquad b = [1 \quad -3].$$

Check your results by substituting for one variable in terms of the other by use of the constraint.

4. Show that if A is positive-definite, A^n is also positive-definite for all positive integral n. If A is negative-definite, is A^n necessarily negative-definite?

REVIEW

R7

Semipositive and Dominant Diagonal Matrices

This review is concerned with two related topics that are used in different places in the economic analysis. The properties of semipositive matrices (Sections R7.2 and R7.3) are essential for Chapter 6, while the properties of dominant diagonal matrices (Section R7.4) are essential for Chapter 12. Familiarity with the content of Reviews R1 through R3 and R5 is assumed. Reviews R4 and R6 are not essential prerequisites. The proofs of the results set out in Sections R7.3 and R7.4 are given in Section R7.5 and may be omitted on first reading.

R7.1 INTRODUCTION[1]

Following the terminology for vectors, we define a *positive* (strictly positive, if emphasis seems called for) matrix as a matrix all of whose elements are positive. Similarly, we define a *nonnegative* matrix as a matrix consisting entirely of nonnegative elements.

We also introduce the idea of a *semipositive* matrix. An important property of semipositive vectors is that the product of a semipositive vector and a positive vector is always positive. We shall analogously define a semipositive

[1] For a discussion that brings together both semipositive and dominant diagonal properties, see McKenzie [3]. Other references are given later in the Review.

matrix in such a way that the product of a semipositive matrix and a positive row or column vector is a positive vector. This leads to the following definition.

A matrix is semipositive if all its rows are semipositive vectors and all its columns are also semipositive vectors.[2]

Our interest will be mainly in semipositive *square* matrices, but the definition is applicable to any type of matrix.

We now define a second type of matrix in which we are interested:

A square matrix A of order $n \times n$ has a dominant diagonal if there exist n positive numbers d_j such that

$$d_j\,|a_{jj}| > \sum_{i \neq j} d_i\,|a_{ij}| \qquad j = 1, \ldots, n.$$

That is, A is a dominant diagonal matrix if positive weights can be found such that absolute value of every weighted diagonal element exceeds the weighted sum of the absolute values of the off-diagonal elements in the same column. It is essential that the same weights apply to every column.

The above definition is a generalization by McKenzie of the more usual definition, that a matrix has a dominant diagonal if the absolute value of every diagonal element exceeds the simple sum of the absolute values of the off-diagonal elements in the same column (or row).[3]

The weights d_j in the more general definition permit us, in effect, to use "excess" dominance in one column to offset lack of dominance in another. If D is the diagonal matrix $[d_j]$ and A has a dominant diagonal on the generalized definition then DA has a dominant diagonal in the usual sense.

Since we shall be concerned with real matrices, we have regarded $|a_{ij}|$ as the absolute value of a_{ij}. We shall, however, be investigating matrices with complex diagonal elements, in which cases $|a_{jj}|$ will be the *modulus* of a_{jj}.

On the surface, there may seem no relationship between semipositive and dominant diagonal matrices. Beneath the surface, there are, however strong connections between the properties of the two classes of matrices and many of the properties can be proved jointly. This is the reason they are treated together in this review.

We shall first discuss the concept of *indecomposability* which is important in relation to semipositive matrices. Then we shall list, without proof, the

[2] This requires that the matrix has at least one nonzero element in every row *and* every column. A semipositive diagonal matrix has all its diagonal elements positive.

[3] We shall call such a matrix a *Hadamard matrix*. See Section R7.5.

properties of semipositive matrices that are important in economics. Then we shall list the important properties of dominant diagonal matrices, also without proof. Finally, we shall give the proofs for both classes of matrices.

R7.2 INDECOMPOSABILITY[4]

A positive matrix is a positive matrix, but the properties of a semi-positive matrix depends to a considerable extent on its *structure*.

To appreciate what is meant by structure in this context, consider a system of linear equations which is a representation of some "system" in the substantive sense. Certain of the variables may play no part in some aspects of the operation of the system and will be absent from the relevant equations. The relevance or irrelevance of certain variables to the operation of a particular part of the system is a structural property of the system.

The matrix of the equation system derived from the "real" system will contain a zero in the (i, j)th place if the jth variable is absent from the ith equation. We shall refer to the location of zeros in the matrix resulting from this cause as part of the structure of the matrix. In some contexts it may be appropriate to differentiate between a zero element in the matrix that represents part of the underlying structure and an element which simply happens to have the numerical value zero. In general, we shall assume that the zero pattern is essentially structural.

We wish, before proceeding further, to make another kind of distinction between systems which is of an even more fundamental kind. Consider two familiar types of equation system,

$$Ax = b,$$

$$Ax = \lambda x.$$

In the first type, an ordinary linear equation system, the indexing of the variables is entirely independent of the indexing of the equations. We can choose any variable to be the first and independently choose any equation to be the first equation, so that *the row index is independent of the column index.*

In the second system, that associated with characteristic roots and vectors, the right-hand side of the ith equation is λx_i. If we reorder the variables, then we must reorder the equations in the same way. In this case, *the row index is determined by the column index.*

[4] The topic of indecomposability is covered in all elementary discussions of input-output analysis. See, for example, Dorfman, Samuelson, and Solow, Allen [2], or Yamane.

The following structural definitions apply only to systems in which the row index is determined by the column index.

The square matrix A is said to be *decomposable* if we can permute the indices in such a way to give a matrix of the form

$$\begin{bmatrix} A_{11} & A_{12} \\ 0 & A_{22} \end{bmatrix},$$

where A_{11} and A_{22} are *square*, but not necessarily of the same order.

A_{11}, A_{12}, and A_{22} may be further decomposable. If A_{12} is a zero matrix and continued decomposition of A_{11}, A_{22} together with suitable index permutation leads finally to a matrix of the form

$$\begin{bmatrix} A_{11} & 0 & 0 & \cdots & 0 \\ 0 & A_{22} & 0 & \cdots & 0 \\ \cdot & \cdot & \cdot & \cdots & \cdot \\ 0 & 0 & 0 & \cdots & A_{kk} \end{bmatrix},$$

where the matrices $A_{11} \cdots A_{kk}$ are square and indecomposable, not necessarily of the same order, we refer to the matrix as *completely decomposable.*

If a matrix cannot be decomposed, even partially, it is *indecomposable* and this is the type of matrix in which we are chiefly interested. The terms *reducible, irreducible* are sometimes used in place of decomposable, indecomposable.

To understand what is implied by decomposability, consider the equation system $Ax = kx$, where A is decomposable. Let us suppose that, after suitable permutation if necessary, we obtain a matrix in the form

$$\begin{bmatrix} A_{11} & A_{12} \\ 0 & A_{22} \end{bmatrix},$$

where A_{11} is a $k \times k$ matrix and A_{22} is a $(n - k) \times (n - k)$ matrix.

If we partition the vector x into x_1, containing the first k components, and x_2 containing the remaining $n - k$, then we can write the system as:

$$\begin{bmatrix} A_{11} & A_{12} \\ 0 & A_{22} \end{bmatrix} \begin{bmatrix} x_1 \\ x_2 \end{bmatrix} = k \begin{bmatrix} x_1 \\ x_2 \end{bmatrix},$$

which becomes

$$\begin{aligned} A_{11}x_1 + A_{12}x_2 &= kx_1 \\ A_{22}x_2 &= kx_2. \end{aligned}$$

Thus there is a complete subsystem $A_{22}x_2 = kx_2$ which can be solved quite independently from the rest of the system. Given x_2 from this subsystem, the other subsystem can then be solved. Thus the variables x_1 depend on the variables x_2, but there is no dependence in the reverse direction. If A_{11} and A_{22} were indecomposable and A_{12} a zero matrix, then the system would be completely decomposable. In this case the variables x_1 would be completely independent of the variables x_2 and vice versa. On the other hand, if A were indecomposable, the variables x_1 would depend on the variables x_2 and vice versa.

Thus complete decomposability implies independent subgroups of variables, decomposability implies, at most, one-way dependence $x_2 \rightarrow x_1$, and indecomposability implies two-way dependence $x_2 \rightleftarrows x_1$.

A positive matrix is necessarily indecomposable, so the problem of decomposability arises only for semipositive matrices.

The term *indecomposable* will also be used in the book with reference to systems which have the fundamental property that no subgroup of the system is independent of the rest, but in which the system is not represented by a single square matrix as it is here. The more general idea of indecomposability appears in Section R8.8 of the next Review, and in Chapter 10, Sections 10.3 and 10.5.

R7.3 PROPERTIES OF SEMIPOSITIVE SQUARE MATRICES[5]

As stated earlier in the chapter, the leading properties of positive and semipositive matrices which are useful in the analysis of linear economic models are set out here without proof. Section R7.5 contains the proofs.

Many properties hold for semipositive indecomposable matrices (and thus necessarily for positive matrices), so positive matrices will only be mentioned when they have some property not possessed by indecomposable semipositive matrices. Some properties of importance are possessed by decomposable semipositive matrices. These are, except for one special property, weaker versions of properties possessed by all semipositive matrices.

[5] Lists of the properties of such matrices are given in various sources, including Dorfman, Samuelson, and Solow and Karlin [1]. The list given here includes all that are most relevant for economic analysis. One result, the *Hawkins-Simon condition*, is not given since its content is covered by other results. This condition played an important part in the early analysis of Leontief type models.

An elementary analysis, with proofs, is given in Yamane.

If A is a semipositive matrix it has, among its characteristic roots, one particular root λ^* *(which we shall refer to as its dominant root), with an associated characteristic vector* x^**, such that:*

(*a*) λ^* *is real and nonnegative;*

(*b*) *No other characteristic root has a modulus exceeding* λ^**;*

(*c*) x^* *is a nonnegative vector;*

(*d*) *For all* $\mu > \lambda^*$, $\mu I - A$ *is nonsingular and* $(\mu I - A)^{-1}$ *is a semipositive matrix.*

If, in addition, A is indecomposable, Conditions (*a*), (*c*), (*d*) *can be strengthened to:*

(*a*.1) λ^* *is positive and is not a repeated root;*

(*c*.2) x^* *is a strictly positive vector;*

(*d*.2) $(\mu I - A)^{-1}$ *is a strictly positive matrix.*

If A is a strictly positive matrix, condition (*b*) *can be strengthened to:*

(*b*.2) λ^* *exceeds the modulus of every other root.*

Properties (a.1), (b.2), (c.2), proved for positive matrices only, are the content of *Perron's theorem*. Properties (a.1), (b), (c.2), together with some other results not relevant to our interests here, and proved for semipositive indecomposable matrices, constitute the *Frobenius theorem*. Semipositive, indecomposable matrices are sometimes referred to as *Frobenius matrices* and λ^* as the *Frobenius* (or *Perron*) *root*.

The following additional results, derived from various sources, certainly hold for semipositive indecomposable matrices:

(*e*) *There is no nonnegative characteristic vector other than* x^**;*

(*f*) *If s is the smallest, and S the largest, of the row sums of A* (*that is,* $s = \min_i \sum_j a_{ij}$, $S = \max_i \sum_j a_{ij}$), *then* $s < \lambda^* < S$, *unless* $s = S$ *in which case* $s = \lambda^* = S$. *The same kind of relationship holds for column sums;*

(*g*) *The only nonnegative solution of either of the inequality systems* $Ax \leqq \lambda^* x$, $Ax \geqq \lambda^* x$ *is* x^*, *and the only nonnegative solution of either of the inequality systems* $Ax \leqq kx$, $Ax \geqq kx$ *where AT LEAST ONE EQUALITY HOLDS is* $k = \lambda^*$, $x = x^*$;

(*h*) *If A, B are both semipositive indecomposable and of the same order then* $A \gg B$ *implies* $\lambda_A^* > \lambda_B^*$;

(*i*) *For any choice of indices i, j, there is some power* $r \leqq n$ *of A such that the* (i, j)*th element of* A^r *is positive.*

Finally, if A is completely decomposable:

(*j*) x^* *is strictly positive, if and only if each of the diagonal submatrices has the same value for its dominant root, which is then the value of the dominant root of the whole matrix.*

R7.4 PROPERTIES OF DOMINANT DIAGONAL MATRICES[6]

In Section R7.1 we defined a matrix A as having a dominant diagonal if there exist positive numbers d_i such that

$$d_j\,|a_{jj}| > \sum_{i \neq j} d_i\,|a_{ij}| \qquad (\text{all } j).$$

This implies that the weighted modulus of the diagonal element is greater than the weighted sum of the absolute values of the other elements in the same *column*. We could have chosen to make the defining property

$$|a_{ii}|\,d_i > \sum_{j \neq 1} |a_{ij}|\,d_j \qquad (\text{all } i),$$

which would define *row dominance* as compared with *column dominance*. Obviously A^T has row dominance if A has column dominance.

Since the important properties of dominant diagonal matrices relate to singularity and the characteristic roots, which are the same for A, A^T, it does not matter whether we show row or column dominance. The column form is most appropriate in economics, since we shall have occasion to show the existence of diagonal dominance by using prices (typically a row vector) as weights corresponding to the d_i's.

Since $d_i > 0$, all i, $d_i\,|a_{ij}| = |d_i a_{ij}|$, so that if $B = DA$ where D is the diagonal matrix of the d_i's, diagonal dominance of A in our sense implies

$$|b_{jj}| > \sum_{i \neq j} |b_{ij}| \qquad (\text{all } j).$$

This unweighted form of the definition of dominance is the original definition used by Hadamard.[7] We shall refer to a matrix like B as a *Hadamard matrix* (the term is sometimes restricted to the case in which $b_{jj} > 0$, all j). Thus if A is a dominant diagonal matrix, DA is a Hadamard matrix for some diagonal matrix D with positive diagonal elements.

The fundamental property from which all other properties of dominant diagonal matrices are derived is:

(*a*) ***Generalized Hadamard Theorem***
If A has a dominant diagonal, it is nonsingular.

The dominant diagonal property we shall find especially useful is:

(*b*) *If A has a dominant negative diagonal, all its characteristic roots have negative real parts.*

[6] See McKenzie [3].

[7] For entry into the mathematical literature, commence with Brauer.

Another useful property is:

(*c*) *If A has column dominance, no characteristic root has modulus exceeding the largest column sum of absolute values* ($\max_j \sum_i |a_{ij}|$), *and if it has row dominance no characteristic root has modulus exceeding the largest row sum of absolute values.*

The above properties are consequences of the existence of a dominant diagonal property that are useful in economic applications. The remaining properties, some of which are trivial, are primarily of use in ascertaining whether a matrix might have a dominant diagonal.

(*d*) *A has a dominant diagonal only if the Hadamard inequality* $|a_{jj}| > \sum_{i \neq j} |a_{ij}|$ *holds for at least one j.*

(*e*) *If A has a dominant diagonal, every principal submatrix of A has a dominant diagonal.*

(*f*) *A has a dominant diagonal if there exist positive* d_i*'s such that*

$$d_j |a_{jj}| \geqq \sum_{i \neq j} d_i |a_{ij}|,$$

with strict inequality for at least one j.

Proofs of the above properties are given, along with proofs of the results for semipositive matrices, in the next section.

R7.5 PROOFS[8]

Hadamard's Theorem

We commence with dominant diagonal matrices and first prove:
If B is a Hadamard matrix, it is nonsingular.

Suppose, to the contrary, that B is singular. Then $yB = 0$ has a solution

[8] For dominant diagonal matrices and the extension of their properties to semipositive matrices, see McKenzie [3].

Karlin [1] gives a fairly extensive discussion of semipositive matrices. Bellman [3] discusses positive, but not semipositive matrices.

The most extensive treatment of semipositive matrices is in Gantmacher [1] or [2]. In the economics literature, see Debreu and Herstein and the articles by Wong and Woodbury in Morgenstern.

The proofs given here, which cover all the fundamental results given in Section R7.3 for semipositive matrices, but omit some of the minor results, are algebraic, along the general lines of Gantmacher. Fixed point theorems can be used for some proofs.

$y \neq 0$. Noting that $|y_i|\,|a_{ij}| \geqq y_i a_{ij}$, each of the equations $\sum_i y_i b_{ij} = 0$ implies the inequality

$$\sum |y_i|\,|b_{ij}| \geqq 0,$$

so that

$$|y_j|\,|b_{jj}| \leqq \sum_{i \neq j} |y_i|\,|b_{ij}| \qquad (\text{all } j).$$

Let k be the index for which $|y_k|$ is a maximum. Then $|y_i| \leqq |y_k|$, all i, so that

$$|y_k|\,|b_{kk}| \leqq \sum_{i \neq k} |y_k|\,|b_{ik}|.$$

Giving

$$|b_{kk}| \leqq \sum_{i \neq k} |b_{ik}| \qquad (\text{since } |y_k| \neq 0)$$

in contradiction of the Hadamard property. Thus $yB = 0$ has no nontrivial solution and B is therefore nonsingular.

If A has a dominant diagonal, then $B = DA$ is a Hadamard matrix for the diagonal matrix D with positive diagonal. But $\det B = (\det D)(\det A)$. Since $\det B$, $\det D \neq 0$, $\det A \neq 0$, so A is nonsingular, proving (a) of Section R7.4.

Now consider the matrix $A - \lambda I$. By definition, λ is a characteristic root of A, if and only if $A - \lambda I$ is singular. Thus we can use the Hadamard theorem to investigate the roots of A.

To prove (c) of Section R7.4, suppose to the contrary that $|\lambda| > \max_j \sum_i |a_{ij}|$. Then clearly $|\lambda| > |a_{ij}|$, all j and

$$\begin{aligned} |a_{ij} - \lambda| &\geqq |\lambda| - |a_{jj}| \qquad (\text{all } j) \\ &> \sum_i |a_{ij}| - |a_{jj}| \\ &> \sum_{i \neq j} |a_{ij}|, \end{aligned}$$

so that $A - \lambda I$ is a Hadamard matrix and nonsingular, and λ cannot be a characteristic root, proving the result.

Now suppose A has a dominant negative diagonal. Then DA is a Hadamard matrix, also with a negative diagonal. Further, the roots of DA are positive multiples of the roots of A. Now consider the Hadamard matrix $B = DA$ and suppose it has some root λ such that $\operatorname{Re}(\lambda) \geqq 0$. Then

$$\begin{aligned} |b_{jj} - \lambda| &= |\lambda - b_{jj}| \\ &\geqq |\operatorname{Re}(\lambda - b_{jj})| \\ &\geqq \operatorname{Re}(\lambda) + |b_{jj}| \qquad (\text{since } b_{jj} > 0) \\ &\geqq |b_{jj}|. \end{aligned}$$

But B is a Hadamard matrix, so $B - \lambda I$ is also a Hadamard matrix and nonsingular. Thus λ cannot be a root of B if $\text{Re}(\lambda) \geqq 0$, and the roots of B must have negative real parts. But then the roots of A must also have negative real parts, proving (b) of Section R7.4.

Proofs of (d), (e), and (f) of Section R7.4 are trivial. A principal submatrix of A retains the relevant diagonal elements but loses off-diagonal elements, hence (e). Putting $d = \min_i d_i$ in the definition of the dominant diagonal, then cancelling, gives (d). Result (f) follows by making a small variation in d_i for the i corresponding to the strict inequality.

Turning now to semipositive matrices, we shall first prove:

Fundamental Lemma for Semipositive Indecomposable Matrices
Let $Z(x)$ denote the number of zero components of the vector x. Then, if A is semipositive indecomposable and x is semipositive, $Z[(I + A)x] < Z(x)$.

Put $y = (I + A)x$. Since $I + A \geqslant 0$, it is clear that $Z(y) \leqq Z(x)$. We need only rule out the equality case. Suppose the equality did hold then, since we must have $y \geqq x$ (since $I + A \geqq I$), the zeros of y must occupy the same positions as the zeros of x. We shall suppose the zeros to occupy the last r places, so that we can partition the vectors, y, x

$$y = \begin{bmatrix} \hat{y} \\ 0 \end{bmatrix} \qquad x = \begin{bmatrix} \hat{x} \\ 0 \end{bmatrix},$$

where $\hat{x}, \hat{y} \gg 0$. Partition the matrix $I + A$ conformably with the partition of x, y. Then we have

$$\begin{bmatrix} \hat{y} \\ 0 \end{bmatrix} = \begin{bmatrix} I + A_{11} & A_{12} \\ A_{21} & I + A_{22} \end{bmatrix} \begin{bmatrix} \hat{x} \\ 0 \end{bmatrix}.$$

Multiplying out, this gives

$$\hat{y} = (I + A_{11})\hat{x}$$
$$0 = A_{21}\hat{x}.$$

Since $\hat{x} \gg 0$, this implies $A_{21} = 0$. But this contradicts the assumption of indecomposability, so that we cannot have $Z(y) = Z(x)$ and the lemma is proved.

We can now prove the following:

If A is semipositive indecomposable of order n, then $(I + A)^{n-1} \gg 0$.

If we take an arbitrary semipositive vector x and apply the fundamental lemma repeatedly, it is clear that we must have eliminated all zero components in $(I + A)^{n-1}x$. Since x is arbitrary, we must have $(I + A)^{n-1} \gg 0$.

By expanding $(I + A)^{n-1}$ binomially, it is obvious that a positive (i, j)th element must occur in at least one of the terms of the expansion for every i, j, proving Result (i) of Section R7.3.

We shall now define the following number associated with the semipositive indecomposable matrix A:

$$r(v^*) = \max_{v \in V} \min_i \frac{(Ax)_i}{x_i},$$

where $V = \{v \mid v \geqq 0, \sum v_i = 1\}$ and $(Ax)_i$ means the ith component of the vector Ax. v^* is the vector for which the maximum occurs.

We shall first derive some important properties of $r(v^*)$, then show that it is equal to λ^*, the special root that plays such an important role in the properties of Section R7.3.

Since $A, v \geqq 0$, $r(v^*) \geqq 0$. We shall show that it is not zero by taking the vector $u \in V$, all of whose components are equal to $1/n$.

Define

$$\begin{aligned} r(u) &= \min_i \frac{(Au)_i}{u_i} \\ &= \min_i \sum_j a_{ij} \\ &> 0 \qquad \text{since } A \text{ is semipositive.} \end{aligned}$$

Obviously $r(v^*) \geqq r(u)$, so that $r(v^*) > 0$.[9]

By definition, we have $r(v^*) = \min_i (Av^*)_i/v_i^*$. We shall now show that $r(v^*)$ is the largest number satisfying the inequality

$$Av^* \geqq rv^*.$$

For the kth component of the inequality we have

$$(Av^*)_k \geqq rv_k^*$$

$$\frac{(Av^*)_k}{v_k^*} \geqq r.$$

If $r = r(v^*)$, the equality holds for some k, so that no number greater than $r(v^*)$ can satisfy it.

The above property of $r(v^*)$ implies that

$$[Av^* - r(v^*)\, v^*] \geqq 0.$$

Suppose the equality does not hold, then $[Av^* - r(v^*)v^*]$ is a semipositive

[9] Note that we have shown in the process that $r(v^*) \geqq \min_i \sum_j a_{ij}$, a result we shall use later.

vector. Multiplying by the strictly positive matrix $(I + A)^{n-1}$ we have, necessarily

$$(I + A)^{n-1}[Av^* - r(v^*)\, v^*] \gg 0,$$

or,

$$Ay - r(v^*)\, y \gg 0,$$

where $y = (I + A)^{n-1}v^*$ [expansion of $(I + A)^{n-1}A$ will convince the reader that $(I + A)^{n-1}A = A(I + A)^{n-1}$].

But the inequality $r(v^*)y \ll Ay$ contradicts the maximum property of $r(v^*)$, so that we must have the equality in the relationship $Av^* \geqq r(v^*)$. Thus $r(v^*)$ must be a characteristic root of A, and v^* a characteristic vector.

Since $r(v^*)$ is a root of A, $[1 + r(v^*)]^{n-1}$ is a root of $(I + A)^{n-1}$ and v^* is a characteristic vector of both A, $(I + A)^{n-1}$. Using these relationships and the strict positivity of $(I + A)^{n-1}$ it follows that

$$[1 + r(v^*)]^{n-1}v^* = (I + A)^{n-1}v^* \gg 0.$$

Since $r(v^*) > 0$, this implies $v^* \gg 0$.

Now consider any other characteristic root, λ, of A, with associated characteristic vector v. Then

$$\lambda v = Av.$$

Taking moduli, we obtain

$$|\lambda|\, v^+ \leqq Av^+,$$

where v^+ is the vector of the moduli of the components of v. The inequality is derived from the relationship $|a|\,|b| \leqq |ab|$, since the left-hand side is in the form $|a|\,|b|$ but the right-hand side is in the form $|ab|$ since A is non-negative and real.

Since we have already found $r(v^*)$ to be the largest number satisfying an inequality of the above kind, it follows that

$$|\lambda| \leqq r(v^*)$$

for every root of A. Thus we can now identify $r(v^*)$ as the *dominant root* of A. It is the root λ^* of Section R7.3, and we shall identify it as such from now on.

At this stage we have proved that λ^* is the dominant root, that it is positive, and that its associated characteristic vector is strictly positive, all on the assumption of indecomposability. These are Results (a), (b), and (c) of Section R7.3, except that we have not shown that λ^* cannot be a repeated root. Since we did not discuss repeated roots in Review R5, this proof will be omitted.

We shall now prove that no other root has an associated nonnegative characteristic vector. Suppose the root λ did have a characteristic vector v that was nonnegative. Then we have the two equations

$$Av^* = \lambda^* v^* \qquad Av = \lambda v.$$

Transposing the second and multiplying by x^* we obtain

$$v'A'v^* = \lambda v'v^*.$$

Multiplying the first by v', we obtain

$$v'Av^* = \lambda^* v'v^*.$$

Since the quadratic expressions $v'Av^*$, $v'A'v^*$ are equal, we must have $(\lambda^* - \lambda)v'v^* = 0$. Since $\lambda^* \neq \lambda$ we must have $v'v^* = 0$. Since $v^* \gg 0$, this is not possible if v is nonnegative, proving the result, which is (e) of Section R7.3.

Now consider the matrix $(\mu I - A)^{-1} = \mu^{-1}(I - \mu^{-1}A)^{-1}$, for $\mu > \lambda^*$. Obviously the dominant root of $\mu^{-1}A$ is $\lambda^*/\mu < 1$, so that the matrix can be expanded as a convergent infinite series:

$$(I - \mu^{-1}A)^{-1} = I + \mu^{-1}A + \mu^{-2}A^2 + \cdots.$$

But we have already shown that, for any i, j, the (i, j)th element of A^r is positive for some $r < n$. Since $\mu > 0$, there is a positive entry in every place in the matrix $(I - \mu^{-1}A)^{-1}$, so that $(\mu I - A)^{-1} \gg 0$ for all $\mu > \lambda^*$. This is Result (d) of Section R7.3.

We have already shown, in the process of our analysis, that

$$\lambda^* \geqq \min_i \sum_j a_{ij}.$$

Now from the dominant diagonal analysis we have seen that any number k such that

$$|a_{jj} - k| \geqq \sum_{i \neq j} |a_{ij}|$$

with strict inequality for some j, cannot be a characteristic root of any matrix A. Thus k cannot be the dominant root for the semipositive indecomposable matrix A if

$$k - a_{jj} \geqq \sum_{i \neq j} a_{ij} \qquad \text{(some strict inequality)}$$

$$k \geqq \sum_i a_{ij} \qquad \text{(some strict inequality)}.$$

Thus we must have

$$\lambda^* < \max_i \sum_j a_{ij} \qquad \text{(unless all sums are equal)}$$

with an equivalent relationship for the row sums. This is Result (f) of Section R7.3.

The other results (g) and (h) are simple corollaries of the main results already proved, or of other results proved as lemmas.

EXERCISES

1. Which of the two matrices below is indecomposable?

$$\begin{bmatrix} 2 & 3 & 0 \\ 2 & 1 & 1 \\ 1 & 0 & 4 \end{bmatrix} \qquad \begin{bmatrix} 1 & 0 & 1 \\ 3 & 2 & 2 \\ 1 & 0 & 4 \end{bmatrix}$$

Show, by direct calculation, that $(I + A)^2 \gg 0$ for the indecomposable matrix, but not for the other.

2. Calculate the characteristic roots and vectors of the following matrices, and show that they conform to the results of Section R7.3. (Note that one matrix is decomposable):

$$\begin{bmatrix} 1 & 2 \\ 4 & 3 \end{bmatrix} \qquad \begin{bmatrix} 0 & 4 \\ 2 & 1 \end{bmatrix} \qquad \begin{bmatrix} 1 & 2 \\ 0 & 4 \end{bmatrix}$$

3. Choose a value of b to give a dominant negative diagonal in the following matrix, then show, by direct calculation, that its roots have negative real parts:

$$\begin{bmatrix} b & 2 \\ -4 & -1 \end{bmatrix}$$

Find a semipositive diagonal matrix D such that DA is a Hadamard matrix.

REVIEW

R8

Continuous Functions

The material contained in Sections R8.1 through R8.6 is basic to all nonlinear analysis. General familiarity with the content of Reviews R1 through R6 is assumed, but a first reading can be given with only a broad background in these matters. Sections R8.7 and R8.8 are of a more specialized nature. Section R8.8 assumes familiarity with Review R7.

R8.1 INTRODUCTION[1]

It is assumed that the reader is familiar with the ordinary calculus properties of the function of a single variable. We shall be concerned here with the properties of functions of several variables, including several calculus techniques. Since calculus of functions of several variables employs the same general type of analysis as that for functions of a single variable, many results will be given here without proof.

We are concerned here with functions of the general form $f(x)$, where x is a vector rather than a scalar. It is more appropriate to the purposes of this

[1] The material of Sections R8.1 through R8.4 is standard and widely available in advanced calculus texts, not necessarily in the same form as here. Among suitable mathematical texts are Courant and Buck. The same material is also available, in a simplified form, in Allen [1].

The contents of Sections R8.5 through R8.8 are more specialized. References are given at the relevant point in the text.

book to consider x as a vector of n components rather than to consider f as a function of n different variables, although the two approaches are equivalent. We can regard $f(x)$ as the rule for a mapping from R^n into R. If emphasis is needed, the ordinary function $f(x)$ can be described as a *scalar function of a vector*.

If we have m such functions $f^i(x)$, all functions of the same vector x, it is often convenient to regard the numbers $y_i = f^i(x)$ as components of a vector and write the single relationship

$$y = F(x),$$

which is to be interpretated as simply a shorthand statement for the m relationships

$$y_i = f^i(x) \qquad i = 1, \ldots, m$$

in the same way the matrix equation $y = Ax$ represents the m linear functional relationships $y_i = A_i x$.

We shall speak of $F(x)$ as a *vector-valued function*, or simply a *vector function*, and refer to $f^i(x)$ as the *ith component of* $F(x)$.

If y is an m vector and x an n vector, $F(x)$ is a rule for a *point-to-point mapping from* R^n *to* R^m. We may have $n = 1$, so that $F(x)$ is a *vector function of a scalar*, a type of function we shall be concerned with in differential and difference equations (Review R10).

In this review, we are primarily interested in functions of vectors. These can be taken to be *scalar functions* unless they are specified to be vector functions.

Examples

(a) $y = ax_1^2 - be^{x_2}$ is a scalar valued function of the vector (x_1, x_2).

(b) $y_1 = ax^2$, $y_2 = bx^3$ are two functions of x, or alternatively, a vector function of a scalar.

(c) $y_1 = a_1x_1 - b_1x_2^2$, $y_2 = a_2x_1^3 - b_2x_2 + cx_3^2$ constitute a vector function of a vector, giving a mapping from R^3 into R^2.

A scalar function $f(x)$ of a vector x is *continuous* at x_0 if, given any number $\varepsilon > 0$, however small, it is possible to find $\delta(x_0, \varepsilon) > 0$ such that, for all x satisfying $|x - x_0| < \delta$, the inequality $|f(x) - f(x_0)| < \varepsilon$ is true. The norm $|x - x_0|$ is usually taken to be the Euclidean norm, but other norms are possible.

A function is continuous over a region or set if it is continuous at every point x_0 in the set. If is *uniformly continuous* if the number δ in the continuity definition can be made independent of x_0.

We shall consider a vector function to be continuous if all its components are continuous.

If $f(x)$ is a continuous function, this fact is often stated by saying "$f \in C$" or "f is of class C". We shall take continuous to imply uniformly continuous.

R8.2 DERIVATIVES AND DIFFERENTIALS

In the simple calculus of functions of a single variable, the derivative of $f(x)$ is defined as

$$df(x)/dx = \lim_{h\to 0} \frac{f(x+h) - f(x)}{h},$$

if the limit exists. This definition cannot directly be applied to functions of a vector, since h would then be a vector and division by h would be an undefined operation.

We can, however, regard $f(x)$ as a function of only one component of x, say x_j, treating the other components as constants. We then define

$$\lim_{h\to 0} \frac{f(x_1, \ldots, x_j + h, \ldots) - f(x_1, \ldots, x_i, \ldots)}{h},$$

(if it exists) as *the partial derivative of $f(x)$ with respect to x_j*. It is calculated by treating every component of x other than x_j as constant and taking the ordinary derivative with respect to x_j. Various notations are used for the partial derivative, of which we shall use the following at various times:

$$\partial f(x)/\partial x_j,\ D_j f(x), f_j(x).$$

Where no ambiguity is likely to arise, the simple subscript notation has an attractive simplicity, often with the argument omitted to give the particularly simple form f_j. To avoid any confusion with this notation, members of a set of related functions should always be indexed with a *superscript*, as $f^i(x)$.

A function of an n-vector has n such partial derivatives, one for each component of x. The n partial derivatives may themselves be regarded as the components of a vector. This vector is called the *gradient vector* of the function f and is written

$$\text{grad} f \quad \text{or} \quad \nabla f \quad \text{(the latter read "del } f\text{")}.$$

Usually x will be written as a column vector, and ∇f will then be written as a *row* vector, for reasons that will become clear.

If $f(x)$ is continuous and has continuous first order partial derivatives f_j it is said to be $\in C^1$.

Now any partial derivative f_j is a function of the vector x and, if it is suitably continuous, we can take partial derivatives of f_j with respect to x_j or *any* component of x. Thus we obtain *second order partial derivatives.*

$$\frac{\partial}{\partial x_i}\left(\frac{\partial f(x)}{\partial x_j}\right) = \frac{\partial^2 f(x)}{\partial x_i \partial x_j}; \qquad D_i[D_j f(x)] = D_{ij} f(x); \qquad f_{ij}(x).^2$$

There are n^2 partial derivatives of the second order. Since each partial derivative has two indexes, it can be regarded as the element of a matrix:

$$H(x) = [f_{ij}(x)].$$

The matrix $H(x)$ is called the *Hessian matrix* and must be evaluated at each point x.

Example. If $f(x) = ax_1^3 + bx_1x_2 + cx_2^2$, then

$$\nabla f = [3ax_1^2 + bx_2, bx_1 + 2cx_2]$$

$$H = \begin{bmatrix} 6ax_1 & b \\ b & 2c \end{bmatrix}.$$

If $f(x)$ is continuous and has continuous first and second order partial derivatives it is said to be $\in C^2$. It can be assumed that all continuous functions in economics that are otherwise unspecified will have this degree of continuity.

Since the second order partial derivatives are also functions of x, we can define third and higher order derivatives, for example $D_i[D_{jk}f(x)] = D_{ijk}f(x)$.

We can state without proof the following well-known fundamental theorem on partial derivatives:

If $f(x) \in C^2, f_{ji} = f_{ij}$.[3]

As a consequence of this result, *the Hessian is symmetric*, a property that is crucial to the use of the Hessian matrix in analysis.

In addition to the partial derivative, we can also define another type of derivative that is analogous to the simple derivative of a function of a single variable.

Let v be an arbitrary vector which we take to be normalized ($v'v = 1$). We then define:

$$D_v f(x) = \lim_{h \to 0} \frac{f(x + hv) - f(x)}{h},$$

[2] In calculating f_{ij} we treat x_i as constant to obtain f_j, then treat x_j as constant in calculating $D_i(f_j)$.

[3] There is an analogous result for higher order derivatives.

where h is a scalar, as *the derivative of $f(x)$ in the direction v*, assuming the limit exists. $D_v f(x)$ is a *directional derivative.*

Whereas there are n partial derivatives, there is a single directional derivative, given v. On the other hand, there is a directional derivative defined for every normalized vector v.[4]

The directional and partial derivatives are related by the following fundamental relationship (which implies that the partial derivative is the derivative in the direction of the coordinate vector giving $v_i = 1$, $v_j = 0$, $j \neq i$).

If $f(x) \in C^1$, $D_v f(x) = \nabla f \cdot v \;(= \sum f_j v_j)$.

As an illustration of typical analysis in the area under discussion, we shall give an outline proof of the above result.

Consider the simple case $n = 2$, so that we have $f(x_1, x_2)$. Define the expression,

$$g(h) = f(x + hv) = f(x_1 + hv_1, x_2 + hv_2). \tag{R8.2.1}$$

$g(h)$ is a function of a single variable and will be continuous if $f(x)$ is continuous. We use the ordinary mean value theorem:

$$g(h) - g(0) = hg'(\theta h) \qquad 0 \leqq \theta \leqq 1. \tag{R8.2.2}$$

From the definition of $g(h)$ we have

$$\begin{aligned} g(h) - g(0) &= f(x_1 + hv_1, x_2 + hv_2) - f(x_1, x_2) \\ &= f(x_1 + hv_1, x_2 + hv_2) - f(x_1, x_2 + hv_2) \\ &\qquad + f(x_1, x_2 + hv_2) - f(x_1, x_2). \end{aligned}$$

We now consider the first two terms on the right-hand side as functions of x_1 only and use the mean value theorem, and also use the mean value theorem on the last two terms consider as functions of x_2 only, to obtain

$$\begin{aligned} g(h) - g(0) &= hv_1 D_1 f(x_1 + \mu_1 hv_1, x_2 + hv_2) \\ &\quad + hv_2 D_2 f(x_1, x_2 + \mu_2 hv_2) \qquad 0 \leqq \mu_1, \mu_2 \leqq 1. \end{aligned}$$

Substituting in (R8.2.2) and canceling h ($\neq 0$), we obtain:

$$g'(\theta h) = v_1 D_1 f(x_1 + \mu_1 hv_1, x_2 + hv_2) + v_2 D_2 f(x_1, x_2 + \mu_2 hv_2).$$

If now $h \to 0$, the left-hand side approaches $g'(0)$, the right-hand side approaches $v_1 D_1 f(x_1, x_2) + v_2 D_2 f(x_1, x_2)$. Assuming the limits exist, we have, ultimately:

$$g'(0) = v_1 f_1 + v_2 f_2. \tag{R8.2.3}$$

[4] If the directional derivative is defined for all v, $f(x)$ is said to be *differentiable.* This is a weaker property than being $\in C^1$, but not of interest in most economic applications.

From the definition of $g(h)$ and the definition of the relevant derivatives, it is obvious that $g'(0) = D_v f(x)$. Thus we have proved the result for a vector of two components. The general proof is a simple extension.

Consider the directional derivative at a point x. If x is fixed, then we can consider $D_v f$ to be a function of v. Now we define

$$\phi(h, v) = hD_v f(x)$$

as the *differential* of $f(x)$ at the point x. The differential is a *function*, not a number.

From the previous result we have

$$\phi(h, v) = h\nabla f \cdot v = \sum f_j(hv_j).$$

The ordinary mean value theorem gives

$$f(x + hv) - f(x) = hD_v f(x + \theta hv) \qquad 0 \leqq \theta \leqq 1,$$

so that *the differential can be regarded as an approximation to the change in $f(x)$ for a small movement h in the direction v.*[5]

The vector hv is a vector with Euclidean norm h, and can therefore be regarded as a "small" vector. We usually write the vector hv in the form dx, meaning the vector with components $dx_j = hv_j$. Because of the relationship given above, we also usually write $\phi(dx)$ as df. Thus we have the differential in its common form

$$df = \nabla f \cdot dx = \sum f_j \, dx_j.$$

The importance of the differential is that it is a *linear* function of dx with the following properties:

(1) df is an approximation for $f(x + dx) - f(x)$;
(2) $\lim_{h \to 0} (1/h)\, df = D_v f$ where v is a normalized vector

which is proportional to dx.

It is sometimes convenient to refer to the direction dx, meaning the relevant direction vector v, and to use the notation $|dx| \to 0$ to mean $h \to 0$.

In a context in which manipulation of differentials occurs, followed by a limiting process in order to obtain a derivative, it may be useful to employ the notations Δf, Δx for the variables of the differential in order to prevent confusion with the use of df in the expression for the derivative.

[5] From ordinary calculus, the difference between $D_v f(x + \theta hv)$ and $D_v f(x)$ is of order h^2.

R8.3 SOME MAPPING RELATIONSHIPS

If D denotes the derivative with respect to a single variable, we have, from elementary calculus

$$D(cy) = cDy,$$
$$D(y + z) = Dy + Dz,$$

so that the derivative is a linear operator, and the mapping $y \rightarrow Dy$ is a linear mapping. Since partial derivatives have the properties of ordinary derivatives, this linearity holds for them too. It holds also for directional derivatives, since these are linear combinations of partial derivatives. The differential is defined as a linear function, so that derivatives and differentials are associated with linear mappings.

Various well-known results can be derived from this general property.

(a) Compound Functions

Consider the function $f(x)$, where x is a vector valued function of z denoted by $x_j = \phi^j(z)$. x is of order n, z of order m. The differentials of f, ϕ^j are

$$df = \sum f_j \, dx_j$$
$$dx_j = \sum \phi_r^j \, dz_r \qquad j = 1, \ldots, n.$$

Since the relationships are linear, we can immediately write down the differential of f, now regarded as a function of dz,

$$df = \sum_{j=1}^{n} \sum_{j=1}^{m} f_j \phi_r^j \, dz_r.$$

If we denote the matrix $[\phi_r^j]$ by M, we can write this in matrix-vector form,

$$df = \nabla f \cdot M \cdot dz.$$

Choose the vector $dz = he^r$, where e^r is the rth coordinate vector in R^m ($e_r^r = 1$; $e_s^r = 0$, $s \neq r$). Then we have

$$\frac{df}{h} = \sum_j f_j \phi_r^j.$$

Taking the limit as $h \rightarrow 0$, we obtain the well-known rule for taking the partial derivatives of a compound function,

$$\frac{\partial f}{\partial z_r} = \sum_j \frac{\partial f}{\partial x_j} \frac{\partial x_j}{\partial z_r}.$$

If z is a scalar, we have the *chain rule*,

$$\frac{df}{dz} = \sum_j \frac{\partial f}{\partial x_j}\frac{dx_j}{dz}.$$

(b) Gradients and Contours

The values of x for which $f(x) = c$ is a *contour* or *level surface* of the function $f(x)$. Since df is an approximation for $f(x + dx) - f(x)$, the direction dx for which $df = 0$ is a direction *along the contour*. If $df = 0$, we have

$$df = \nabla f \cdot dx = 0,$$

so that *the gradient vector is orthogonal to every direction along the contour.* In geometric terms, the gradient vector is the *normal* to the contour.

A scalar function $F(x)$ of the vector x is said to be a *positive monotonic transformation* of some other function $f(x)$ of the same vector if $F(x)$ can be expressed as $F(x) = \phi[f(x)]$, where ϕ is the scalar function of the scalar $f(x)$ such that $\phi' = d\phi/df(x) > 0$.

It follows that $F_i(x) = \phi' f_i(x)$ and hence that $\nabla F = \phi' \nabla f$. Thus $\nabla F \cdot dx = 0$ if and only if $\nabla f \cdot dx = 0$, so that any contour of $f(x)$ is also a contour of $F(x)$ and vice versa, although the labeling of the contours may be different.

(c) The Mapping from R^n into R^n

Consider the vector valued function $y = F(x)$, where y, x are both vectors in R^n. Each component of $F(x)$ has a differential

$$dy_i = \sum_j \frac{\partial y_i}{\partial x_j} dx_j.$$

The matrix $M = [\partial y_i / \partial x_j]$ (whose rows are the gradient vectors ∇F^i of the components of F) is the *Jacobian matrix* of the mapping, or sometimes the *differential of the mapping* (written dF). The determinant $J = \det M$ is called simply the *Jacobian*.

Thus we can write the *vector differential*

$$dy = M\,dx.$$

If $J \neq 0$, we can solve for dx in terms of dy,

$$dx = M^{-1}\,dy.$$

The mapping $dy \rightarrow dx$ is obviously unique under these circumstances. This implies that every point in the neighborhood of y maps into a unique point in the neighborhood of x. It is intuitive, and can be proved rigorously, that this implies the existence of a unique inverse mapping $y \to x$ in the neighborhood of all points for which $J \neq 0$.

Thus we have the well-known result that a non-vanishing Jacobian is necessary and sufficient for the existence of a unique inverse to a continuous point-to-point mapping from R^n into R^n.

(d) Implicit Function Theorem

Consider n implicit functions $F^i = 0, i = 1, \ldots, n$, each a function of the same $n + m$ variables. We divide the variables arbitrarily into two vectors, one of n components (x), the other of m components (y). Thus the functions will be considered in the form $F^i(x, y)$.

We shall show that, if certain conditions are satisfied, there exist m functions

$$y_i = \phi^i(x)$$

that is, *that any m of the variables can be expressed as functions of the remaining n*. This constitutes the main result of the *implicit function theorem* and gives the warrant for the elimination of variables between equations other than linear equations.

We define the following $n + m$ variables z:

$$\begin{aligned} z_i &= x_i \qquad i = 1, \ldots, n \\ &= F^{i-n}(x, y) \qquad i = n + 1, \ldots, n + m. \end{aligned}$$[6]

Thus $x, y \to z$ is a mapping from R^{n+m} into R^{n+m}. If we denote the $m \times n$ matrix $[\partial F^i/\partial x_j]$ by M_1 and the $m \times m$ matrix $[\partial F^i/\partial y_j]$ by M_2, the Jacobian matrix of the mapping is easily seen to have the form:

$$M = \begin{bmatrix} I & 0 \\ M_1 & M_2 \end{bmatrix},$$

(where I is of order $n \times n$, $[0]$ is of order $n \times m$). We have $\det M = \det M_2$, so the Jacobian does not vanish if M_2 is nonsingular, whatever the rank of M_1. Assuming $\det M_2 \neq 0$, M is invertible and so the inverse mapping $z \to x, y$ exists and is unique.

Denote the components of the inverse mapping by

$$\begin{aligned} x_i &= f^i(z) \qquad i = 1, \ldots, n, \\ y_k &= f^{k+n}(z) \qquad k = 1, \ldots, m. \end{aligned}$$

Since we have $z_i = x_i$ (by construction), $f^i(z) = z_1$, $i = 1, \ldots, n$. Now $z_j = F^j(x, y)$ for $j = n + 1, \ldots, n + m$. But, for x, y which satisfy the

[6] The first n components of z equal the components of x, the remainder equal the components of the vector valued function F.

implicit function relationships, these z_j are zero. We already have $z_i = x_i$ for $i \leqq n$, so that the required functions

$$y_i = \phi^i(x)$$

are identified as the functions

$$y_i = f^{i+n}(z_1, \ldots, z_n, 0, \ldots, 0).$$

Apart from suitable continuity (and the functions ϕ^i will possess similar continuity) the main requirements, both of which were used in the proof, are:

(1) the existence of solutions x, y to the n implicit functional relationships over the region under consideration.
(2) that $\det M_2 = \det [\partial F^i/\partial y_j] \neq 0$.

The theorem is no guarantee of the possibility of *explicit* solution, but the partial derivatives of ϕ^i may be found by taking the derivatives of the implicit functions in the following way:

$$\frac{\partial}{\partial x_j} F^i(x) = F^i_j + \sum \frac{\partial F^i}{\partial y_k} \cdot \frac{\partial \phi^k}{\partial x_j} = 0 \qquad i = 1, \ldots, n.$$

If we denote the vector $[F^i_j]$ by D_jF and the vector $[\partial\phi^k/\partial x_j]$ by $D_j\phi$, then the above relationships can be put in the matrix-vector form,

$$D_jF + M_2 \cdot D_j\phi = 0,$$

giving the required vector of partial derivatives

$$D_j\phi = -M_2^{-1} \cdot D_jF.$$

(e) Comparative Statics

A common problem in economics is to be given m relationships among n $(> m)$ variables. $n - m$ of the variables are given or exogenous, and a solution is found for the remaining m endogenous variables. This solution is usually identified as an equilibrium.

We are interested in how the solution (or equilibrium) values of the endogenous variables change with small variations in the exogenous variables (often called parameters). This problem (*comparative statics*) is suitable for the use of the differential. If we suppose the relationships are expressed implicitly by $F^i(x) = 0$, $i = 1, \ldots, m$, we have

$$dF^i = \sum F^i_j \, dx_j = 0 \qquad i = 1, \ldots, m.$$

Denoting the matrix $[F^i_j]$ by M, the variations dx in the variables (the notation does not distinguish endogenous from exogenous variables) are related to each other by

$$M\,dx = 0.$$

This system can be analyzed using linear theory.[7]

R8.4 MAXIMA AND MINIMA[8]

Taylor's theorem, for a function of a vector, can be expressed in several forms. We shall find the following form suitable for our purposes:

Taylor's Theorem
If $f(x) \in C^2$, then

$$f(x + v) = f(x) + \nabla f(x) \cdot v + \frac{1}{2!} v'Hv,$$

where the Hessian H is evaluated at some point $x + \theta v$, $0 \leqq \theta \leqq 1$.[9]

The proof is not given, being widely available. It is easily derived from Taylor's theorem for a function of a single variable.

We can now give the fundamental theorem for the existence of a maximum or minimum for $f(x)$.

If $f(x) \in C^2$, it has a maximum (minimum) over a convex set S at the interior point x^* if and only if $\nabla f(x^*) = 0$ and H is negative (positive) semidefinite everywhere in S.

If $f(x)$ has a maximum at x^*, then $f(x) \leqq f(x^*)$ for all $x \in S$. Since S is convex we can express any $x \in S$ in the form $x = x^* + v$. From Taylor's theorem,

$$f(x^* + v) \leqq f(x^*) \qquad \text{implies} \qquad \nabla f(x^*) + \frac{1}{2!} v'Hv \leqq 0.$$

[7] Space precluded a full discussion of comparative static analysis in this book. The traditional demand theory analysis (Chapter 8, Section 8.2) is representative of neoclassical methods. For qualitative comparative static methods, see Lancaster [1] and [2].

[8] A knowledge of the properties of quadratic forms (Review R6, Sections R6.1 through R6.2) is desirable for this and the next section.

[9] Formally, this is the Lagrange form of Taylor's theorem with $n = 2$.

Since this is true for all v such that $x^* + v \in S$, it must be true for hv, where $h > 0$ is a small number. For hv the inequality becomes

$$h\nabla f(x^*) \cdot v + \frac{1}{2!} h^2 v' Hv \leqq 0,$$

or

$$f(x^*) \cdot v + \frac{1}{2!} hv' Hv \leqq 0.$$

If $h \to 0$, this implies

$$\nabla f(x^*) \cdot v \leqq 0.$$

But, since x^* is an *interior* point of S (this is crucial), then $x - hv \in S$, if $x + hv \in S$, for sufficiently small h. Using the same argument as above, this implies

$$-\nabla f(x^*) \cdot v \leqq 0.$$

The two inequalities together imply

$$\nabla f(x^*) \cdot v = 0.$$

Such a point, at which all the partial derivatives vanish, is called a *critical point* of f.

If $\nabla f(x^*) = 0$, the inequality

$$f(x^* + v) \leqq f(x^*)$$

then implies

$$v' Hv \leqq 0.$$

Since this must be true for all v (subject to some bound on the norm of v, which does not affect the result), H must be negative semidefinite, and it must be so everywhere in S since it is evaluated at $x + \theta v$.

The corresponding result for the minimum follows in a similar manner from the inequality

$$f(x^* + v) \geqq f(x^*).$$

If we define S to be the open neighborhood of x^*, then we need only consider H evaluated at x^*. This is the most usual from in which we meet the above conditions, giving a *local* maximum or minimum.

The assumption that x^* is interior to S is essential to the above analysis. The analysis of maxima and minima at boundary points of a set is a problem in constrained optima, the subject matter of Chapters 2 through 5.

By using the strong inequality $f(x^* + v) < f(x^*)$, we conclude that:

The maximum (minimum) is unique if H is negative (positive) definite.

R8.5 CONVEX AND CONCAVE FUNCTIONS[10]

A function $f(x)$ is said to be *convex* over the convex set $S \subset R^n$ if, for any two points $x^1, x^2 \in S$,

$$f[\lambda x^1 + (1 - \lambda) x^2] \leqq \lambda f(x^1) + (1 - \lambda) f(x^2) \qquad 0 \leqq \lambda \leqq 1.$$

It is *strictly convex* if the strict inequality holds for all λ such that $0 < \lambda < 1$ and $x^1 \neq x^2$.

A function $f(x)$ is *concave* or *strictly concave* if $[-f(x)]$ is convex or strictly convex.

A linear function is *both convex and concave*, but not *strictly* convex or concave. A function is *locally* convex or concave at x^* if the set S is the neighborhood of x^*.[11]

Consider the sum $f(x) = \sum f^j(x)$ of a number of convex functions, defined over the same convex set S. We have

$$\begin{aligned} f[\lambda x^1 + (1 - \lambda)x^2] &= \sum f^j[\lambda x^1 + (1 - \lambda)x^2], \\ &\leqq \sum [\lambda f^j(x^1) + (1 - \lambda) f^j(x^2)], \\ &\leqq \lambda f(x^1) + (1 - \lambda) f(x^2), \end{aligned}$$

so that *the sum of convex functions is convex* (and the sum of concave functions is concave). Since $cf(x)$ is obviously convex if $f(x)$ is convex and $c > 0$, *any positive linear combination of convex (concave) functions is convex (concave).*

An important property of convex and concave functions is the following:

If $f(x)$ is convex the set $V = \{x \mid f(x) \leqq f(x^*)\}$ is convex. If $f(x)$ is concave the set $V = \{x \mid f(x) \geqq f(x^*)\}$ is convex, (where x^* is any point $\in S$).

We shall prove the result for convex $f(x)$. Consider any x^1, x^2 in the set V. Since $f(x^1), f(x^2) \leqq f(x^*)$ and $f(x)$ is convex

$$\begin{aligned} f[\lambda x^1 + (1 - \lambda)x^2] &\leqq \lambda f(x^1) + (1 - \lambda) f(x^2) \\ &\leqq \lambda f(x^*) + (1 - \lambda) f(x^*) \\ &\leqq f(x^*), \end{aligned}$$

so that the point $\lambda x^1 + (1 - \lambda) x^2 \in V$, proving the result.

[10] Convex and concave functions are of special interest in optimizing and game theory and are not usually discussed in calculus texts. A useful reference is Hadley [3] (Sections 3.10 through 3.12). Karlin [1] contains a list of most properties of these functions (Appendix B), but without proofs, and an advanced discussion of extensions of the properties in Chapter 7.

[11] It is essential in the definition that S be a convex set, since we require that $\lambda x^1 + (1 - \lambda)x^2$ be in S if x^1, x^2 are. S may, of course, be R^n in which case the function is *globally* convex or concave.

It is important, especially in economic models, to note that the above property is *necessary but not sufficient* for the function $f(x)$ to be convex or concave, as the case may be.

We can state this property in another way:

The contour $f(x) = c$ is the upper boundary of a convex set if $f(x)$ is convex and the lower boundary if $f(x)$ is concave.

Typical functions in economics (such as the neoclassical production and utility functions) are such that $f(x) = c$ is the *lower* boundary of a convex set. We can refer to such functions as *concave-contoured.* They are not necessarily concave functions. A neoclassical production function with increasing returns to scale is concave-contoured, but not a concave function. Convexity or concavity of $f(x)$ depends both on the contours and on the way $f(x)$ changes from contour to contour. For this reason, the utility function is concave-contoured, but we cannot know whether it is a concave function or not without a cardinal utility index. For a further discussion of this point, see Chapter 8, Section 8.5.

A stronger convex set property is the following:

If x is defined on the convex set X in R^n, and if y denotes the vector in R^{n+1} given by $[x:\zeta]$ where ζ is a scalar, then $f(x)$ is a convex function, if and only if the set $\bar{S} = \{y \mid x \in X, \zeta \geqq f(x)\}$ is convex, and a concave function if and only if the set $\underline{S} = \{y \mid x \in X, \zeta \leqq f(x)\}$ is convex.

To prove this, assume $f(x)$ is convex and consider the vectors $y^1 = [x^1:\zeta^1]$, $y^2 = [x^2;\zeta^2]$ where $\zeta^1 \geqq f(x^1)$, $\zeta^2 \geqq f(x^2)$.

Then

$$\begin{aligned} y^* = \lambda y^1 + (1-\lambda)\, y^2 &= [\lambda x^1 + (1-\lambda)\, x^2 : \lambda\zeta^1 + (1-\lambda)\zeta^2] \\ &= [x^*:\zeta]. \end{aligned}$$

Now $x^* \in X$ (since X is convex) and

$$\begin{aligned} f(x^*) &= f[\lambda x^1 + (1-\lambda)x^2] \\ &\leqq \lambda f(x^1) + (1-\lambda) f(x^2) \qquad \text{[convexity of } f(x)]. \end{aligned}$$

But

$$\begin{aligned} \zeta^1 \geqq f(x^1),\ \zeta^2 \geqq f(x^2),\ \text{so that} \\ \zeta^* &= \lambda f(x^1) + (1-\lambda)\, f(x^2) \\ &\geqq f(x^*), \end{aligned}$$

and $y^* = (x^*:\zeta^*) \in \bar{S}$.

To prove sufficiency, suppose $y^* \in \bar{S}$, so that $\zeta^* \geqq f(x^*)$. If the hypothesis is true for all y, it is true for $\zeta^1 = f(x^1)$, $\zeta^2 = f(x^2)$. In this case,

$$f(x^*) \leqq \zeta^* = \lambda f(x^1) + (1 - \lambda) f(x^2),$$

completing the proof.

Another result of importance is the following:

$f(x)$ is convex, if and only if, for all $x, x^* \in X$

$$f(x) - f(x^*) \geqq \nabla f^* \cdot (x - x^*)$$

with the inequality reversed for concavity and strict for strict convexity or concavity.

To prove this, write $\lambda x + (1 - \lambda)x^*$ in the form $x^* + \lambda(x - x^*)$. Then if $f(x)$ is convex we have

$$f[x^* + \lambda(x - x^*)] \leqq f(x^*) + \lambda[f(x) - f(x^*)].$$

Using Taylor's theorem, the left-hand side can be expanded,

$$f[x^* + \lambda(x - x^*)] = f(x^*) + \lambda \nabla f[x^* + \lambda\theta(x - x^*)] \cdot (x - x^*),$$

$(0 \leqq \theta \leqq 1)$. Substituting in the inequality and dividing out $\lambda\ (>0)$ we obtain the relationship

$$\nabla f[x^* + \lambda\theta(x - x^*)] \cdot (x - x^*) \leqq f(x) - f(x^*),$$

which gives the result on letting $\lambda \to 0$. The other results follow easily.

The following related result will be used in Chapter 8:

If $f(x)$ is convex (concave), its Hessian is positive (negative) semidefinite. If $f(x)$ is strictly convex (concave), its Hessian is positive (negative) definite.

For the proof, assume $f(x)$ is convex and use the second order Taylor expansion around x^*:

$$f(x^* + v) = f(x^*) + \nabla f^* \cdot v + (1/2!)v'Hv,$$

where H is evaluated at some point between x^*, $x^* + v$.

From the previous result, we have immediately,

$$(1/2!)v'Hv = f(x^* + v) - f(x^*) - \nabla f^* \cdot v \geqq 0.$$

Since this is true for all v, H is positive semidefinite. The associated results follow immediately.

From the analysis of Section R8.4, it then follows immediately that

if a convex (concave) function has a critical point, it has a minimum (maximum) at that point.

The association between convex functions and minima, and between concave functions and maxima, is the reason for our interest in them. The association extends to constrained maxima and minima, even though the function may have no critical points. See Chapter 2 for a discussion of the constrained optimum case.

A function $f(z)$, in which the variables z are divided into two groups x, y so that we write $f(z) = f(x, y)$, may be concave when considered as a function of the vector x (with y constant) and convex as a function of y (with x constant). At a critical point of such a function, we will have a maximum with respect to x and a minimum with respect to y. Such a point is called a *saddle point* of $f(x, y)$. Functions with saddle points are important in the theory of games (not treated in this book) and in optimizing theory. Their place in optimizing theory is discussed in Chapter 5.

R8.6 HOMOGENEOUS AND HOMOTHETIC FUNCTIONS[12]

Homogeneous functions are of considerable importance in mathematical economics, especially in neoclassical production theory. Homothetic functions, which are of a more general kind, are used from time to time in various contexts.

A scalar-valued function $f(x)$ is said to be homogeneous of degree ρ if it satisfies the relationship

$$f(tx) = t^{\rho} f(x),$$

for all numbers $t > 0$ and all vectors x for which $f(x)$ is defined. It is essential for the definition that $f(x)$ be defined over a cone, since we require that $f(tx)$ be defined if $f(x)$ is defined.

The function is *positively* homogeneous if the relationship holds for $x \geqq 0$, $t > 0$. It is *homogeneous of the first degree*, or *linearly homogeneous* if $\rho = 1$, and *homogeneous of degree zero* if $\rho = 0$, implying that $f(tx) = f(x)$. Typically a function that is homogeneous of degree zero will be an implicit function $F(x) = 0$. It is obvious that, if we solve explicitly for, say, the nth variable as $x_n = f(x_1, \ldots, x_{n-1})$, the function f is homogeneous of the *first* degree.

Homogeneous functions are a special case of a general class of functions satisfying the relationship $f(tx) = \phi(t) \cdot f(x)$. The more general functions are usually referred to as *homothetic.*

[12] Homogeneous functions are a traditional interest of economists and are treated in texts on mathematics for economists. See Allen [1] for a fairly detailed treatment in simpler terms than that given here.

An important property of homogeneous functions is obtained if we put $t = 1/x_j$, where x_j is any component of x. We then have

$$f\left(\frac{1}{x_j}x\right) = \left(\frac{1}{x_j}\right)^{\rho} f(x)$$

which gives

$$\begin{aligned} f(x) &= x_j^{\rho} f(v_1, \ldots, v_{j-1}, 1, v_{j+1}, \ldots, v_n), \\ &= x_j^{\rho}\, \phi(v) \end{aligned}$$

where v is an $(n - 1)$-vector of components x_i/x_j, $i \neq j$. This reduction of a homogeneous function is often useful.

The most important property of homogeneous functions is, however,

Euler's Theorem

If $f(x)$ is homogeneous of degree ρ, then the relationship $\rho f(x) = \nabla f(x) \cdot x$ $(= \sum f_j x_j)$ is satisfied for all x.

To prove this we take the derivatives of both sides of the defining relationship with respect to t, considering x constant. This gives

$$\sum f_j(tx)\, x_j = \rho t^{\rho-1} f(x).$$

The result then follows by putting $t = 1$.

By employing the same analysis for a homothetic function we obtain the *generalized Euler relationship*,

$$\phi'(1) \cdot f(x) = \nabla f(x) \cdot x.$$

If we take the partial derivative of the defining relationship for the homogeneous function with respect to the jth variable, we obtain:

$$t D_j f(tx) = t^{\rho} D_j f(x)$$

or,

$$D_j f(tx) = t^{\rho-1} D_j f(x),$$

so that $D_j f(x)$ is homogeneous of degree $\rho - 1$.

By taking derivatives of this homogeneous function, we then conclude:

The nth order partial derivatives of a homogeneous function of degree ρ are themselves homogeneous functions of degree $\rho - n$, provided, of course, the function is of class C^n.

Finally, since the Euler relationship holds for all x, we can take partial derivatives of both sides, to obtain

$$\rho f_i = f_i + \sum f_{ij} x_j \qquad i = 1, \ldots, n$$

or

$$(\rho - 1) f_i = \sum f_{ij} x_j \quad i = 1, \ldots, n.$$

Multiplying each equation by x_i and summing, we have

$$(\rho - 1) \sum f_i x_i = \sum \sum f_{ij} x_i x_j.$$

Putting this in matrix form, we can give the following result; since $\sum f_i x_i = \rho f(x)$ from Euler's theorem:

If $f(x)$ is homogeneous of degree ρ, then

$$\rho(\rho - 1) f(x) = x'Hx,$$

where H is the Hessian at the point x.

R8.7 THE BROUWER FIXED POINT THEOREM[13]

A vector-valued function of a vector defines a point-to-point mapping. The mapping $x \to F(x)$ is continuous if $F(x)$ is continuous, that is, if all the components of $F(x)$ are continuous. Many of the properties of continuous point-to-point mappings can therefore be determined by calculus methods, as in the preceding section.

For some purposes, however, calculus methods are not applicable. In the discussion of the implicit function theorem in the preceding section, it was assumed that solutions to the functional relationships existed, and we investigated properties in the neighborhood of these solutions. Calculus could not tell us whether solutions did or did not exist.

A powerful tool for proving the existence of solutions (when they do exist!) is the following:

Brouwer Fixed Point Theorem
If $F(x)$ defines a continuous point-to-point mapping of a compact convex set S into itself, there is some point $x^* \in S$ such that $x^* = F(x^*)$.[14]

Proof of the Brouwer theorem requires topological methods beyond the scope of this book. We can, however give a true proof for a set $S \subset R$, and give a heuristic demonstration with something of a topological atmosphere for $S \subset R^2$.

[13] Every mathematically inclined economist should be familiar with the statement of both the Brouwer and the Kakutani fixed point theorems. The latter is discussed in Section R9.5 of the next review.

Proofs of these theorems beyond the heuristic level requires a background in topological methods. For the reader interested in further exploration, Lefschetz is probably the most relevant introduction and includes a proof of the Brouwer theorem. A simple description of the Brouwer theorem is given in Courant and Robbins, and also in Dorfman, Samuelson, and Solow.

[14] x^* is said to be mapped into itself and is called a *fixed point.*

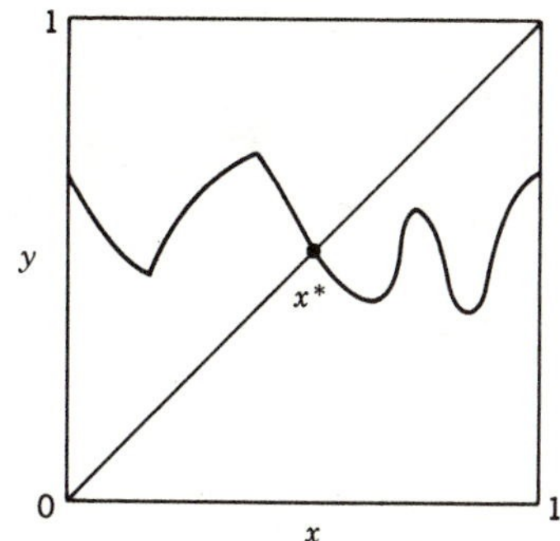

Figure R8–1 Brouwer fixed point theorem in R^1.

Consider $S \subset R^1$. Then, if S is compact and convex it must be some closed interval, say [0, 1]. Then the range of $F(x)$, now a scalar function of a scalar, must lie within [0, 1]. We can draw the graph of the mapping in two dimensions as in Figure R8–1, which shows x on the horizontal axis and $y = F(x)$ on the vertical axis. Since $F(x)$ is defined for all $x \in S$, its graph must cross the square from the left-hand boundary to the right-hand boundary. Since $F(x)$ is continuous, the graph must have no breaks in it and so must intersect the diagonal somewhere. But the diagonal is the set of points $y = x$, so there must be some point x^* such that $x^* = F(x^*)$. The graph in the diagram has been drawn with deliberate irregularity to emphasize that the only requirement on $F(x)$ is continuity (no breaks in the graph).

For $S \subset R^2$, we can argue in the following way. Let S be the compact convex set depicted in Figure R8–2a and T be the image of S under the mapping. Since the mapping is continuous, T must be compact (but not necessarily convex), and $T \subset S$.

Suppose for simplicity that T is a proper subset of S, as shown in the diagram. Make an arbitrary division of S into two subsets such that T is entirely contained in one of them. Consider the subset which does not contain T (shaded in the diagram). This has some image set in T, for example the shaded area of T in the diagram. Since the shaded region of T is the image of a set which contains no point of T, it obviously contains no fixed point.

Now discard both shaded areas. We now have the sets S_1, T_1 of Figure R8–2b in which $T_1 \subset S_1$ and T_1 contains the image of every point in S_1. Since the relationship between S_1, T_1 is of the same kind as the relation between S and T, we can carry out the partitioning and discarding process on S_1, T_1, then on the sets S_2, T_2 derived from them, and so on.

At no stage do we ever discard any points in the remaining piece of S which are covered by the remaining piece of T. Eventually we have only a

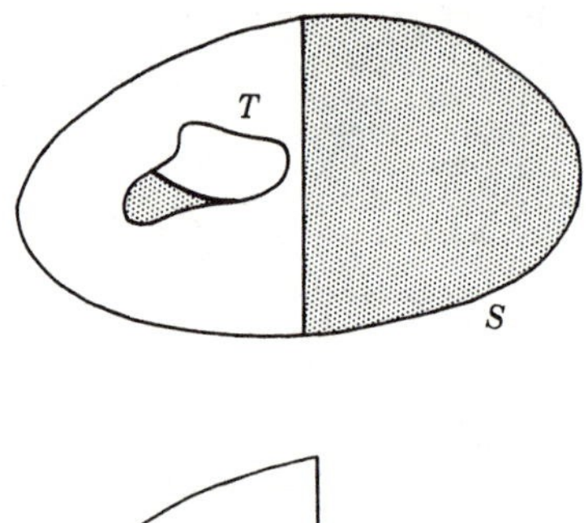

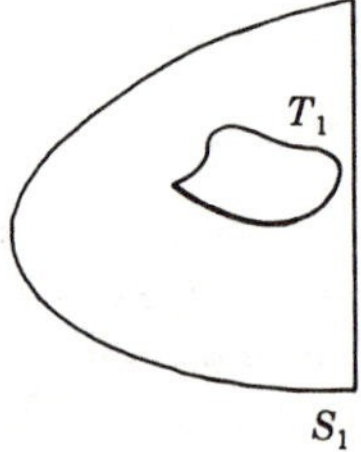

Figure R8–2 Brouwer fixed point theorem in R^2.

neighborhood of some point $x^* \in S$ and the image of this neighborhood in T, which lies within the neighborhood in S. In the limit, we have $x^* \in T$, giving the fixed point.

The Brouwer theorem is, of course, a sufficient condition. A mapping may give a point x^* such that $x^* = F(x^*)$ without satisfying the conditions of the theorem.

We shall use the Brouwer theorem later in the next section. The Kakutani theorem (Section R9.5) is a generalization of the Brouwer and this provides the essential key to problems of the existence of equilibrium in a model of the economy (Chapter 9).

R8.8 LINEAR HOMOGENEOUS VECTOR-VALUED FUNCTIONS[15]

An important class of functions in the theory of growth with neoclassical production functions is of the kind

$$y = F(x),$$

where $x, y \in R^n$ and each component $F^i(x)$ is homogeneous of the first degree.

[15] The material of this section is quite specialized and required only for growth theory (Chapter 10, Section 10.5). It is a generalization of the results for semipositive matrices and should be read after Review R7.

The original analysis of growth using these methods was Solow and Samuelson, but the subject has been developed by Morishima. See Morishima [1] (Appendix) for the most extensive treatment of the topic. For mathematical generalization, see Karlin [2].

We shall assume that $F(x)$ is a *positive* function ($x \geqq 0$ implies $F(x) \geqq 0$) and a *nondecreasing* function [$x^* \geqq x$ implies $F(x^*) \geqq F(x)$].

Thus the mapping defined by F:

(a) maps from R^n into R^n;
(b) maps $x \geqq 0$ into $F(x) \geqq 0$;
(c) maps $x^* \geqq x$ into $F(x^*) \geqq F(x)$;
(d) maps tx into $tF(x)$.

One mapping that has all these properties is that defined by $y = Ax$, where A is a semipositive square matrix. The function $F(x)$ can be regarded as a generalization of the linear function defined by a semipositive square matrix. We are, in fact, interested in just those properties of $F(x)$ that are generalizations of the properties of semipositive matrices set out in Review R7.

In the theory of semipositive matrices, the idea of indecomposability is a prominent one, which greatly strengthens many of the results. We shall make an analogous assumption here, for the same reasons.

$F(x)$ is *decomposable* if it is possible to find some submapping $\hat{x} \in R^m \to \hat{y} \in R^m$ which is independent of the components of $\hat{x}$ which are not included in x. That is, it is decomposable if there exists some set S of indices such that $x_i^* = x_i$ implies $F^i(x^*) = F^i(x)$ for all $i \in S$, whatever relations hold between x_j^*, x_j for $j \notin S$.

$F(x)$ is *indecomposable* if it is not decomposable. Now if $x_i^* = x_i$, $i \in S$ and $x_j^* > x_j$, $j \notin S$ we have $F^i(x^*) \geqq F^i(x)$ for all i as a result of the nondecreasing property. Indecomposability then implies $F^i(x^*) > F^i(x)$ for at least one $i \in S$.

We shall now state and prove the results in which we are chiefly interested, which are obvious generalizations of the Perron-Frobenius results for matrices:

(a) There exists at least one number $\lambda \geqq 0$ associated with a linearly normalized vector $v \geqq 0$, $\sum v_i = 1$, such that

$$\lambda v = F(v).$$

(b) If $F(x)$ is indecomposable there is only one λ, v satisfying (a) and, in addition, $\lambda > 0$ and $v \gg 0$.[16]

The existence proof is simple if we draw upon the Brouwer fixed point theorem (Section R8.7). Consider the set,

$$Y = \{y \mid y \in R^n, y \geqq 0, \textstyle\sum y_i = 1\}$$

[16] v can be regarded as a nonlinear characteristic vector or eigenvector of $F(x)$.

and the mapping $T(y)$ defined by

$$T(y) = \frac{1}{1 + \sum F^i(y)} [y + F(y)].$$

From the general properties of y, F it is clear that $T(y) \geqq 0$ and, from the construction, that $\sum T^i(y) = 1$, so that $T(y) \in Y$. Thus $T(y)$ defines continuous a point-to-point mapping of Y into itself. From the Brouwer theorem it follows that there is some y such that $y = T(y)$.

Put $v = y$, $\lambda = \sum F^i(y) = \sum F^i(v)$ and we have

$$v = \frac{1}{1 + \lambda} [v + F(v)],$$

giving

$$\lambda v = F(v),$$

with $\lambda \geqq 0$, $v \geqq 0$ by construction. Thus we have proved (a) and also identified the value of λ in terms of $F(v)$.

Now we shall assume indecomposability and prove (b). Suppose that we do not have $v \gg 0$. Let S be the set of indices for which $v_i = 0$, so that $v_i > 0$ for $i \notin S$. Consider tv, with $t > 1$. We have $F^i(tv) = tF^i(v)$, all i. For $i \in S$ we have $v_i = tv_i = 0$, and for $i \notin S$ we have $tv_i > v_i$, so that, if $F(x)$ is indecomposable we must have $F^i(tv) > F^i(v)$ for some $i \in S$. But $F^i(v) = \lambda v_i = 0$ for all $i \in S$, so that $F^i(tv) = tF^i(v) = 0$ for all $i \in S$, contradicting the previous result. Thus the set S must be empty and $v \gg 0$.

Suppose now that we had $\lambda = 0$. From the existence proof we have $\lambda = \sum F^i(v)$ and $F^i(v) \geqq 0$, all i, so that $\lambda = 0$ implies $F^i(v) = 0$, all i. Now choose u such that $u_1 = v_1$ and $v_i > u_i \geqq 0$, $i = 2, \ldots, n$ (possible since $v \gg 0$). Then if $F(v)$ is indecomposable we must have $F^1(u) < F^1(v)$. But $F^1(v) = \lambda v_1 = 0$ and F is a positive function so this is impossible. Thus we must have $\lambda > 0$.

Finally we must prove uniqueness. Suppose that there are two numbers λ, λ^* and two vectors v, v^*. Identify λ^* as the greater of λ, λ^*, if they differ, so that $\lambda^* \geqq \lambda$.

Define

$$\alpha = \max_i (v_i^*/v_i),$$

and let S be the set of indices for which $v_i^*/v_j = \alpha$. Then $v_i^*/v_i < \alpha$ for $i \notin S$. Thus $\alpha v_i = v_i^*$, $i \in S$ and $\alpha v_i > v_i^*$, $i \notin S$. From the indecomposability assumption we must have $F^i(\alpha v) > F^i(v^*)$ for at least one $i \in S$. Consider this to be $i = 1$. Then we have

$$\begin{aligned} F^1(\alpha v) &= \alpha F^1(v) && \text{(homogeneity)}, \\ &= \alpha \lambda v_1 && \text{(definition of } v) \end{aligned}$$

But $F^1(v^*) = \lambda^* v_i^*$, so that $F^1(\alpha v) > F^1(v^*)$ implies $\alpha \lambda v_1 > \lambda^* v_i^*$. Since

$\alpha v_i = v_i^*$ for all $i \in S$, we have $v_1 = v_j^*$, giving $\lambda^* < \lambda$, in contradiction of the original ordering of λ^*, λ. Thus we cannot have two different positive λ's, hence we cannot have two different positive v's, and the uniqueness property is proved.

EXERCISES

1. For the function $f(x) = 5x_1^3 - 2x_1x_2 + 3x_2^2$ calculate:

a. ∇f.

b. the directional derivative in the direction $\frac{1}{3}, \frac{2}{3}$;

c. the Hessian matrix.

2. Find the critical points of the following functions, and determine whether each critical point is a maximum or a minimum:

a. $f(x) = 4x_1^2 - x_1x_2 + x_2^2$;

b. $f(x) = -x_1^3 - x_2^3 + 3x_1 + 3x_2$;

c. $f(x) = x_1^2x_2 - x_2$.

3. Show by using any property from Section R8.5 that the function $f(x) = x_1^2 - 2x_1x_2 + 3x_2^2$ is strictly convex, and check that it shows all the other relevant properties set out in that Section.

4. Show that the function of Exercise **3** is also homogeneous of degree 2, and check that it shows all the properties of homogeneous functions set out in Section R8.6.

5. Show that a function which is homogeneous of the first degree has a Hessian which is singular.

6. Show that $f(x) = x_1x_2$ is concave-contoured for $x_1, x_2 \geqq 0$, but not a concave function.

REVIEW

R9

Point-to-Set Mappings

The material in this review is essential for Chapter 9 and for the discussion of the minimax theorem in Chapter 5. For the rest of the book it is desirable but not essential. It is assumed that the reader is familiar with Review R1, convex sets and general optimizing theory. Section R8.7 of the previous review (on the Brouwer fixed point theorem) should have been read.

R9.1 INTRODUCTION[1]

Point-to-set mappings, in which a point maps into a set rather than a single point, are particularly associated with optimizing models. These mappings may also be referred to as set-valued functions or correspondences.

Consider the simple linear program illustrated in Figure R9–1. The shaded area is the feasible set K. The objective function is px, with $p = [x_1, x_2] \geqslant 0$. For each p, the optimal x vector will be some point on the

[1] The material of this section is required only for Chapter 9 (General Equilibrium) and for Section 5.6 of Chapter 5. The section contains a heuristic treatment of some topological matters. At this level of treatment, the author can find little supplementary reading to recommend. Debreu [1] (Sections 1.7 and 1.10) contains an elliptic treatment of the same material, and Dorfman, Samuelson, and Solow touches lightly on some of it. Karlin [1] (Appendix C) also contains some discussion. Introductory topology texts (including Lefschetz, recommended for most purposes) do not deal with the matters discussed here.

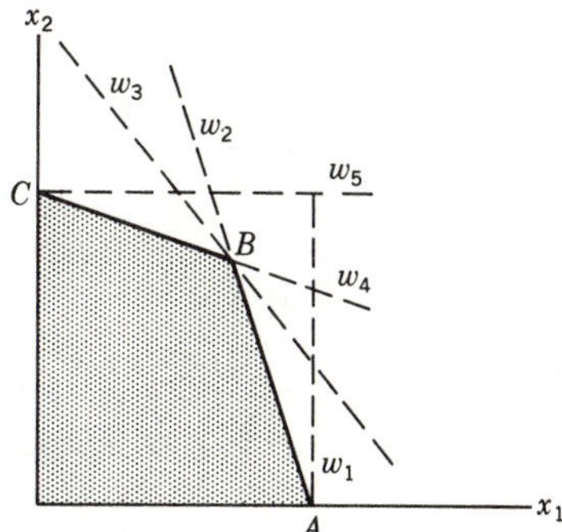

Figure R9–1 Point-to-set mappings.

boundary ABC. Since the optimal point is the same for p, λp ($\lambda > 0$), we shall suppose p to be linearly normalized so that $p_1 + p_2 = 1$.

If $p = [1, 0]$, the level line (in more than two dimensions it would be a hyperplane) of the objective function, w_1, will be vertical and the optimal point will be A. As p varies from $[1, 0]$ to $[0, 1]$ the level lines swing around through $w_2, w_3, \ldots$ to w_5, and the optimal x vector will change. Starting from A, this point will remain optimal until w swings to w_2, coincident with AB. For w_2, A, B and all points in the segment AB are optimal. For a further change in p, B alone becomes optimal and remains so until w reaches w_4, when any point on BC is optimal. The optimal point then becomes C alone, and so remains for the remaining values of p.

If we define:

$$S(p) = \{x^* \mid px^* \geqq px, \qquad \text{all } x, x^* \in K\},$$

then it is clear that the mapping $p \to S(p)$ is a point-to-set mapping. It happens that $S(p)$ contains but a single point for all but two values of p, but unless $S(p)$ contains a single point for *every* p, we must treat it as a point-to-set mapping.

Mappings of the above kind represent the typical occurrence of point-to-set mappings in economics.

We note that the points A, B, C in the diagram are optimal for more than one price vector, so that, if we define:

$$S'(x) = \{p^* \mid p^*x \geqq px; \qquad p, p^* \in P, \qquad x \in K'\},$$

where P is the set of normalized semipositive price vectors and K' the subset of K represented by AB and BC, then the mapping $x \to S'(x)$, which is inverse to $p \to S(p)$ is also point-to-set.

We are interested in this review in the continuity properties of point-to-set mappings of the above kind, and in the existence of the fixed point property for a set subject to these mappings. The continuity properties are most easily investigated through the graph of the mapping, to which we now turn.

R9.2 THE GRAPH OF A MAPPING

The graph of a simple function of a single variable is a useful analytical tool, as well as a visual aid, because some properties that are difficult to express explicitly in algebraic or other terms can be defined in terms of the properties of the graph. This characteristic has been well used in economics.

We shall generalize from the graph of $y = f(x)$ in the following way. Suppose $f(x)$ is a single-valued function of a single variable, and let S, T be subsets of real numbers. S will be the domain of f and T its range. Then $S \times T$ is the set of real number pairs $x \in S$, $y \in T$. If we take an arbitrary pair x, y then either $y = f(x)$ or $y \neq f(x)$, so that the pairs (x, y) for which $y = f(x)$ form a *subset* of $S \times T$.

The subset of $S \times T$ *containing the pairs* $[x, f(x)]$ *is defined as the graph of the mapping* $f(x)$. In the simple case, the line or curve by which $y = f(x)$ is represented in the two-dimensional diagram is regarded as a *subset* of R^2 rather than as a line or curve in the geometric sense.

This definition of a graph is applicable to point-to-point mappings of any kind. If x is in R^n and the point $y = T(x)$ is in R^m, then the graph of the mapping $x \to T(x)$ is a subset of R^{n+m}.

Furthermore, we can immediately generalize for point-to-set mappings. The mapping from $x \in S$ to $Y \subset T$ has a graph which is the subset of $S \times T$ containing the pairs which now consist of a point x and each point in its image set Y.

Figure R9–2 illustrates the graph of a point-to-set mapping for a simple

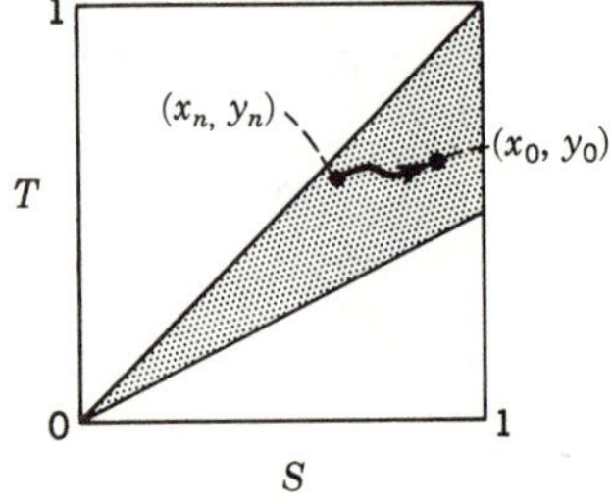

Figure R9–2 Graph of a mapping.

example. Let S, T be the real number closed intervals [0, 1] and the mapping rule be

$$Y = \{y \mid \tfrac{1}{2}x \leqq y \leqq x, \qquad x \in S\},$$

where Y is the image set of x. Then the graph is the set represented by the shaded area in the diagram.

Our interest in the graph is in its properties as a *set*, such as whether it is closed or open.

The term graph in the techniques known as *graph theory*[2] is related to the definition we have given, but the ultimate interest is different. Graph theory is primarily concerned with *finite graphs*, that is, with the graphs of mappings from one finite set to another. Usually the context is that of a point-to-point mapping from S into itself. Since the number of points is finite, we can suppose each $x \in S$ to be linked by a *path* to its image x'. Then x' itself may have an image x'', and this path can be supposed to be drawn. Then graph theory is concerned with the relationship of all the paths which connect a finite number of points with each other.

We shall also use the term *path*, but in the case of the infinite graphs with which we are concerned here, a path merely means any sequence $x_0 \to x_n$ paired with its image sequence $y_0 \to y_n$ (for a point-to-point mapping), or with *any* sequence $y_0 \to y_n$ in a point-to-set mapping which is such that $y_i \in Y(x_i)$ where $Y(x_i)$ is the image set of x_i. It is obvious that a sequence of the latter kind is represented by any path in the geometric sense that lies wholly in the graph. In Figure R9–2, such a path is illustrated from (x_n, y_n) to (x_0, y_0).

R9.3 CONTINUITY

For an ordinary function $f(x)$, continuity at x_0 is defined in terms of the relationship between $f(x)$ and $f(x_0)$ on the one hand and x and x_0 on the other. In heuristic terms, $f(x)$ is continuous at x_0 if $f(x) \to f(x_0)$ as $x \to x_0$. The graph of the function will be connected if it is continuous at all points.

[2] *Graph Theory* has important and growing applications in the structural analysis of economic systems and models. Such matters as matrix decomposability (Section R7.2) and the structure of certain programming and game problems are easily amenable to graph theoretic treatment. Lady has generalized some of the author's structural observations concerning qualitative solutions by using these techniques.

Several books on graph theory are available, some useful only as light reading. The author recommends Berge.

It is obvious that this approach to continuity cannot be directly applied to point-to-set mappings where unique values of $f(x), f(x_0)$ do not exist. Instead, we define the following terms:

A point-to-set mapping is said to be *upper semicontinuous* at x_0 if every sequence in which $x \to x_0$ has a limit point which lies in the image set of x_0.

A point-to-set mapping is said to be *lower semicontinuous* at x_0 if every point in the image set of x_0 is the limit point of some sequence such that each y lies in the image set of its associated x.

A point-to-set mapping is *continuous* if it is both upper and lower semicontinuous.

A mapping is upper or lower semicontinuous or continuous over a set S if the relevant property holds for all $x \in S$.

An equivalent description of the continuity properties is the following. Let S be the set of limit points of all sequence paths in the graph of the mapping as $x \to x_0$, and let $Y(x_0)$ be the image set of x_0. Then the mapping is upper semicontinuous at x_0 if $S \subset Y(x_0)$ and lower semicontinuous if $Y(x_0) \subset S$. If it is both upper and lower semicontinuous at x_0, then $S = Y(x_0)$, according with the intuitive idea of a continuous mapping.

These continuity ideas can be illustrated in terms of the graph of a simple point-to-set mapping fom one set of real numbers into another.

Figure R9–3 gives such an illustration. For $x < 1$, the mapping rule is the same as in Figure R9–2. For $x = 1$, we have $0 \leqq y \leqq 1$ in diagram (a) and $3/4 \leqq y \leqq 1$ in diagram (b). The image set of $x = 1$ is shown by the heavy vertical line in both cases. The arrows on the boundaries of the part of the graph for $x < 1$ indicate that these boundaries do not "reach" the line $x = 1$. The set S of limit points of all sequence paths in the graph for which $x \to x_0$ is indicated in each diagram.

It is obvious from the diagram that $S \subset Y(x_0)$ in (a), so that the mapping of that diagram is upper semicontinuous. On the other hand, $Y(x_0) \not\subset S$, so that it is not lower semicontinuous. In terms of the first set of definitions given, (a) is upper semicontinuous because every sequence path with $x \to x_0$ and y in the image set of its corresponding x must end in $Y(x_0)$, but it is not lower semicontinuous because a point like B in $Y(x_0)$ has no approach path to it such that each y lies in the image set of its appropriate x.

Diagram (b), on the other hand, represents a lower semicontinuous mapping because $Y(x_0) \subset S$ but it is not upper semicontinuous because $S \not\subset Y(x_0)$. In terms of sequences, we can find a sequence path lying in the graph to every point (like C) in $Y(x_0)$ (lower semicontinuous) but there exist sequence paths from points inside the graph whose limit points (like D) are not in $Y(x_0)$ (hence not upper semicontinuous).

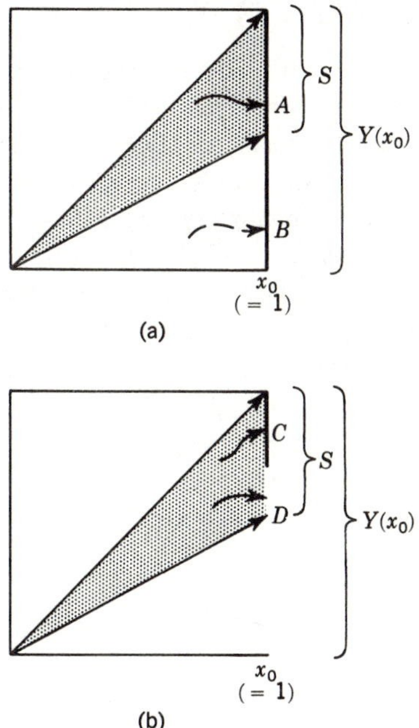

Figure R9–3 Upper and lower semicontinuous mappings.

In Figure R9–2, we have $S = Y(x_0) = \{y \mid \frac{1}{2} \leqq y \leqq 1\}$, so the mapping in that figure is continuous.

It is obvious from the illustrations and also from the formal definitions that *a mapping is upper semicontinuous if and only if its graph is closed.* Every sequence path in the graph must have as limit point an interior point or a boundary point, and if all boundary points are in the set the upper semicontinuous property is assured.

Upper semicontinuity is a very important property since it is characteristic of many point-to-set mappings in economics and it is also a sufficient degree of continuity for use of the Kakutani fixed point theorem (Section R9.5). Usually it is easiest to show the existence of upper semicontinuity by showing that the graph is closed rather than by investigating sequences as such.

We are not usually interested in lower semicontinuous mappings in their own right. We investigate lower semicontinuity in order to show that a mapping, already shown to be upper semicontinuous, is, in fact, continuous. Establishing lower semicontinuity involves demonstrating the existence of appropriate approach paths to all points in $Y(x_0)$. This may require considerable ingenuity.

If we take the above continuity definitions for a point-to-set mapping and apply them to the special case in which the image set of each x contains a single point, we see that the three concepts upper semicontinuous, lower semicontinuous, continuous are exactly equivalent, and that they coincide with the ordinary definition of continuity for a point-to-point mapping.

Because of this, it seems appropriate to eliminate some potential terminological confusion. A *function* $f(x)$ is said to be *upper semicontinuous*[3] at x_0 if $f(x) - f(x_0) < \varepsilon$ for x sufficiently close to x_0, although the distance $f(x_0) - f(x)$ need not be small. [Lower semicontinuity for $f(x)$ is equivalent to upper semicontinuity for $-f(x)$].

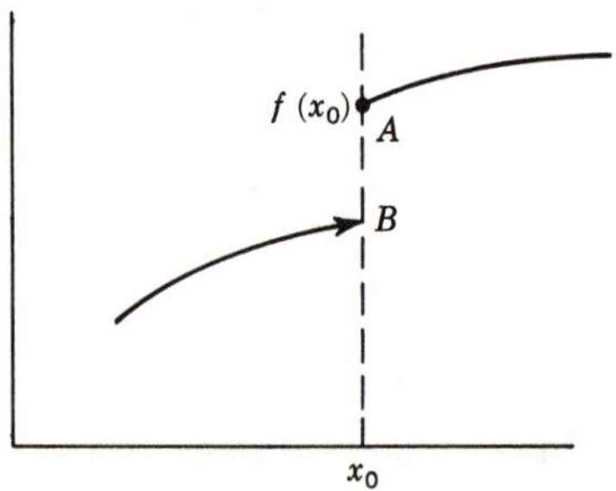

Figure R9–4 Semicontinuous functions.

Figure R9–4 illustrates the graph of an *upper semicontinuous function*. It can be seen that the function approaches $f(x_0)$ (the point A) from above in a continuous manner, but not from below. The lower point of the graph has limit point B, corresponding to the set S in the point-to-set mapping. $Y(x_0)$ is the single point A. Clearly we have neither $S \subset Y(x_0)$ nor $Y(x_0) \subset S$, so that the function is *discontinuous* if considered as a point-to-set mapping. On the other hand the point-to-set mapping $x \to Y(x)$ where $Y(x) = \{y \mid 0 \leqq y \leqq f(x)\}$ *is* upper semicontinuous.

The terms *outer* and *inner* semicontinuous would seem to be a better choice for describing the continuity properties of the point-to-set mapping, since the criterion is that of a certain set being contained in or containing another.

The following two properties of compound mappings are of importance and follow almost immediately from the definitions:

[3] This is the most widely used sense of the term "upper semicontinuous" in the general mathematics literature.

(a) If $x \to y = f(x)$ is a continuous point-to-point mapping, and $y \to Z(y)$ is an upper semicontinuous point-to-set mapping, then the point-to-set mapping defined by $x \to Z[f(x)]$ is also upper semicontinuous.

(b) If $x_1 \to \phi_1(x_1)$, $x_2 \to \phi_2(x_2)$ are upper semicontinuous point-to-set mappings, then the mapping $x \to \phi(x)$, where $x = (x_1, x_2)$ and $\phi(x) = \phi_1 \times \phi_2$, is also upper semicontinuous. This can be extended to cover the Cartesian product of any number of point-to-set mappings.

In both the above statements, "lower semicontinuous" can be inserted in place of upper semicontinuous.

R9.4 CONTINUITY PROPERTIES OF OPTIMAL SOLUTIONS

The standard cases in which point-to-set mappings and their continuity properties are relevant in economic analysis are all concerned with optimal solutions. The continuity properties of typical mappings which arise from optimization and parameter variation are given from the general theorem set out here.

Let u, v be two vectors in R^n and $u \to K(u)$ be a mapping which defines a set $K(u) \subset R^n$ for every u in some set $S \subset R^n$. Let $f(u, v)$ be a single valued function of u, v.

Define the set

$$V(u) = \{v \mid f(u, v) \text{ maximized for } v \in K(u)\},$$

and the function

$$\phi(u) = \max_{v \in K(u)} f(u, v).$$

In a typical economic context, u may be a price vector, v a consumption or production vector. $K(u)$ will be the feasible set for some choice situation—typically $K(u)$ will be a constant set for production, but will vary with u for consumption. $f(u, v)$ is the objective or valuation function—typically profits (hence a function of u, v) in the production case, or utility (a function of v only) for consumption. $V(u)$ is then the set of optimal production or consumption vectors associated with u, and $\phi(u)$ is the optimum level of profits or utility.

We can now state[4]:

If (a) the mapping $u \to K(u)$ is a continuous mapping and $K(u)$ is a compact convex set;

(b) the function $f(u, v)$ is continuous and concave in v for every u;

[4] A more general version of the theorem is given in Debreu [1]. (Section 1.8k).

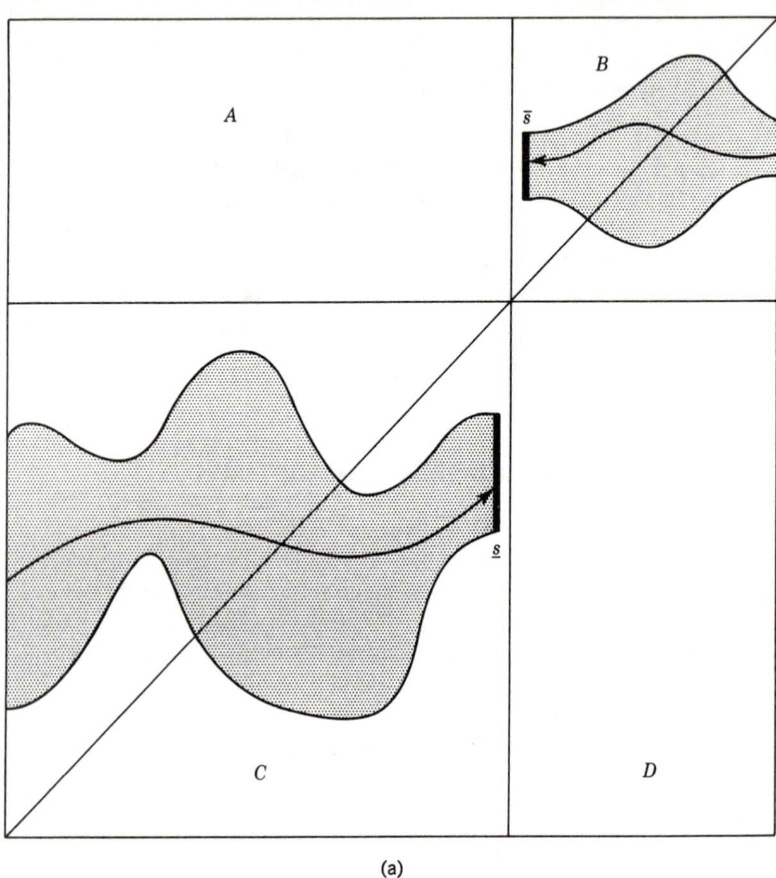

(a)

then (i) the mapping $u \to V(u)$ is upper semicontinuous and $V(u)$ is a compact convex set

(ii) the function $\phi(u)$ is continuous.

We shall not give a rigorous proof, but can note the following points. From general optimizing theory, the concavity of $f(u, v)$ and the convexity of $K(u)$ guarantee that $V(u)$ is either a point or a convex set. If we consider any sequence $u \to u^*$, $V \in V(u)$, the continuity of $f(u, v)$ and of the mapping $u \to K(u)$ guarantee that the limit point v^* is $\in V(u^*)$, establishing upper semicontinuity.

Now consider $\phi(u)$. Given u, $\phi(u) = f[u, \text{any } v \in V(u)]$. If u is in the neighborhood of u^*, $v \in V(u)$ is in the neighborhood of some $v^* \in V(u^*)$ [from the upper semicontinuity of $u \to V(u)$]. It follows that $\phi(u) = f[u, \text{any } v \in V(u)]$ is a continuous function of u.

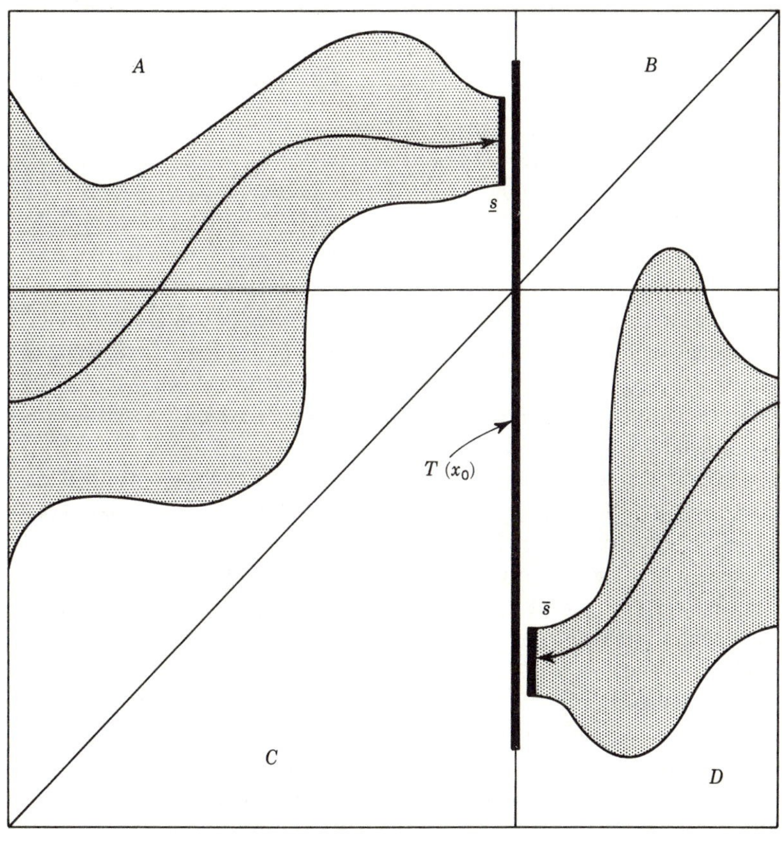

(b)

Figure R9–5 Kakutani fixed point theorem in R^1.

R9.5 THE KAKUTANI FIXED POINT THEOREM[5]

The Brouwer fixed point theorem was discussed earlier (Section R8.7). The following is a generalization of the Brouwer theorem which is of great importance in general equilibrium analysis and other contexts requiring existence proofs.

[5] The theorem is given in Kakutani, with later generalizations in Eilenberg and Montgomery and Begle. What is presented here is Kakutani's theorem 1, corollary. Kakutani's theorem 2 is concerned with the existence of a common point of two mappings and provides an easy proof of the Minimax theorem. It is closely related to a theorem of Von Neumann.

Statements of the theorem, in the form given here, appear without proof in various economic references, including Karlin [1], Debreu [1], Dorfman, Samuelson, and Solow, and Morishima [1].

Kakutani Fixed Point Theorem

If $x \to T(x)$ defines an upper semicontinuous point-to-set mapping of a compact convex set S into itself such that each $T(x)$ is compact and convex, there is some $x^* \in S$ such that $x^* \in T(x^*)$.

As in the case of the Brouwer theorem, we can give a proof for the case $S \subset R^1$ which illustrates the role of the various conditions of the theorem.

Consider, for simplicity, a point-to-set mapping which is continuous everywhere except at x_0, where it is upper semicontinuous only. Since S is compact and convex, it must be some closed interval which we shall take to be [0, 1]. The graph of the mapping is then in $S \times S$, which is a unit square as in Figure R9–5.

The set $S \times S$ has been divided into four closed subsets by the divisions $x = x_0$ and $y = T(x) = x_0$. Denote by $\underline{s}$ the set of limit points of sequences $x \to x_0$, $y \in T(x)$ for $x \leqq x_0$ and by $\bar{s}$ the equivalent set for $x \geqq x_0$. Since the mapping is continuous except at x_0, it is lower semicontinuous except at x_0. Thus every point in $\underline{s}$ has an approach path in the graph from $0 \to x_0$ and every point in $\bar{s}$ has an approach path $1 \to x_0$.

It is obvious that if $\underline{s} \subset C$ or $\bar{s} \subset B$, an approach path must intersect the diagonal somewhere in C or B, as shown in (a) of the figure, giving a fixed point. In this case, the nature of the mapping at x_0 is irrelevant.

Suppose this is not the case, so that $\underline{s} \subset A$ and $\bar{s} \subset D$, as shown in (b) of the figure. Since the mapping is upper semicontinuous, $\underline{s} \subset T(x_0)$ and $\bar{s} \subset T(x_0)$. But $T(x_0)$ is convex, by hypothesis, so must be a connected interval in this case. Then the diagonal must pass through $T(x_0)$, giving a fixed point.

A similar argument can be given for each point at which the mapping is not continuous.

EXERCISES

1. For $x \in S$, where $S = \{x \mid -1 \leqq x \leqq 1\}$ we define three point-to-set mappings as follows:

For $x \neq 0$, all mappings are the same, giving

$$-1 \leqq y \leqq -\tfrac{1}{2}(x+1) \qquad \text{for} \qquad 0 < x \leqq 1$$

$$-\tfrac{1}{2}(x-1) \leqq y \leqq 1 \qquad \text{for} \qquad -1 \leqq x < 0.$$

At $x = 0$ the three mappings give, respectively:

a. $-1 < y < 1$;

b. $-1 \leqq y \leqq 1$;

c. $-1 \leqq y \leqq -\frac{1}{2}$ and $\frac{1}{2} \leqq y \leqq 1$.

Show that there is a fixed point for two of the mappings. Of the three mappings:

d. One satisfies all the conditions of the Kakutani theorem;

e. one is upper semicontinuous but does not satisfy another condition of the theorem (which one?);

f. one is neither upper nor lower semicontinuous.

Determine which of the mappings has the properties in **d**, **e**, and **f**, respectively. (Hint: Draw the graph in R^2.)

REVIEW

R10

Linear Differential and Difference Equations

Specific use is made of the material in this review only in Chapter 12, but its content (like that of all previous reviews) should be regarded as standard equipment for the analytical economist. Familiarity with the content of Reviews R1, R2, and R5 is assumed. The reader who has read the sections on complex numbers, vectors and characteristic roots only lightly should now study them in detail.

R10.1 PRELIMINARY REMARKS[1]

Since many dynamic economic models can be specified either as differential equations or as difference equations, it is convenient to discuss the two together rather than as separate subjects. This brings out both the similarities and differences of the two types of equation.

[1] Differential equations is a large field in mathematics with a vast literature. We touch on only one corner of the field. Difference equations are more specialized, although the computer has resulted in a rapid increase in interest in them.

An elementary introduction to differential and difference equations (including the vector equation) is given in Yamane, and a more extensive discussion with economic applications is given in Baumol [1]. Allen [2] contains an extensive discussion of the linear scalar equation of the n'th order, but not of the vector form with which this Review is mainly concerned.

We consider an *unspecified* function $y(t)$ of the single independent variable t. The independent variable in typical differential or difference equations in economics will almost always be identified with time, so we write it as t rather than as the more conventional x.

We shall denote the derivatives $dy(t)/dt, d^2y(t)/dt^2, \ldots$, by $Dy(t)$, $D^2y(t), \ldots, D^ny(t)$ for convenience here. Sometimes it is useful to view D^r as a *linear operator* but, here we shall consider only the properties of $D^ry(t)$ as an rth order derivative. It is important not to confuse, for example, D^2y and $(Dy)^2$; the first is the second derivative of y, the second is the square of the first derivative.

Now consider $y(t)$ at different values of t separated by *equal finite intervals*, h. We write these $y(t), y(t + h), y(t + 2h), \ldots$. By choosing suitable units for t we can always make $h = 1$, and we shall assume this to be done unless it is otherwise clear from the context. In economic models we shall usually identify this interval with a time period.

We define

$$\Delta y(t) = y(t + 1) - y(t)$$

as the *first difference* of $y(t)$. [Sometimes it is also called a *forward* difference, as contrasted with the *backward* difference $y(t) - y(t - 1)$.]

Now

$$\begin{aligned}\Delta[\Delta y(t)] &= \Delta y(t + 1) - \Delta y(t)\\ &= [y(t + 2) - y(t + 1)] - [y(t + 1) - y(t)]\\ &= y(t + 2) - 2y(t + 1) + y(t).\end{aligned}$$

We write $\Delta[\Delta y(t)]$ as $\Delta^2y(t)$ and refer to it as the *second difference*. In an inductive fashion we can define the *n'th difference* as

$$\begin{aligned}\Delta^n y(t) &= \Delta^{n-1}y(t + 1) - \Delta^{n-1}y(t)\\ &= y(t + n) - ny(t + n-1) + \cdots + (-1)^n y(t),\end{aligned}$$

where the expansion on the right-hand side is binomial. For our immediate purposes we merely note that $\Delta^n y(t)$ is a linear combination of $y(t), y(t + 1), \ldots, y(t + n)$.

Outside the economics literature, Goldberg provided an adequate coverage of difference equations, including the vector form. Vector differential equations are discussed in Bellman [3] (Chapter 10) and Bellman [1] (more advanced). Among standard differential equations texts, the author recommends Kaplan.

The importance of *inequalities* in modern economic models has considerably reduced the role of differential and difference equations in dynamic formulations.

The relationship between D and Δ is given by the mean value theorem,

$$Dy(t + \theta h) = \frac{1}{h}[y(t + h) - y(t)] \qquad 0 \leqq \theta \leqq 1$$
$$= \frac{1}{h}\Delta y(t).$$

$(1/h)\Delta y(t)$ is therefore equal to the derivative at some point between t and $t + h$ and $(1/h)\Delta y(t) \to Dy(t)$ as $h \to 0$. For numerical solutions, this approximation of a derivative by a difference is of considerable use.

An important property of a difference, shared by a derivative, is that it removes a constant. If $y(t)$ has the form $f(t) + a$, $\Delta y(t) = f(t + 1) - f(t)$.

We are now ready to define differential and difference equations:

(a) A relationship of the form $f[t, y(t), Dy(t), \ldots, D^n y(t)] = 0$ is a *differential equation of the nth order.*

(b) A relationship of the form $f[t, y(t), \Delta y(t), \ldots, \Delta^n y(t)] = 0$ is a *difference equation of the nth order.*

A difference equation can always be expressed in the alternative form

(c) $f[t, y(t), y(t + 1), \ldots, y(t + n)] = 0,$

where the order of the equation is the difference between the earliest and latest dates attached to variables appearing within the relationship.[2]

If f is a *polynomial*, the *degree* of the equation is the power of the *highest order* derivative or difference. The order and degree of an equation should never be confused.

Examples

(a) $D^3y(t) + [D^2y(t)]^2\, dy(t) - 2y^5 = 0$ is a *third order* differential equation of the *first degree.*

(b) $[\Delta y(t)]^2 - ky(t) + t^7 = 0$ is a *first order* difference equation of the *second degree.*

(c) $y(t + 5) - ky(t) = 0$ is apparently a *fifth order* difference equation. In this case, however, changing the units of t enable us to put it in the form of a *first order* equation. If values of y appeared for periods between t and $t + 5$, the change to a first order equation could not be made.

If the relationship takes the special form

(1) $a_0(t)D^ny(t) + a_1(t)D^{n-1}y(t) + \cdots + a_ny(t) = \phi(t);$

(2) $a_0(t)\Delta^ny(t) + a_1(t)\Delta^{n-1}y(t) + \ldots + a_ny(t) = \phi(t),$

[2] A relationship of the form $f[t, y(t), \ldots, D^ry(t), \ldots, \Delta^sy(t)] = 0$ is a *mixed difference-differential equation.* Although such equations sometimes appear to be a good specification for an economic model, their solution is extraordinarily difficult.

where the a's can be functions of t, but not of y or its derivatives, the equation is a *linear differential (or difference) equation of the nth order*. If the a's are constants, it is a *linear differential (or difference) equation with constant coefficients*. If $\phi(t) = 0$ it is a *homogeneous* linear equation.

We shall be almost entirely concerned with linear equations having constant coefficients, and so shall usually refer to these simply as "linear equations."

In the case of linear equations, it turns out, as we shall see later, that it is as easy to discuss equations in which $y(t)$ is a *vector-valued function of t* and the a's are matrices A, as it is to discuss the equation in which $y(t)$ is scalar valued and the a's are simple numbers. We shall refer to such equations as *vector* difference or differential equations. If a distinction is necessary, we shall refer to the ordinary equation as a *scalar* equation.

Differential and difference equations of the kind discussed so far, in which y is a scalar- or vector-valued function of the single independent variable t are called *ordinary* differential or difference equations. If y is a function of several variables partial derivatives are necessarily involved in forming differential equations (and the equivalent, partial differences, in forming difference equations), and such equations are *partial* differential or difference equations.

Equations involving partial derivatives abound in neoclassical mathematical economics. These are *not*, in general, partial differential equations. In the typical neoclassical relationship involving partial derivatives we are assumed (in principle) to be given a function [say $u(x)$ as in consumer theory] and required to find the *point* at which the partial derivatives of this known function satisfy the relationships.

We would have a differential equation only if we were given a set of relationships involving the derivatives and were required to find the *function* whose derivatives satisfied these relationships. Problems of the kind, what type of utility function gives a certain type of demand curve, involve partial differential equations.

The solution of a differential or difference equation is a *function*, not a *point*. In some cases (numerical solution) the function may be specified only by a list of the points in its graph, but it is the function, not the individual points, which constitute the solution.

R10.2 SOLUTIONS

We now formally define a *solution* of the differential equation

$$f[t, y(t), Dy(t), \ldots, D^n y(t)] = 0$$

as *a function $y(t)$ which, when inserted in f above, makes the relationship $f = 0$ an identity*. The solution of a difference equation is defined analogously.

Considerable insight into the nature of solutions is obtained by considering a reverse process. Let $y(t) = f(t, c_1, \ldots, c_n)$ (no relationship between this f and that above) be an arbitrary function of t involving n constants $c_1, \ldots, c_n$. These constants may be incorporated in any way in the function, as algebraic coefficients, exponents, and so on.

Take the function at any arbitrary value of t and consider the derivatives of $y(t)$ at this point up to the nth order. Each derivative is a function of t and of the constants $c_1, \ldots, c_n$. We obtain the following $n + 1$ relationships:

$$
\begin{aligned}
y(t) &= f(t, c_1, \ldots, c_n), \\
Dy(t) &= f_1(t, c_1, \ldots, c_n), \\
&\cdots\cdots\cdots \\
D^n y(t) &= f_n(t, c_1, \ldots, c_n).
\end{aligned}
$$

Assuming no complications arise, such as singularity of the Jacobean, we can take the first n of these equations and solve for the constants $c_1, \ldots, c_n$ in terms of $y(t)$ and the first $n - 1$ derivatives to obtain relationships of the form

$$c_i = \phi_i[t, y(t), Dy(t), \ldots, D^{n-1}y(t)].$$

If these are now inserted in the equation

$$D^n y(t) = f_n(t, c_1, \ldots, c_n),$$

we obtain the *nth order differential equation*:

$$D^n y(t) - F[t, y(t), Dy(t), \ldots, D^{n-1}y(t)] = 0.$$

An exactly analogous process can be used to eliminate the n constants by differencing and to obtain an nth order difference equation.

The original function $f(t, c_1, \ldots, c_n)$ is called the *primitive* of the differential or difference equation which is obtained from it in the above way. Thus solving a differential or difference equation can be regarded as searching for a primitive from which the equation could have been derived. Since a primitive with n arbitrary constants can be reduced to a differential or difference equation of the nth order, we have the following *expectation*:

A differential or difference equation of the nth order will have a solution containing n abitrary constants.[3]

[3] The equation may contain any number of constants itself. The arbitrary constants in the solution are additional constants which do not appear in the equation itself. Another way of looking at the matter is to consider that solving an nth order differential equation is equivalent to n successive integrations, each of which introduces an arbitrary constant of integration.

Now let us consider a related solution problem. We are given an nth order differential equation and n *initial conditions* the actual values, y_0, $y'_0, y''_0, \ldots, y_0^{(n-1)}$ of some arbitrary function of t and its first $n - 1$ derivatives at $t = 0$. Can we find a solution of the differential equation *satisfying the initial conditions*, that is, a function which is both a solution of the differential equation and whose value and the value of whose first $n - 1$ derivatives equal $y_0, \ldots, y_0^{(n-1)}$ at $t = 0$?

We noted, in deriving the differential equation from the primitive, that we could express the arbitrary constants in the form

$$c_i = \phi_i[t, y(t), \ldots, D^{n-1}y(t)] \qquad i = 1, \ldots, n.$$

Putting $t = 0$ and inserting the values $y_0, y'_0, \ldots$ in the above relationships, we obtain the appropriate values of $c_1, \ldots, c_n$ to satisfy the initial conditions. Thus, given a solution of the differential equation, we can satisfy the initial conditions by making a suitable choice of the arbitrary constants.

An analogous argument can be given for the difference equation. The initial conditions for a difference equation are usually given as the values of $y(0), y(1), \ldots, y(n - 1)$, rather than as $y(0)$ and the first $n - 1$ differences. Either specification can be derived directly from the other.

Thus we have the following expectation:

A solution can be found for an nth order differential or difference equation which will satisfy the n initial conditions $y(0)$, $Dy(0), \ldots, D^{n-1}y(0)$ *(for the differential equation) or* $y(0), y(1), \ldots, y(n - 1)$ *(for the difference equation).*

For the difference equation, justification for this expectation is straightforward, as a consequence of the following:

If a difference equation of the nth order can be written in the explicit form $\Delta^n y(t) = F(t, y(t), \Delta y(t), \ldots, \Delta^{n-1}y(t)]$, *where F is single valued, a solution which satisfies the initial conditions* $y(0), y(1), \ldots, y(n - 1)$ *can always be found by iteration.*

If the equation can be written in the form given, it can necessarily also be written in the form

$$y(t + n) = f[t, y(t), y(t + 1), \ldots, y(t + n - 1)].$$

By putting $t = 0$ and inserting the initial values in the right-hand side we immediately obtain $y(n)$. We can then put $t = 1$, insert the values $y(1), \ldots, y(n)$ in the right-hand side and obtain $y(n + 1)$, and so on.

By approximating a differential equation by such a difference equation, numerical solutions of differential equations can be obtained, given the initial conditions. Much of the recent growth of interest in difference equations has been due to the ability of computers to handle iterative processes of the required kind. The economist's interest in difference equations has not been due to this so much as to the natural suitability of difference equations to the specification of many dynamic processes in economics and to the possibility of using inequalities together with finite differences. Inequalities and continuous changes do not go well together.

The approximation approach can also be used as one method of proving the following[4]:

Fundamental Existence Theorem for Solutions of Differential Equations *If* $F(z_1, \ldots, z_{n+1})$ *is a continuous function of the* $n + 1$ *variables* $z_1, \ldots, z_{n+1}$ *and has continuous first partial derivatives, then the differential equation*

$$D^n y(t) = F[t, y(t), Dy(t), \ldots, D^{n-1}y(t)],$$

where z_1 *has been replaced by* t, z_2 *by* $y(t)$ *and* z_i *by* $D^{i-2}y(t)$ *has a solution* $y = f(t)$ *which satisfies arbitrary initial conditions* $f(t_0) = y_0$, $Df(t_0) = y_0', \ldots, D^{n-1}f(t_0) = y_0^{(n-1)}$. *Furthermore this solution is unique. (This holds for the region in which F is continuous.)*

R10.3 THE FIRST ORDER LINEAR SCALAR EQUATION

We shall examine linear scalar equations of the first order, with constant coefficients, of the following type:

$$(1) \quad Dy(t) - \lambda y(t) = \phi(t),$$

$$(2) \quad \Delta y(t) - \lambda y(t) = \phi(t).$$

First we make a very important observation. If $f_1(t), f_2(t)$ are any two different solutions of (1), then, if $y(t) = f_1(t) + f_2(t)$,

$$\begin{aligned} Dy(t) - y(t) &= [Df_1(t) - \lambda f_1(t)] - [Df_2(t) - \lambda f_2(t)] \\ &= \phi(t) - \phi(t) \\ &= 0, \end{aligned}$$

so that $y(t)$ is a solution of the related *homogeneous equation*

$$Dy(t) - \lambda y(t) = 0. \tag{3}$$

[4] Proofs are available in Kaplan and standard differential equations texts.

Thus if $f(t)$ is *any* solution of (1), called a *particular solution*, the set $Y(t)$ of all solutions of (1) is given by

$$Y(t) = f(t) + \{y(t)\},$$

where $\{y(t)\}$ is the set of all solutions to the homogeneous equation (3). (The similarity to the relationship between the solutions of nonhomogeneous and related homogeneous linear *algebraic* equations will be noted.)

Any function $y(t)$ which is a member of $\{y(t)\}$ is called the *complementary function* to the particular solution. The set $\{y(t)\}$ obviously has the fundamental linear properties

$$ky(t) \in \{y(t)\} \quad \text{if} \quad y(t) \in \{y(t)\},$$

$$y_1(t) + y_2(t) \in \{y(t)\} \quad \text{if} \quad y_1(t), y_2(t) \in \{y(t)\}.$$

Thus we see that the solution technique for equations like (1) and (2) (the above analysis obviously holds for the difference equation as well) consists in finding any particular solution, then adding to it the set of solutions of the associated homogeneous equation.

Let us, therefore, turn to examine the solutions of the homogeneous equation $Dy(t) - \lambda y(t) = 0$. Dropping the explicit notation that t is the independent variable, we can write this as

$$\frac{dy}{dt} = \lambda y.$$

Separating the variables, we obtain

$$\begin{aligned} \lambda t &= \int \frac{dy}{y} + c \\ &= \log y + c. \end{aligned}$$

Taking exponents, this becomes

$$y = ke^{\lambda t}, \qquad \text{where } k = e^c.$$

Since it is a constant of integration, c, hence k, is the *arbitrary constant* that we expect to find in the solution of the first order equation.

Thus the *general solution* to (1) has the form

$$y(t) = f(t) + ke^{\lambda t},$$

where $f(t)$ is a particular solution and k an arbitrary constant. It is obvious that the set $\{ke^{\lambda t}$, all $k\}$ is the general solution set of the homogeneous equation.

Now consider the homogeneous form of the *difference* equation:

$$\Delta y(t) - \lambda y(t) = 0. \tag{4}$$

If we write this in the alternate form

$$y(t + 1) = (1 + \lambda)\, y(t),$$

we can solve by iteration to obtain

$$y(t) = k(1 + \lambda)^t \qquad \text{where } k = y(0) \text{ is arbitrary.}$$

The general solution to (2) is, therefore,

$$y(t) = f(t) + k(1 + \lambda)^t.$$

We can summarize as follows:

The general solutions of (1) and (2) have the form

$$y(t) = f(t) + ke^{\lambda t}, \tag{5}$$

$$y(t) = f(t) + k(1 + \lambda)^t, \tag{6}$$

where $f(t)$ is a particular solution of the relevant equation and k an arbitrary constant.

At this stage we shall not concern ourselves with the general problem of finding a particular solution, but consider only the simple case in which $\phi(t) = c$ (a constant).

For the differential equation, (1) becomes

$$Dy(t) - \lambda y(t) = c. \tag{7}$$

It is obvious that $y(t) = y^* =$ constant will be a solution for appropriate choice of y^*. This is easily seen to be $y^* = -c/\lambda$, so that $f(t) = y^* = -c/\lambda$ is a particular solution of (7).

For the difference equation (2) we have

$$\Delta y(t) - \lambda y(t) = c. \tag{8}$$

It is obvious that $y^* = -c/\lambda$ is also a particular solution of (8), so we can write the general solutions of (7) and (8), respectively, as

$$y(t) = y^* + ke^{\lambda t}, \tag{9}$$

$$y(t) = y^* + k(1 + \lambda)^t. \tag{10}$$

We can now fit these solutions to the arbitrary initial condition $y(0) = y^0$. Putting $t = 0$ in (9) we obtain

$$y(0) = y^0 = y^* + k,$$

giving

$$k = y^0 - y^*,$$

with the same result from putting $t = 0$ in (10).

Thus we have the complete solution for the equations which satisfy arbitrary initial conditions:

The solution of $Dy(t) - \lambda y(t) = c$ *subject to* $y(0) = y^0$ *is*

$$y(t) = y^* + (y^0 - y^*)e^{\lambda t}. \tag{11}$$

The solution of $\Delta y(t) - y(t) = c$ *subject to* $y(0) = y^0$ *is*

$$y(t) = y^* + (y^0 - y^*)(1 + \lambda)^t, \tag{12}$$

where $y^* = -c/\lambda$ *in both cases.*

A particular solution of a differential or difference equation which, like y^*, is constant is an *equilibrium solution* [if $\phi(t)$ is not constant, an equilibrium solution in this sense will not usually exist]. Every linear differential or difference equation with constant coefficients and $\phi(t) = \text{constant}$ has an equilibrium solution. If the equation is homogeneous the equilibrium solution is always $y^* = 0$.

If, for all y^0 in the region $|y^* - y^0| \leqq K$, every solution of the equation approaches y^* as $t \to \infty$, the equation is said to be *stable* in the region. If K is unlimited, the equation is *globally stable*, if K is specified as "small" the equation is *locally stable.*

The behavior of any solution (11) or (12) relative to t depends on the *dynamic term* $(y^* - y^0)e^{\lambda t}$ or $(y^* - y^0)(1 + \lambda)^t$. We note that, if y^0 is already the equilibrium y^*, the dynamic term vanishes, so that a dynamic path which commences as an equilibrium path remains so. However, if $y^0 \neq y^*$ (however small the difference) we must consider the dynamic term whose influence over t is determined by the behavior of $e^{\lambda t}$ or $(1 + \lambda)^t$.

For the *differential equation*, there are only two modes of behavior (we assume $\lambda \neq 0$):

(a) If $\lambda < 0$, $e^{\lambda t} \to 0$ as $t \to \infty$ and the equation is stable. In this case it is globally stable.

(b) If $\lambda > 0$, $e^{\lambda t}$ increases without limit as $t \to \infty$. However small $y^* - y^0$ may be, the dynamic term ultimately becomes very large, so that $y(t) - y^*$ becomes very large and continues to increase. The equation is unstable.

In the case of the differential equation $e^{\lambda t} \geqq 0$ for all λ, t so the sign of $y(t) - y^*$ is always the sign of $y^0 - y^*$ and $y(t)$ remains on the side of y^* from which it started.

For the *difference equation*, there is more variety of behavior.

(a) If $|1 + \lambda| < 1$, then $(1 + \lambda)^t \to 0$ as $t \to \infty$ and the equation is stable, as in the differential equation with $\lambda < 0$. Here, however, the term $(1 + \lambda)^t$ can approach zero in either of two ways, depending on the sign of $(1 + \lambda)$:
   (i) If $0 \leqq 1 + \lambda < 1$ $(-1 \leqq \lambda < 0)$, then $(1 + \lambda)^t \geqq 0$ and $y(t)$ remains on the same side of y^* as y^0.
   (ii) If $-1 < 1 + \lambda < 0$ $(-2 < \lambda < -1)$, then $(1 + \lambda)^t$ and $(1 + \lambda)^{t+1}$ have opposite signs and $y(t)$ *alternates* from one side of y^* to the other, with the absolute distance $|y(t) - y^*|$ decreasing with t.[5]
(b) If $|1 + \lambda| > 1$, $(|1 + \lambda|)^t$ increases as t increases, and the equation is unstable. Again, there are two modes of stability. If $1 + \lambda > 1$ $(\lambda > 0)$, $(1 + \lambda)^t$ increases steadily with t, whereas, if $1 + \lambda < -1$ $(\lambda < -2)$, the distance $|y(t) - y^*|$ increases steadily, but with $y(t)$ alternating from one side of y^* to the other.

We can compare and contrast the differential and difference equation by summarizing the nature of the solutions of the two for different values of λ.

(a) If $\lambda > 0$, both equations are unstable.
(b) If $-2 < \lambda < 0$, both equations are stable, although for $-2 < \lambda < -1$, the solution of the difference equation converges to y^* in an alternating fashion.
(c) If $\lambda < -2$, the differential equation is stable, but the difference equation is unstable.

Thus whereas $\lambda < 0$ is both necessary and sufficient for stability of the differential equation, it is *necessary but not sufficient* for stability of the difference equation. For the latter, $|1 + \lambda| < 1$ is necessary and sufficient.

Stability in the technical sense here may not always coincide with stability in the substantive sense for an economic model. See Chapter 12, Section 12.1.

R10.4 COMPLEX SOLUTIONS

In a first order linear equation with real coefficients, λ is necessarily real. In order to examine the nature of solutions of difference and differential equations further, we need to examine the second order homogeneous equation.

Consider the homogeneous linear differential equation:

$$D^2y(t) + a_1Dy(t) + a_2y(t) = 0.$$

[5] An *alternation*, with $(1 + \lambda)^t$ alternately positive and negative should not be confused with a sinusoidal movement or *oscillation*, such as we shall meet in the next section.

Since $e^{\lambda t}$ was a solution of the first order equation, let us try such a solution here. Inserting $e^{\lambda t}$ in the equation, we obtain

$$(\lambda^2 + a_1\lambda + a_2)\, e^{\lambda t} = 0.$$

Now $e^{\lambda t} \neq 0$, so that $e^{\lambda_i t}$ is a solution, if and only if λ_i is a root of the *auxiliary equation*

$$\lambda^2 + a_1\lambda + a_2 = 0.^6$$

Similarly we can show that $(1 + \lambda_i)^t$ is a solution of the second order homogeneous difference equation

$$\Delta^2 y(t) + a_1 \Delta y(t) + a_2 y(t) = 0,$$

if and only if λ_i is a root of an auxiliary equation of exactly the same form as that of the differential equation.

Due to the linearity of the solution sets, it is obvious that, if λ_1, λ_2 are the two roots of the auxiliary equation, then $y(t) = c_1 e^{\lambda_1 t} + c_2 e^{\lambda_2 t}$ and $y(t) = c_1(1 + \lambda_1)^t + c_2(1 + \lambda_2)^t$ are solutions of the differential and difference equations, respectively, for any arbitrary constants c_1, c_2.

Since the auxiliary equation is quadratic, it may have either real or complex roots. If the roots are real, the behavior of the solution $y(t)$ is simply some weighted sum of the types of behavior described in the previous section.

If the roots are complex then, since the coefficients a_1, a_2 are taken to be real, the two roots must be complex conjugates which we shall write λ, λ^*, and both $e^{\lambda t}$ and $(1 + \lambda)^t$ will be complex numbers.

We shall now interest ourselves exclusively in the complex root case. Consider the solution of the differential equation at $t = 0$. We have

$$y(0) = c_1 + c_2.$$

Now $y(0)$ is a real number, so that $c_1 + c_2$ is real. Thus the imaginary parts of c_1, c_2 must be opposite in sign and numerically equal. Write $c_1 = \alpha_1 + i\beta$, $c_2 = \alpha_2 - i\beta$.

Consider the first derivative of $y(t)$ at $t = 0$. We have

$$Dy(0) = c_1\lambda + c_2\lambda^*.$$

$Dy(0)$ is real, so we must have $\operatorname{Im}(c_1\lambda + c_2\lambda^*) = 0$.

[6] The roots of the above algebraic equation are also called, by transference, the roots of the corresponding differential equation.

Put $\lambda = a + ib$, then $\lambda^* = a - ib$. Using the values for c_1, c_2, we have

$$c_1\lambda + c_2\lambda^* = (\alpha_1 + i\beta)(a + ib) + (\alpha_2 - i\beta)(a - ib),$$

$$\text{Im}(c_1\lambda + c_2\lambda^*) = (\beta a + b\alpha_1) - (\beta a + b\alpha_2),$$

$$= 0, \text{ if and only if } \alpha_1 = \alpha_2 = \alpha.$$

Thus we have $c_1 = \alpha + i\beta$, $c_2 = \alpha - i\beta$. That is, if $c_1 = c$, $c_2 = c^*$.[7]

Thus we have the very important result that, for a real equation, *all complex roots will be in conjugate pairs and the associated constants will also be conjugates.*

We can now write the solution for the equation, in the complex root case (for the differential equation) as

$$y(t) = ce^{\lambda t} + c^*e^{\lambda^* t}$$
$$= ce^{\lambda t} + (ce^{\lambda t})^*.$$

Now the sum of a complex number and its conjugate is equal to twice the real part, so that

$$y(t) = 2\text{Re}\,(ce^{\lambda t}).$$

Put $\lambda = \alpha + i\omega$, then

$$e^{\lambda t} = e^{(\alpha + i\omega)t}$$
$$= e^{\alpha t}e^{i\omega t}.$$

Now put c in the polar form $c = \frac{1}{2}ke^{i\phi}$. Then we have

$$ce^{\lambda t} = \tfrac{1}{2}ke^{\alpha t}e^{i(\omega t + \phi)}$$
$$= \tfrac{1}{2}ke^{\alpha t}[\cos(\omega t + \phi) + i\sin(\omega t + \phi)],$$

so that

$$y(t) = 2\text{Re}\,(ce^{\lambda t})$$
$$= ke^{\alpha t}\cos(\omega t + \phi),$$

where $\alpha = \text{Re}(\lambda)$ and $\omega = \text{Im}(\lambda)$. ϕ depends only on the arbitrary constant.

For the difference equation we have complex conjugate roots which, by the same type of argument as that used for the differential equation, will be associated with conjugate constants. Thus we will have

$$y(t) = 2\text{Re}\,[c(1 + \lambda)^t].$$

In this case write the complex number $1 + \lambda$ in the polar form $\rho e^{i\omega}$ where $\rho = |1 + \lambda|$ so that $(1 + \lambda)^t = \rho^t e^{i\omega t}$.

[7] Although, given c, we have c^*, there are still *two* arbitrary constants. These are the real and imaginary parts of c.

If we write c in the same form as in the differential equation case, we have

$$c(1 + \lambda)^t = \tfrac{1}{2}k\rho^t e^{i(\omega t + \phi)},$$

and, in the same way as in the differential equation case, we obtain

$$y(t) = k\rho^t \cos(\omega t + \phi),$$

where $\rho = |1 + \lambda|$ and $\tan \omega = \text{Im}(1 + \lambda)/\text{Re}(1 + \lambda)$.

Thus we can summarize our results for complex roots:

If a real homogeneous differential or difference equation has a complex root λ, it has another complex root which is conjugate to the first, and the part of the solution associated with the two conjugate roots can be written in the form

$$y(t) = ke^{\alpha t} \cos(\omega t + \phi) \qquad \textit{(differential equation);}$$

$$y(t) = k\rho^t \cos(\omega t + \phi) \qquad \textit{(difference equation),}$$

where $\alpha = \text{Re}(\lambda)$, $\rho = |1 + \lambda|$. *In both cases ω depends on* $\text{Im}(\lambda)$ *and $\omega = 0$, if* $\text{Im}(\lambda) = 0$. *The arbitrary constants are k, ϕ.*

If we had $\alpha = 0$ or $\rho = 1$, the solutions would both take the form $y(t) = k \cos(\omega t + \phi)$, the path of a sine curve. The maximum values of $y(t)$ would be k, $-k$, the *amplitude* of the *oscillation*. Since the constant ϕ has the same effect as shifting the time origin, it gives the *phase* of the oscillation. Since $\cos(2\pi + x) = \cos x$, the curve repeats, at t', its behavior at t whenever $\omega(t' - t) = 2\pi$, so that $t' - t = 2\pi/\omega$ is the *period* of the oscillation.

If $\alpha < 0$ or $\rho < 1$, the amplitude of the path is k at $t = 0$, at other values of t it is given by $ke^{\alpha t}$ or $k\rho^t$ and decreases as t increases. Such a path, a *damped oscillation*, is clearly stable.

If $\alpha > 0$ or $\rho > 1$, the amplitude of the oscillation increases with t (sometimes the oscillation is said to be antidamped). Such a path is obviously unstable.

Thus we can state the stability conditions for solutions with complex roots:

For the *differential equation* $\alpha = \text{Re}(\lambda) < 0$ is both necessary and sufficient for stability, coinciding with the condition $\lambda < 0$ for real roots.

For the *difference equation*, $\rho = |1 + \lambda| < 1$ is necessary and sufficient. Now $\rho^2 = [1 + \text{Re}(\lambda)]^2 + [\text{Im}(\lambda)]^2$, so that $\text{Re}(\lambda) < 0$ is *necessary*, but *not sufficient*, for stability.

R10.5 THE FIRST ORDER VECTOR EQUATION

We shall now turn to equations of the kind,

$$Dy(t) - Ay(t) = 0, \tag{1}$$

$$\Delta y(t) - Ay(t) = 0, \tag{2}$$

where $y(t)$ is an n vector, A an $n \times n$ matrix and $Dy(t)$, $\Delta y(t)$ are the vectors with typical components $Dy_i(t)$, $\Delta y_i(t)$. We shall assume throughout that A is nonsingular with distinct nonzero characteristic roots. The justification for considering matrices which appear in economic models always to have distinct characteristic roots has been made in Section R5.5.

The above equations can be regarded as a vector equation, or as a system of simultaneous scalar equations. Our approach here is to consider them as vector equations.

Let us first consider the difference equation (2). This can be written in the alternate form,

$$y(t + 1) - y(t) - Ay(t) = 0,$$

or,

$$y(t + 1) = (I + A)\, y(t). \tag{R10.5.1}$$

Taking an arbitrary $y(0)$, this equation can be solved by iteration:

$$y(1) = (I + A)\, y(0),$$

$$y(2) = (I + A)\, y(1) = (I + A)^2\, y(0),$$

$$\cdot \quad \cdot \quad \cdot \quad \cdot \quad \cdot \quad \cdot \quad \cdot \quad \cdot \quad \cdot \quad \cdot \quad \cdot$$

$$y(t) = (I + A)^t\, y(0). \tag{R10.5.2}$$

This is already the solution in a form analogous to that for the scalar equation, but it is convenient to manipulate it into a somewhat different form.

Since A has distinct characteristic roots, and the roots of $I + A$ are $1 + \lambda_i$, where λ_i is a root of A, we can diagonalize $(I + A)^t$ and we shall do so in the following way:

$$V^{-1}(I + A)^t U^{-1} = \Lambda^t,$$

giving

$$(I + A)^t = V\Lambda^t U,$$

where V is the matrix of column characteristic vectors (right-hand eigenvectors) of A (A, $I + A$ have the same characteristic vectors), U is the matrix of row characteristic vectors (left-hand eigenvectors) of A, and Λ is the diagonal matrix with typical element $1 + \lambda_i$, where λ_i is a characteristic root of A.

Substituting in the solution (R10.5.2), we obtain

$$y(t) = V\Lambda^t U y(0).$$

Make the substitution $k = Uy(0)$. Since the initial vector $y(0)$ is arbitrary, k is an arbitrary vector. Then

$$y(t) = V\Lambda^t k.$$

Now consider the matrix $V\Lambda^t$. Since the diagonal matrix Λ^t postmultiplies the matrix V, it multiplies the jth column of V by the jth diagonal element $(1 + \lambda_j)^t$. The jth column of V is v^j, the jth characteristic column vector of A, so that the jth column of $V\Lambda^t$ is the vector $(1 + \lambda_j)^t v^j$.

Thus the solution of (2) can be written in the form

$$y(t) = \sum_j k_j(1 + \lambda_j)^t v^j. \qquad \text{(R10.5.3)}$$

The solution vector $y(t)$ is thus a linear combination of the n linearly independent vectors v^j, with weights which depend on t. Each of the weights $k_j(1 + \lambda_j)^t$ is analogous to a solution of the first order scalar equation, if λ_j is real.

If λ_j is complex, then it forms a conjugate pair with some other root (since A is real), which we can take to be λ_{j+1}. The associated characteristic vectors are also complex conjugates, and, by arguing in the same way as we did for the scalar case, we conclude that the associated constants k_j, k_{j+1} will also be conjugate.

Thus we will have, as a component of the complete solution, the sum

$$k_j(1 + \lambda_j)^t v^j + k_{j+1}(1 + \lambda_{j+1})^t v^{j+1} = k(1 + \lambda)^t v + k^*[(1 + \lambda)^t]^* v^*,$$

where $k = k_j$, $\lambda = \lambda_j$, $v = v^j$.

If we put $v = u + iw$, where u is the vector of the real parts of the components of v, and w is the vector of the imaginary parts, we can show by similar methods to those of the previous section, that the relevant part of the solution can be put in the form

$$k\rho^t\{[\cos(\omega t + \phi)]u - [\sin(\omega t + \phi)]w\}. \qquad \text{(R10.5.4)}$$

where $\rho = |1 + \lambda|$.

Thus the solution of the vector equation is the weighted sum of the characteristic vectors (or of vectors derived from the characteristic vectors if complex) and these vectors (which are linearly independent) can be regarded as a basis. Part of the weight associated with each vector is derived from the arbitrary constant and part, $(1 + \lambda_j)^t$ is a function of t. As t increases, the part played by v^j in the solution depends on the behavior of $(1 + \lambda_j)^t$ as t increases. It is obvious that the characteristic vector associated with the particular root having the largest modulus will dominate the solution when t is very large, whatever the initial conditions, and that the characteristic vector associated with a root giving $|1 + \lambda| < 1$ will have negligible influence on the solution as t becomes large. Thus the solution will approach closer and closer to the proportions of the vector associated with the root of largest modulus as t increases.

Now consider the existence of an equilibrium solution of

$$\Delta y(t) - Ay(t) = c. \tag{2a}$$

If we try the constant vector y^*, we easily see that this is a particular solution if

$$y^* = -A^{-1}c. \tag{R10.5.5}$$

Such a solution always exists if A is nonsingular, which we shall assume to be the case.

Thus the vector equation will, in general, possess an *equilibrium solution* (which will be $y^* = 0$ in the homogeneous case) like the scalar equation. It is obvious that the Equation (2a) will converge to y^*, that is, it will be *stable* if *all* the terms of the solution (R11.5.3) of the homogeneous equation approach zero as $t \to \infty$. The behavior of each individual term (or pair of terms, in the case of complex roots) is similar to that of the first or second degree scalar equation, as discussed in Sections R11.3 and R11.4, so that:

The difference equation (2) *is stable, if and only if every root of A satisfies the condition* $|1 + \lambda_j| < 1$.

The exact description of the solution behavior of the vector equation is complicated. $y(t)$ is the linear combination of basis vectors with weights varying over time. Some weights may be increasing or decreasing steadily, some alternating, some oscillating. Usually we shall be interested only in general properties, such as which characteristic vector dominates, whether the solution converges to an equilibrium, or under what circumstances the solution will be a scalar expansion along a path of fixed proportions.

Now let us turn to the vector *differential* equation (1). Since $e^{\lambda t}$ was a solution of the equivalent scalar equation, we may expect that the matrix equivalent e^{tA} is a solution of the vector equation. Let us examine this possibility.

From the definition of e^{tA} we have:

$$e^{tA} = I + tA + (1/2!)t^2A^2 + (1/3!)t^3A^3 + \cdots.$$

Taking the derivative with respect to t, we obtain:

$$\begin{aligned} De^{tA} &= A + tA^2 + (1/2!)\,t^2A^3 + \cdots \\ &= A(I + tA + (1/2!)\,t^2A^2 + \cdots) \\ &= Ae^{tA}. \end{aligned}$$

Put $y(t) = e^{tA}z$, where z is an undetermined constant vector, and substitute in (1). This gives:

$$\begin{aligned} De^{tA}z - Ae^{tA}z &= Ae^{tA}z - Ae^{tA}z \\ &\equiv 0, \end{aligned}$$

so that $y(t) = e^{tA}z$ is, indeed, a solution of (1) for all vectors z. If we put $t = 0$, we identify z as $y(0)$, so that we can write the general solution of the homogeneous vector differential equation as

$$y(t) = e^{tA}y(0). \tag{R10.5.6}$$

This is obviously related to the solution of the scalar differential equation in exactly the same way the solution (R10.5.2) of the vector difference equation is related to the solution of its equivalent scalar equation.

We can diagonalize, as in the difference equation case, to obtain

$$y(t) = Ve^{\Lambda t}Uy(0).$$

Putting $Uy(0) = k$, as before, and noting that $e^{\Lambda t}$ is the diagonal matrix with typical element $e^{\lambda_j t}$, we can put the solution in the form

$$y(t) = \sum_j k_j e^{\lambda_j t} v^j. \tag{R10.5.7}$$

The analysis of this solution is precisely analogous to that of the difference equation solution. Among other things we therefore conclude that:

The differential equation (1) is stable, if and only if every root of A satisfies the condition $\mathrm{Re}\,(\lambda_j) < 0$.

A matrix all of whose roots have negative real parts is often called a *stable matrix* or *stability matrix* as a consequence of the above statement.

Note that such a matrix is stable only for the *differential* equation. The condition is *necessary but not sufficient* for the stability of the difference equation, as we saw earlier in the scalar case.

Let us now consider a point of some importance in relation to growth theory.

If we consider the homogeneous equation with the initial condition $y(0)$ given, we immediately obtain the constant vector k as $k = Uy(0)$. (This was the definition of k.)

Now suppose $y(0)$ is proportional to the jth characteristic vector, which we necessarily assume to be real. Without loss of generality we can suppose $y(0) = v^j$, since a constant of proportionality will make no difference to the analysis.

Consider $k = Uv^j$ in the form

$$k_i = u_i v^j \qquad i = 1, \ldots, n.$$

Each row u_i of U is a characteristic row vector. Now the characteristic row and column vectors are orthonormal (Section R5.10) so that $u_i v^j = 0$ if $i \neq j$, and $u_j v^j = 1$. Thus the constant vector k consists entirely of zeros except for $k_j = 1$. This means that the solutions of the differential and difference equations will have the special form:

$$y(t) = e^{\lambda_j t} y(0) \qquad \text{(differential equation)},$$

$$y(t) = (1 + \lambda_j)^t y(0) \qquad \text{(difference equation)}.$$

Thus we can assert the following result, of importance in growth theory:

The behavior of the solution of a first order vector differential or difference equation will be determined solely by a particular characteristic root λ_j, if and only if the initial conditions are represented by a vector proportional to the characteristic vector v^j associated with λ_j. In this case, the solution vector $y(t)$ will remain proportional to $y(0)$ for all values of t.

R10.6 REDUCTION TO FIRST ORDER VECTOR EQUATION

Consider the nth order linear homogeneous scalar equation with constant coefficients,

$$D^n y(t) + a_1 D^{n-1} y(t) + \cdots + a_n y(t) = 0,$$

which we shall put in the form,

$$D^n y(t) = -a_1 D^{n-1} y(t) - \cdots - a_n y(t).$$

Now define n variables $z_i(t)$ as follows:

$$\begin{aligned}
z_1(t) &= y(t), \\
z_2(t) &= Dy(t) = Dz_1(t), \\
z_3(t) &= D^2y(t) = Dz_2(t), \\
&\cdots\cdots\cdots \\
z_n(t) &= D^{n-1}y(t) = Dz_{n-1}(t).
\end{aligned}$$

Then the above differential equation is equivalent to the n equations,

$$\begin{aligned}
Dz_1(t) &= z_2(t), \\
Dz_2(t) &= z_3(t), \\
&\cdots\cdots\cdots, \\
Dz_{n-1}(t) &= z_n(t), \\
Dz_n(t) &= -a_n z_1(t) - a_{n-1} z_2(t) - \cdots - a_1 z_n(t).
\end{aligned}$$ [8]

If we regard the z_i as components of the vector $z(t)$, we can write the above equations as

$$Dz(t) = Az(t),$$

or

$$Dz(t) - Az(t) = 0,$$

where A is the matrix

$$\begin{bmatrix}
0 & 1 & 0 & \cdots & 0 & 0 \\
0 & 0 & 1 & \cdots & 0 & 0 \\
\cdot & \cdot & \cdot & \cdots & \cdot & \cdot \\
0 & 0 & 0 & \cdots & 0 & 1 \\
-a_n & -a_{n-1} & \cdot & \cdots & -a_2 & -a_1
\end{bmatrix},$$

or

$$\left[\begin{array}{c|c} 0 & I \\ \hline \multicolumn{2}{c}{-\hat{a}} \end{array}\right].$$

In the partitioned matrix, a is the (row) vector of the coefficients of the original scalar equation, I is of order $(n - 1) \times (n - 1)$, and 0 is a column of $(n - 1)$ zeros.

Thus an nth order linear scalar equation can be converted into a first order linear equation in vectors of order n, so that the theory of the nth order scalar equation parallels that of the first order vector equation.

[8] Obviously $D^n y(t) = D[D^{n-1}y(t)] = Dz_n(t)$. The relationship $z_1(t) = y(t)$ is not used since $y(t)$ is an unspecified function and calling it $z_1(t)$ is merely a change in name.

The reverse process can also be carried out, the vector equation converted into an nth order scalar equation, since any matrix can be transformed into the form given for A.

By analogous argument we can show that a linear *vector* equation of order m can be transformed into a *first order* vector equation in vectors of order mn, where n is the order of the vectors in the original equation.

Although the analysis has been carried out only for the differential equation, exactly the same relationships hold for difference equations.

R10.7 A NOTE ON PARTICULAR SOLUTIONS

For the nonhomogeneous linear differential or difference equation

$$Dy(t) - Ay(t) = F(t),$$

$$\Delta y(t) - Ay(t) = F(t),$$

where $F(t)$ is an arbitrary vector-valued function, a particular solution can always be found, although it may not be easy to obtain it in explicit form.

For economic models we sometimes wish to have, for example, an exogenous time path for consumption in a growth model. Usually we are free to choose some appropriate function. The case $F(t) = c$, which might be an appropriate choice, has already been treated.

It would seem a reasonable conjecture, that turns out to be true, that it should be easy to obtain a particular solution if $F(t)$ has the same general form as the solutions of the homogeneous equation, namely $e^{\beta t}c$ for the differential equation and $(1 + \beta)^t c$ for the difference equation, where c is a constant vector. These forms are, in a sense, "natural" to the equations with which they are associated.

Let us put $y(t) = e^{\beta t}c$ in the differential equation and see what happens. We have

$$\begin{aligned} D(e^{\beta t}c) - e^{\beta t}Ac &= \beta e^{\beta t}c - e^{\beta t}Ac \\ &= e^{\beta t}(\beta I - A)c. \end{aligned}$$

Thus, if $F(t) = e^{\beta t}(\beta I - A)c$, $y(t) = e^{\beta t}c$ is a particular solution. Hence if $F(t) = e^{\beta t}b$, the particular solution will be $y(t) = e^{\beta t}c$, where $b = (\beta I - A)c$. We can, as a consequence, immediately state the following:

If $F(t) = e^{\beta t}b$, the differential equation has a particular solution $y(t) = e^{\beta t}(\beta I - A)^{-1}b$, provided β is not a root of A.

The last condition is necessary to ensure that $(\beta I - A)$ is nonsingular. If β is a root of A, $F(t)$ is redundant since the appropriate solution is already included among solutions of the homogeneous equation.

In the difference equation we can apply the same reasoning to solutions of the form $(1 + \beta)^t$ and reach an analogous conclusion:

If $F(t) = (1 + \beta)^t b$, the difference equation has a particular solution $y(t) = (1+ \beta)^t(\beta I - A)^{-1}b$, provided β is not a root of A.

The above forms for $F(t)$, with their simple particular solutions, will satisfy many requirements in economic models. More complex paths for $F(t)$ can often be approximated by linear combinations of above forms. Since the equations are linear, the particular solution of a linear combination will be a linear combination of the individual particular solutions.

EXERCISES

[For simplicity y is written in place of $y(t)$.]

1. For A $= \begin{bmatrix} -5/2 & 2 \\ -1 & 1/2 \end{bmatrix}$, find

a. the solution of $Dy - Ay = 0$;
b. the solution of $\Delta y - Ay = 0$;

2. The matrix $A = \begin{bmatrix} -1 & 1 \\ -1 & -1 \end{bmatrix}$ has complex roots. Express in both standard form and in terms of trigonometric functions with real parameters, the solutions of

a. $Dy - Ay = 0$;
b. $\Delta y - Ay = 0$.

Investigate the stability of **a** and **b**.

3. Using the method of Section R10.6, express the third order differential equation

$$D^3y - 6D^2y + 11Dy - 6y = 0$$

as a first order vector equation. Given that one root of the system matrix is unity, show that the others are also real and positive.

4. Find the solution of $Dy = Ay$ $\left(\text{where } A = \begin{bmatrix} -3 & 4 \\ -2 & 3 \end{bmatrix}\right)$ which satisfies the initial condition $x(0) = \begin{bmatrix} 1 \\ 2 \end{bmatrix}$.

5. Find the general solution of the equation

$$Dy - Ay = b,$$

where $b = \begin{bmatrix} 1 \\ 1 \end{bmatrix}$ and A is the matrix of Exercise **2**.

REVIEW

R11

Calculus of Variations and Related Topics

This review contains a treatment of the basic principles of calculus of variations, required for Chapter 11, Section 11.6. Brief notes are given on other topics, including Pontryagin's principle, for the information of the reader. Familiarity is assumed with the material of the previous reviews, with optimizing theory generally, and with elementary integral calculus.

R11.1 OPTIMIZING WITH AN INFINITE NUMBER OF VARIABLES

As pointed out in a footnote to Chapter 2, a point can be regarded as a function of its indices, defined by enumeration. If the number of indices becomes very large, we can write the vector in functional form $x(t)$, $t = 1, \ldots, n$, and we can thus regard a function $x(t)$ of the continuous variable t as the infinite-dimensional analog of a point.

Given a point x defined over some set S, a rule associating a real number with each $x \in S$ is a *function*. If $x(t)$ is itself a function which is a member of some class, C, of functions, a rule associating a real number with each function in C is called a *functional*.[1] The definite integral

[1] The term functional, like the term function, may also be used in a wider sense than that given here.

$$J = \int_a^b x(t)\,dt$$

associates a real number, the value of the integral, with every real integrable function $x(t)$, and is thus a functional. The integral is the most important type of functional for our purposes.

Optimizing with an infinite number of variables is thus a problem in which the unknown is a function rather than a point, and the optimand is a functional which will take the form of an integral in the analysis of this review. The classic problem is that handled by the calculus of variations. There are other approaches, related to the results (but not always the methods) of classical calculus of variations, for solving problems not amenable to the latter. In this review, we shall examine the elements of the calculus of variations and also Pontryagin's maximum principle.

As in the case of differential equations, the topics being touched upon are very broad, with a large literature. Only the barest outline is presented, with proof given only for the basic result of the calculus of variations.

R11.2 BASIC CALCULUS OF VARIATIONS[2]

The simplest problem suitable for the calculus of variations [often referred to, in fact, as the *First* (*or Fundamental*) *Problem of the Calculus of Variations*] is that of finding a minimum or maximum of an integral of the kind[3]

$$J = \int_a^b F(x, \dot{x}, t)\,dt,$$

where x is an unknown function of t, but the functional form of F and the endpoints of the path are x given. $x(t)$ may be either a scalar or a vector function. We shall initially take x as a scalar, and assume that $x(t)$ will be of class C^2. The fundamental result for this case is:

The first necessary condition for $x(t)$ to give a maximum or minimum for the integral

$$J = \int_a^b F(x, \dot{x}, t)\,dt$$

[2] For a detailed discussion of classical calculus of variations methods, with many examples, see Forsyth. A more modern (and more advanced) treatment that covers also other topics given in this review is provided by Hestenes. The elements of the calculus of variations is given in most texts on mathematics for physical scientists or engineers, in many advanced calculus texts (for example, Buck), and in Allen [1].

[3] The occurrence of $\dot{x}$ (used here instead of Dx as in the previous Review) as an argument of F means that the value of the integral depends on the *path* from a to b. If $\dot{x}$ does not appear in the integral (or in a side constraint as in the control problem discussed in Section R11.4), the integral depends only on the endpoint values and becomes an ordinary optimizing problem, or trivial.

is that $x(t)$ *satisfies the Euler differential equation*

$$F_x - \frac{d}{dt} F\dot{x} = 0.^{4}$$

A function $x(t)$ satisfying the Euler equation is called an *extremal.*

We shall prove this result by classical (or *Euler-Lagrange*) methods. By the ordinary reasoning for an optimizing problem, it is clear that, if $x(t)$ is optimal, small "variations" in $x(t)$ will increase the integral (assuming we seek a minimum). If x is a point, variations from x means arbitrary movements away from the point. Since $x(t)$ is a function, we consider variations of the kind

$$z(t) = x(t) + \varepsilon\eta(t),$$

where $\eta(t)$ is an arbitrary function of the same class as $x(t)$ (C^2) and ε is a number independent of x, t, η. Since the endpoints of the integration are fixed, we must have $\eta(a) = \eta(b) = 0$. We can visualize $x(t)$ as an arc joining the points a, b. The varied functions $z(t)$ can be visualized as arbitrary smooth deformations of $x(t)$ which leave the endpoints unchanged.

With the varied function, the integral becomes

$$J(z) = \int_a^b F(z, \dot{z}, t)\, dt,$$

where $\dot{z} = \dot{x} + \varepsilon\dot{\eta}$. Let us assume that $x(t)$ is the optimal function and that an arbitrary $\eta(t)$ has been chosen, but ε is variable. Then $J(z)$ is a function of ε only. Noting that we can differentiate through the integral sign and that z, $\dot{z}$ are functions of ε with derivatives η, $\dot{\eta}$, respectively, we have

$$dJ/d\varepsilon = \int_a^b (\eta F_z + \dot{\eta} F_{\dot{z}})\, dt.$$

Since $x(t)$ is optimal, J must reach its optimal value at $\varepsilon = 0$, and we must have $(dJ/d\varepsilon)_{\varepsilon=0} = 0$. When $\varepsilon = 0$, $z = x$, $\dot{z} = \dot{x}$ so that the relationship

$$\int_a^b (\eta F_x + \dot{\eta} F_{\dot{x}})\, dt = 0$$

must be satisfied if $x(t)$ is optimal. Thus a necessary condition for optimality is

$$\begin{aligned} 0 &= \int_a^b (\eta F_x + \dot{\eta} F_{\dot{x}})\, dt \\ &= \int_a^b \eta F_x\, dt + \int_a^b \dot{\eta} F_{\dot{x}}\, dt \\ &= \int_a^b \eta F_x\, dt + [\eta F_{\dot{x}}]_a^b - \int_a^b \eta \frac{d}{dt}(F_{\dot{x}})\, dt. \end{aligned}$$

[4] F_x, $F_{\dot{x}}$ are the partial derivatives of F with respect to x, $\dot{x}$, respectively, where these variables are treated as independent of each other and of t.

The third line is obtained by integrating the second integral in the second line by parts. Now $\eta(a) = \eta(b) = 0$ so the second term in the last line vanishes and the condition reduces to:

$$\int_a^b \eta \left[F_x - \frac{d}{dt} F_{\dot{x}}\right] dt = 0.$$

This relationship must hold for every function η. It is obvious that, if the term in the bracket is, say positive, for some range, we can choose an acceptable function η which is positive over the same range and zero elsewhere, giving a positive value to the integral. Hence the bracketed term must vanish for all t, giving the

Euler equation

$$F_x - \frac{d}{dt} F_{\dot{x}} = 0.$$

If we write out $(d/dt)F_{\dot{x}}$ in full, we obtain the full form of the Euler equation,

$$F_x - F_{\dot{x}t} - \dot{x}F_{x\dot{x}} - \ddot{x}F_{\dot{x}\dot{x}} = 0,$$

which shows it to be a second order nonlinear differential equation. We cannot usually expect to solve such an equation, but there are two special cases of great importance in which the Euler equation can be immediately reduced to the first order:

(a) *F* Does Not Contain *x* Explicitly

In this case $F_x = 0$, so that the Euler equation has an immediate first integral

$$F_{\dot{x}} = C.$$

(b) *F* Does Not Contain *t* Explicitly[5]

This is an important case of common occurrence. The procedure is not immediately obvious. We have

$$\frac{d}{dt} F = \dot{x}F_x + \ddot{x}F_{\dot{x}} \qquad (\text{since } F_t = 0),$$

and

$$\frac{d}{dt}(\dot{x}F_{\dot{x}}) = \ddot{x}F_{\dot{x}} + \dot{x}\frac{d}{dt} F_{\dot{x}},$$

so that

$$\frac{d}{dt}\left[F - \dot{x}F_{\dot{x}}\right] = \dot{x}\left[F_x - \frac{d}{dt} F_{\dot{x}}\right].$$

[5] This is the case that is relevant to Chapter 11, Section 11.6.

Thus the Euler equation implies that the left-hand side is zero, giving an immediate first integral

$$F - \dot{x}F_{\dot{x}} = C.$$

Even if we cannot integrate the first order equations obtained in (a) and (b), we can find out a great deal about the nature of the path $x(t)$, especially if we can solve explicitly for $\dot{x}$. An example of what can be done is given in Chapter 11, Section 11.6.

So far, we have assumed $x(t)$ to be a scalar valued function. It is obvious that, if we fix all but the ith coordinate function at optimal levels, $x_i(t)$ must satisfy the same conditions with respect to variations as if it were the only function to be varied. Thus we have immediately:

If x is a vector-valued function of t with n components, the single Euler equation is replaced by the n equations[6]

$$F_{x_i} - \frac{d}{dt} F_{\dot{x}_i} = 0 \qquad i = 1, \ldots, n.$$

R11.3 EXTENSIONS OF THE BASIC ANALYSIS[7]

(a) Integral Constraints[8]

Suppose we have a problem of the form

$$\min \text{ (or max) } J = \int_a^b F(x, \dot{x}, t)\, dt$$

$$\text{S.T.} \quad I_k = \int_a^b G^k(x, \dot{x}, t)\, dt + c_k \leqq 0 \qquad k = 1, m.$$

It can be shown, by a simple extension of the basic calculus of variations methods, that

(i) The Euler equation (or equations if x is a vector) have the same form as if we were optimizing an unconstrained integral whose integrand was the function

$$\phi(x, \dot{x}, t) = F(x, \dot{x}, t) - \sum \lambda_k G^k(x, \dot{x}, t),$$

where the λ_k are constant Lagrange multipliers.

[6] In continuous time growth models, this form of the Euler equations is the analog of the intertemporal efficiency conditions, given in Chapter 11, Section 11.3. See Samuelson [4].

[7] This section is designed primarily as a guide to these extensions. A topic that seems relevant to the reader's problem should be pursued in the calculus of variations literature.

[8] See Forsyth, Chapter VIII and Hestenes, Chapter 2.

(ii) $\lambda_k = 0$ whenever $I_k < 0$.

The analogy of these results to classical static optimizing results is obvious. Calculus of variations problems with integral constraints are often referred to as *Isoperimetric Problems* (the first such problem tackled being that of finding the figure of maximum area with a given perimeter).

(b) Variable Endpoints[9]

This problem arises if we wish to optimize an integral along a path which terminates somewhere on a given surface, rather than at a particular point. In an economic example, we might wish to minimize the time for the economy to grow to some isovalue surface ($\sum p_i x_i = c$) rather than to some specific configuration.

It can be shown that, if the terminal conditions are of the form $g(x) = 0$, the optimal path $x(t)$ must satisfy the ordinary Euler equations and, in addition, the conditions

$$F_{\dot{x}_i}/g_{x_i} = F_{\dot{x}_j}/g_{x_j} \qquad \text{all } i, j$$

at the terminal time. These conditions are often referred to as *transversality conditions*. The same type of conditions must hold at the inital time if the terminal point is fixed, but the initial point is variable, or if both endpoints are variable.

(c) Paths with Corners[10]

The analysis of the previous section requires that $x(t) \in C^2$, that is, the path $x(t)$ turns smoothly with no corners. By using different methods of proof, paths with corners can be handled. We can avoid $\ddot{x}$ by deriving the Euler equations in *integral form*

$$F_{\dot{x}} = \int_a^T F_x \, dt + c \qquad [\text{all } T \text{ along } x(t)].$$

A path which satisfies the integral form of the Euler equations and, in addition, satisfies the *Weierstrass condition* at every corner:

$$E(x, \dot{x}, u, t) = F(x, u, t) - F(x, \dot{x}, t) - (u - \dot{x})F_{\dot{x}} \geqq 0$$

qualifies as a potential optimal path (such a path is called an *extremaloid*). The Weierstrass condition guarantees that a better path cannot be found by choosing an arbitrary slope u in place of the actual slope $\dot{x}$.

[9] See Forsyth, Chapter I.

[10] Called weak variations in the classical analysis. See Forsyth, Chapter II or, better, Hestenes, Chapter 2.

If a path is optimal, even if it has corners, the functions

$$F_{\dot{x}} \qquad \text{and} \qquad F - \dot{x}F_{\dot{x}}$$

are continuous everywhere along $x(t)$, including the corners.

(d) Second Order and Other Conditions[11]

None of the conditions given will differentiate between a maximum and a minimum. It is usually obvious in the context of the problem whether a proper minimum or maximum exists. Tests which may be relevant in some contexts are given by the *Legendre* and *Jacobi* conditions. The first is analogous to the second order conditions for an ordinary optimum problem, the second is a special condition that rules out, for example, the eastbound circuit of the globe as an optimal path from New York to San Francisco. (Both eastbound and westbound great circle routes will satisfy all the other optimum conditions).

(e) Higher Order Derivatives

If the integrand of the functional J is of the form

$$F(x, Dx, D^2x, \ldots, D^nx, t),$$

the same technique is used as in reducing an nth order differential equation to a first order vector equation (see Section R10.6 of the preceding review). We then obtain F as a function of a vector z and first order derivatives only, giving an ordinary calculus of variations problem.

R11.4 PONTRYAGIN'S PRINCIPLE AND RELATED TOPICS[12]

In the classical calculus of variations formulation, the dynamic specifications are explicitly included in the integrand, which is a function of $\dot{x}$ as well as of x. In this section we shall be concerned with the problem of optimizing an integral of the form

$$\int_a^b F(y)\, dt,$$

[11] See Forsyth, Chapter I and Hestenes, Chapter 3.

[12] No application of Pontryagin's method is given in this book. Some discussion of potential use is given in Chakravarty. See also Paper 1 in Shell. The brief treatment given in this Section is designed merely to introduce the reader to the topic.

For a complete treatment, see Pontryagin and Hestenes. Both are advanced. (The work of Pontryagin and his associates was awarded the 1962 Lenin Prize.)

in which $\dot{y}$ does not appear explicitly, but where the dynamic specifications appear as separate constraints of the form

$$\dot{y}_i = f^i(y) \qquad i = 1, \ldots, n.$$

Some variables may not be subject to dynamic constraints but may be subject to static constraints (that is, any constraints not involving time derivatives).

This general type of problem arises in optimum control theory. It is convenient to divide the variables into two sets, those subject to dynamic constraints, x_i (called *state variables* in control theory) and those subject only to static constraints, u_j (called *control variables*). If we denote the n-vector of state variables by x and the m-vector of control variables by u, we can formally state the problem:

The Control Problem

$$\text{Min } J = \int_a^b F(x, u)\, dt$$

$$\text{S.T.} \quad \dot{x}_i = f^i(x, u) \qquad i = 1, \ldots, n$$

$$u \in U,$$

where U is the feasible set for the control variables. U may be any closed bounded set in R^m. The functions f^i are assumed to have appropriate continuity.

The problem is traditionally set up as a minimum (although the solution method is often called a maximum principle). It is important to note that no relationship is implied between the number of state variables and the number of control variables (n and m), that the control variables may or may not appear in F or any given f^i (if u does not appear in any f^i, the control variables play a trivial part) and that U can be *any* closed bounded set. Thus the restrictions on u can vary from nonnegativity and upper bounds to complex functional relationships $\phi(u) = 0$.

The best-known approach to a problem of this kind is by means of *Pontryagin's Principle*, which can be stated as follows:

Given a control problem of the above kind, we introduce an arbitrary constant k and arbitrary time functions ψ_i, $i = 1, \ldots, n$ (we can regard these as *dynamic multiplier functions*, somewhat analogous to Lagrange multipliers in static problems). We form the *Hamiltonian* function

$$H(\psi, x, u) = kF(x, u) + \sum_i^n \psi_i f^i(x, u).$$

The function H has partial derivatives

$$H_{\psi_i} = f^i$$

$$H_{x_j} = kF_{x_j} + \sum_1^n \psi_i f^i_{x_j}.$$

Then it is a necessary condition for the functions $x(t)$, $u(t)$ to be a solution of the control problem that

(a) *the dynamic multipliers ψ_j satisfy the differential equations:*

$$\dot{\psi}_j = -H_{x_j} \qquad i = 1, \ldots, n.$$

(b) *The functions $u(t)$ satisfy the conditions*

$$H[\psi(t), x(t), u(t)] = \max_{u \in U} H[\psi(t), x(t)].$$

If $x(t)$, $u(t)$ are optimal in the problem, then

(i) $k \leqq 0$
(ii) $H[\psi(t), x(t), u(t)] = 0$ *for all t.*[13]

The above method may be used in economic applications in various ways. It is particularly useful for continuously optimal problems, as compared with terminally optimal problems (see Chapter 11, Section 11.1), and for problems involving linear production conditions with activity switching.

More complex maximum principles can be devised to include integral constraints and static constraints on the state variables.[14]

As in the case of ordinary calculus of variations problems, it may not be possible to obtain solutions of Equations (a), but it will be possible to do so if the functions f^i are linear with constant coefficients, in which case Equations (a) will also be linear with constant coefficients.

[13] Pontryagin's maximum principle proper is given by (a) and (b). See Pontryagin, Chapter 1. The formulation given here is somewhat simplified.

[14] See Hestenes.

REFERENCES

Allen, R. G. D.:

[1] *Mathematical Analysis for Economists*, St. Martin's Press, 1938.

[2] *Mathematical Economics*, St. Martin's Press, 1959.

[3] *Macro-economic Theory: A Mathematical Treatment*, St. Martin's Press 1967.

American Economic Association:

[1] *Readings in Price Theory*, Irwin, 1952.

[2] (with Royal Economic Society) *Surveys of Economic Theory*, 3 vols. St. Martin's Press, 1966.

Arrow, K. J.:

"Alternative Proof of the Substitution Theorem for Leontief Models in the General Case," in *Koopmans* [1].

Arrow, K. J., and Debreu, G.:

"Existence of an Equilibrium for a Competitive Economy," *Econometrica*, **22**, 265–290, 1954.

Arrow, K. J., and Hurwicz, L.:

[1] "On the Stability of Competitive Equilibrium," *Econometrica*, **26**, 522–552, 1958.

[2] "Competitive Stability under Weak Gross Substitutability: the 'Euclidean distance' Approach," *International Economic Review*, **1**, 38–49, 1960.

Arrow, K. J., Block, H. D., and Hurwicz, L.:

"On the Stability of Competitive Equilibrium, II," *Econometrica*, **27**, 82–109, 1959.

Arrow, K. J., Karlin, S., and Suppes, P.:

Eds., *Mathematical Methods in the Social Sciences, 1959*, Stanford University Press, 1960.

Baumol, W. J.:

[1] *Economic Dynamics*, Macmillan, 1951.

[2] *Economic Theory and Operations Analysis*, Prentice-Hall, 1961.

Begle, E. G.:

"A Fixed Point Theorem," *Annals of Mathematics*, **51**, 544–550, 1950.

Bellman, R. E.:

[1] *Stability Theory of Differential Equations*, McGraw-Hill, 1953.

[2] *Dynamic Programming*, Princeton University Press, 1957.

[3] *Introduction to Matrix Analysis*, McGraw-Hill, 1960.

Bellman, R. E., and Dreyfuss, S. E.:

Applied Dynamic Programming, Princeton University Press, 1962.

Berge, C.:

The Theory of Graphs and its Applications, Wiley, 1962.

Bhagwati, J.:

"The Pure Theory of International Trade: A Survey," *American Economic Association* [2], vol. 2.

Blackorby, C.:

Rational Rules for Intertemporal Decision Making, (unpublished dissertation, Johns Hopkins University, 1967).

Brauer, A.:

"Limits for the Characteristic Roots of a Matrix," *Duke Mathematical Journal*, **13**, 387–395, 1946.

Buck, R. C.:

Advanced Calculus, McGraw-Hill, 1956.

Carr, C. R., and Howe, C. W.:

Quantitative Decision Procedures in Management and Economics, McGraw-Hill, 1964.

Chakravarty, S.:

"Alternative Preference Functions in Problems of Investment Planning on the National Level," in *Malinvaud and Bacharach.*

Chiang, A. C.:

Fundamental Methods of Mathematical Economics, McGraw-Hill, 1967.

Chipman, J. S.:

[1] "The Multisector Multiplier," *Econometrica*, **18**, 355–374, 1950.

[2] "A Survey of the Theory of International Trade," 3 parts, *Econometrica* **33**, 477–519, 1965; **33**, 685–760, 1965; **34**, 18–76, 1966.

Cooper, R. N.:

National Economic Policy in an Integrated World, Council on Foreign Relations, 1967.

Corden, W. M.:

Recent Developments in the Theory of International Trade, International Finance Section, Princeton University, Special Paper in International Economics No. 7, 1965.

Courant, R.:

Differential and Integral Calculus, 2 vols., Interscience, 1936 and 1937.

Courant, R. and Robbins, H.:

What is Mathematics?, Oxford University Press, 1941.

Dantzig, G.:

"Maximization of a Linear Function of Variables Subject to Linear Inequalites," in *Koopmans* [1].

Debreu, G.:

[1] *Theory of Value: An Axiomatic Analysis of Economic Equilibrium,* Wiley, 1959, Cowles Monograph 17.

[2] "New Concepts and Techniques for Equilibrium Analysis," *International Economic Review,* **3**, 257–273, 1962.

Debreu, G., and Herstein, I. N.:

"Nonnegative Square Matrices," *Econometrica,* **21**, 597–607, 1953.

Debreu, G., and Scarf, H.:

"A Limit Theorem on the Core of an Economy," *International Economic Review,* **4**, 235–246, 1963.

Dorfman, R., Samuelson, P. A., and Solow, R.:

Linear Programming and Economic Analysis, McGraw-Hill, 1958.

Eilenberg, S., and Montgomery, D.:

"Fixed Point Theorems of Multi-Valued Transformations," *American Journal of Mathematics,* **68**, 214–222, 1946.

Fan, K.:

"On Systems of Linear Inequalities," in *Kuhn and Tucker* [2].

Farrell, M. J., and Hahn, F. H. (for *The Economic Society*):

Problems in the Theory of Optimal Accumulation, Oliver and Boyd, 1967 (This is a hard cover reprint of *Review of Economic Studies,* **34**, 1–151, 1967.

Forsyth, A. R.:

Calculus of Variations, Dover, 1960, originally published in 1926.

Frisch, R.:

Maxima and Minima: Theory and Economic Applications, Rand McNally, 1966.

Gale, D.:

[1] *The Theory of Linear Economic Models*, McGraw-Hill, 1960.

[2] "Convex Polyhedral Cones and Linear Inequalities," in *Koopmans* [1].

[3] "The Law of Supply and Demand," *Mathematica Scandinava*, **3**, 155–169, 1955.

[4] "The Closed Linear Model of Production," in *Kuhn and Tucker.*

Gantmacher, F. R.:

[1] *Applications of the Theory of Matrices*, Interscience, 1959.

[2] *The Theory of Matrices*, 2 vols., Chelsea, 1960.

Gass, S. I.:

Linear Programming: Methods and Applications, McGraw-Hill, 1958, 1964.

Gerstenhaber, M.:

"Theory of Convex Polyhedral Cones," in *Koopmans* [1].

Goldberg, S.:

Introduction to Difference Equations, Wiley, 1961.

Goldman, A. J., and Tucker, A. W.:

[1] "The Theory of Linear Programming," in *Kuhn and Tucker* [2].

[2] "Polyhedral Convex Cones," in *Kuhn and Tucker* [2].

Goodwin, R. M.:

"The Multiplier as Matrix," *Economic Journal*, **59**, 537–555, 1949.

Hadley, G.:

[1] *Linear Algebra*, Addison-Wesley, 1961.

[2] *Linear Programming*, Addison-Wesley, 1962.

[3] *Nonlinear and Dynamic Programming*, Addison-Wesley, 1964.

Hahn, F. H.:

"Gross Substitutes and the Dynamic Stability of General Equilibrium," *Econometrica*, **26**, 169–170, 1958.

Hahn, F. H., and Matthews, R. C.:

"The Theory of Economic Growth: A Survey," in *American Economic Association* [2], Vole 2.

Hancock, H.:

Theory of Maxima and Minima, Dover, 1960, originally published in 1917.

Henderson, J. H., and Quandt, R. E.:

Microeconomic Theory, McGraw-Hill, 1958.

Hestenes, M. R.:

Calculus of Variations and Optimal Control Theory, Wiley, 1966.

Hicks, J. R.:

Value and Capital, Oxford University Press, 1939.

Jorgenson, D. W.:

"On Stability in the Sense of Harrod," *Economica*, **27**, 243–248, 1960.

Kakutani, S.:

"A Generalization of Brouwer's Fixed Point Theorem," *Duke Mathematical Journal*, **8**, 457–458, 1941.

Kaplan, W.:

Ordinary Differential Equations, Addison-Wesley, 1958.

Karlin, S.:

[1] *Mathematical Methods and Theory in Games, Programming, and Economics*, Vol. I, Addison-Wesley, 1959.

[2] "Positive Operators," *Journal of Mathematics and Mechanics*, **8**, 907–937, 1959.

Kemeny, J. G., Morgenstern, O., and Thompson, G. L.:

"A Generalization of the Von Neumann Model of an Expanding Economy," *Econometrica*, **24**, 115–135, 1956.

Koopmans, T. C.:

[1] (Ed) *Activity Analysis of Production and Allocation*, Wiley, 1951, Cowles monograph 13.

[2] *Three Essays on the State of Economic Science*, McGraw-Hill, 1957.

[3] "Analysis of Production as an Efficient Combination of Activities," in *Koopmans* [1].

[4] "Economic Growth at a Maximal Rate," *Quarterly Journal of Economics*, **78**, 355–394, 1964; also in *Malinvaud and Bacharach.*

[5] "Stationary Ordinal Utility and Impatience," *Econometrica*, **28**, 287–309, 1960.

Koopmans, T. C., Diamond, P., and Williamson, R. E.:

"Stationary Utility and Time Perspective," *Econometrica*, **32**, 82–100, 1964.

Kuhn, H. W.:

[1] "A Note on 'the Law of Supply and Demand,'" *Mathematica Scandinava*, **4**, 143–146, 1956.

[2] "On a Theorem of Wald," in *Kuhn and Tucker* [2].

Kuhn, H. W., and Tucker, A. W.:

[1] "Nonlinear programming," in *Neyman.*

[2] Eds., *Linear Inequalities and Related Systems*, Annals of Mathematics Studies No.38, Princteon University Press, 1956.

Lady, G.:

The Structure of Economic Models, unpublished dissertation, Johns Hopkins University, 1967.

Lancaster, K. J.:

[1] "The Theory of Qualitative Linear Systems," *Econometrica*, **33**, 395–408, 1965.

[2] "The Solution of Qualitative Comparative Static Problems," *Quarterly Journal of Economics*, **80**, 278–295, 1966.

[3] "A New Approach to Consumer Theory," *Journal of Political Economy*, **74**, 132–157, 1966.

[4] "Change and Innovation in the Technology of Consumption," *American Economic Review*, (*Papers and Proceedings*), 14–23, May 1966.

Lang, S.:

Linear Algebra, Addison-Wesley, 1966.

La Salle, J., and Lefschetz, S.:

Stability by Liapunov's Direct Method with Applications, Academic Press, 1961.

Lefschetz, S.:

Introduction to Topology, Princeton University Press, 1949.

Leontief, W. W.:

[1] "Quantitative Input and Output Relations in the Economic System of the United States," *Review of Economic Statistics*, **18**, 105–125, 1936.

[2] *The Structure of the American Economy 1919–1929*, Harvard University Press, 1941.

[3] *Input-Output Economics*, Oxford University Press, 1966.

[4] *et al.*, *Studies in the Structure of the American Economy*, Oxford University Press, 1953.

Malinvaud, E., and Bacharach, M.:

Eds., *Activity Analysis in the Theory of Growth and Planning*, St. Martin's Press, 1967.

Metzler, L.:

"A Multiple Region Theory of Income and Trade," *Econometrica*, **18**, 329–354, 1950.

Morgenstern, O.:

Ed., *Economic Activity Analysis*, Wiley, 1954.

Morishima, M.:

[1] *Equilibrium Stability and Growth*, Oxford University Press, 1964.

[2] "Proof of a Turnpike Theorem: the 'No Joint Production Case,'" *Review of Economic Studies*, **28**, 89–97, 1961.

Mosak, J. L.:

General Equilibrium Theory in International Trade, Principia Press, 1944, Cowles Monograph 7.

McKenzie, L.:

[1] "On Equilibrium in Graham's Model of World Trade and Other Competitive Systems," *Econometrica*, **22**, 147–166, 1954.

[2] "On the Existence of General Equilibrium for a Competitive Market," *Econometrica*, **27**, 54–71, 1959.

[3] "Matrices and Economic Theory," in *Arrow Karlin and Suppes.*

[4] "The Dorfman-Samuelson-Solow Turnpike Theorem," *International Economic Review*, **4**, 29–43, 1963.

[5] "Turnpike Theorems for a Generalized Leontief Model," *Econometrica*, **31**, 165–180, 1963.

[6] "Maximal Paths in the Von Neumann Model," in *Malinvaud and Bacharach*.

Mundell, R. A.:

"The Appropriate Use of Monetary and Fiscal Policy for Internal and External Stability," *International Monetary Fund Staff Papers*, **9**, 70–77, 1962.

Neyman, J.:

Ed., *Proceedings of the Second Berkeley Symposium on Mathematical Statistics and Probability*, University of California Press, 1951.

Nikaido, H.:

"Persistence of Continual Growth near the Von Neumann Ray: a Strong Version of the Radner Turnpike Theorem," *Econometrica*, **32**, 151–162, 1964.

Patrick, J.:

Unpublished dissertation, Columbia University, 1968.

Phelps, E. S.:

"Second Essay on the Golden rule of Accumulation," *American Economic Review*, **55**, 793–814, 1965.

Pontryagin, L. S. *et. al.*:

The Mathematical Theory of Optimal Processes, Interscience, 1962.

Radner, R.:

"Paths of Economic Growth That Are Optimal With Regard Only to Final States: A Turnpike Theorem," *Review of Economic Studies*, **28**, 98–104, 1961.

Ramsey, F.:

"A Mathematical Theory of Saving," *Economic Journal*, **38**, 543–559, 1927.

Robbins, L.:

An Essay on the Nature and Significance of Economic Science, Macmillan, London, 1932.

Samuelson, P. A.:

[1] *Foundations of Economic Analysis*, Harvard University Press, 1948.

[2] *Collected Scientific Papers*, Ed. Stiglitz, 2 Vols., MIT Press, 1966.

[3] "A Note on the Pure Theory of Consumer's Behavior and an Addendum," Paper 1 in [2].

[4] "Efficient Paths of Capital Accumulation in Terms of the Calculus of Variations," in *Arrow, Karlin, and Suppes*, Paper 26 in [2].

[5] "Abstract of a Theorem Concerning Substitutability in Open Leontief Models," in *Koopmans* [1], Paper 36 in [2].

Shell, K., (Ed.):

Essays on the Theory of Optimal Economic Growth, MIT Press, 1967.

Slutsky, E. E.:

"On the Theory of the Budget of the Consumer," in *American Economic Association* [1], originally published in 1915.

Solow, R. M., and Samuelson, P. A.:

"Balanced Growth under Constant Returns to Scale," *Econometrica*, **21**, 412–424, 1953, Paper 24 in *Samuelson* [2].

Strotz, R.:

"Myopia and Inconsistency in Dynamic Utility Maximization," *Review of Economic Studies*, **23**, 165–180, 1956.

Tinbergen, J.:

On the Theory of Economic Policy, North Holland, 1952.

Tucker, A. W.:

"Dual Systems of Homogeneous Linear Relations," in *Kuhn and Tucker*.

Uzawa, H.:

"The Stability of Dynamic Processes," *Econometrica*, **29**, 617–631, 1961.

Valentine, F. A.:

Convex Sets, McGraw-Hill, 1964.

Von Neumann, J.:

"A Model of General Economic Equilibrium," *Review of Economic Studies*, **13**, 1–9, 1945.

Wald, A.:

"On Some systems of Equations of Mathematical Economics," *Econometrica*, **19**, 368–403, 1951.

Wong, Y. K.:

Some Mathematical Concepts for Linear Economic Models," in *Morgenstern*.

Woodbury, M. A.:

"Properties of Leontief-Type Input-Output Matrices," in *Morgenstern*.

Yamane, T.:

Mathematics for Economists, Prentice-Hall, 1962.

INDEX

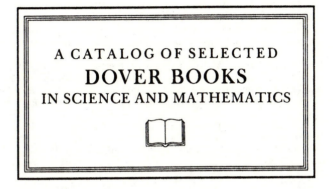

A CATALOG OF SELECTED

DOVER BOOKS

IN SCIENCE AND MATHEMATICS

A CATALOG OF SELECTED
DOVER BOOKS
IN SCIENCE AND MATHEMATICS

ORDINARY DIFFERENTIAL EQUATIONS, Morris Tenenbaum and Harry Pollard. Exhaustive survey of ordinary differential equations for undergraduates in mathematics, engineering, science. Thorough analysis of theorems. Diagrams. Bibliography. Index. 818pp. 5⅜ × 8½. 64940-7 Pa. $16.95

STATISTICAL MECHANICS: Principles and Applications, Terrell L. Hill. Standard text covers fundamentals of statistical mechanics, applications to fluctuation theory, imperfect gases, distribution functions, more. 448pp. 5⅜ × 8½. 65390-0 Pa. $11.95

ORDINARY DIFFERENTIAL EQUATIONS AND STABILITY THEORY: An Introduction, David A. Sánchez. Brief, modern treatment. Linear equation, stability theory for autonomous and nonautonomous systems, etc. 164pp. 5⅜ × 8¼. 63828-6 Pa. $5.95

THIRTY YEARS THAT SHOOK PHYSICS: The Story of Quantum Theory, George Gamow. Lucid, accessible introduction to influential theory of energy and matter. Careful explanations of Dirac's anti-particles, Bohr's model of the atom, much more. 12 plates. Numerous drawings. 240pp. 5⅜ × 8½. 24895-X Pa. $6.95

THEORY OF MATRICES, Sam Perlis. Outstanding text covering rank, nonsingularity and inverses in connection with the development of canonical matrices under the relation of equivalence, and without the intervention of determinants. Includes exercises. 237pp. 5⅜ × 8½. 66810-X Pa. $7.95

GREAT EXPERIMENTS IN PHYSICS: Firsthand Accounts from Galileo to Einstein, edited by Morris H. Shamos. 25 crucial discoveries: Newton's laws of motion, Chadwick's study of the neutron, Hertz on electromagnetic waves, more. Original accounts clearly annotated. 370pp. 5⅜ × 8½. 25346-5 Pa. $10.95

INTRODUCTION TO PARTIAL DIFFERENTIAL EQUATIONS WITH APPLICATIONS, E.C. Zachmanoglou and Dale W. Thoe. Essentials of partial differential equations applied to common problems in engineering and the physical sciences. Problems and answers. 416pp. 5⅜ × 8½. 65251-3 Pa. $10.95

BURNHAM'S CELESTIAL HANDBOOK, Robert Burnham, Jr. Thorough guide to the stars beyond our solar system. Exhaustive treatment. Alphabetical by constellation: Andromeda to Cetus in Vol. 1; Chamaeleon to Orion in Vol. 2; and Pavo to Vulpecula in Vol. 3. Hundreds of illustrations. Index in Vol. 3. 2,000pp. 6⅛ × 9¼. 23567-X, 23568-8, 23673-0 Pa., Three-vol. set $41.85

CHEMICAL MAGIC, Leonard A. Ford. Second Edition, Revised by E. Winston Grundmeier. Over 100 unusual stunts demonstrating cold fire, dust explosions, much more. Text explains scientific principles and stresses safety precautions. 128pp. 5⅜ × 8½. 67628-5 Pa. $5.95

AMATEUR ASTRONOMER'S HANDBOOK, J.B. Sidgwick. Timeless, comprehensive coverage of telescopes, mirrors, lenses, mountings, telescope drives, micrometers, spectroscopes, more. 189 illustrations. 576pp. 5⅜ × 8¼. (Available in U.S. only) 24034-7 Pa. $9.95

GEOMETRY OF COMPLEX NUMBERS, Hans Schwerdtfeger. Illuminating, widely praised book on analytic geometry of circles, the Moebius transformation, and two-dimensional non-Euclidean geometries. 200pp. 5⅜ × 8¼. 63830-8 Pa. $8.95

MECHANICS, J.P. Den Hartog. A classic introductory text or refresher. Hundreds of applications and design problems illuminate fundamentals of trusses, loaded beams and cables, etc. 334 answered problems. 462pp. 5⅜ × 8½. 60754-2 Pa. $9.95

TOPOLOGY, John G. Hocking and Gail S. Young. Superb one-year course in classical topology. Topological spaces and functions, point-set topology, much more. Examples and problems. Bibliography. Index. 384pp. 5⅜ × 8¼. 65676-4 Pa. $9.95

STRENGTH OF MATERIALS, J.P. Den Hartog. Full, clear treatment of basic material (tension, torsion, bending, etc.) plus advanced material on engineering methods, applications. 350 answered problems. 323pp. 5⅜ × 8½. 60755-0 Pa. $8.95

ELEMENTARY CONCEPTS OF TOPOLOGY, Paul Alexandroff. Elegant, intuitive approach to topology from set-theoretic topology to Betti groups; how concepts of topology are useful in math and physics. 25 figures. 57pp. 5⅜ × 8½. 60747-X Pa. $3.50

ADVANCED STRENGTH OF MATERIALS, J.P. Den Hartog. Superbly written advanced text covers torsion, rotating disks, membrane stresses in shells, much more. Many problems and answers. 388pp. 5⅜ × 8½. 65407-9 Pa. $9.95

COMPUTABILITY AND UNSOLVABILITY, Martin Davis. Classic graduate-level introduction to theory of computability, usually referred to as theory of recurrent functions. New preface and appendix. 288pp. 5⅜ × 8½. 61471-9 Pa. $7.95

GENERAL CHEMISTRY, Linus Pauling. Revised 3rd edition of classic first-year text by Nobel laureate. Atomic and molecular structure, quantum mechanics, statistical mechanics, thermodynamics correlated with descriptive chemistry. Problems. 992pp. 5⅜ × 8½. 65622-5 Pa. $19.95

AN INTRODUCTION TO MATRICES, SETS AND GROUPS FOR SCIENCE STUDENTS, G. Stephenson. Concise, readable text introduces sets, groups, and most importantly, matrices to undergraduate students of physics, chemistry, and engineering. Problems. 164pp. 5⅜ × 8½. 65077-4 Pa. $6.95

THE HISTORICAL BACKGROUND OF CHEMISTRY, Henry M. Leicester. Evolution of ideas, not individual biography. Concentrates on formulation of a coherent set of chemical laws. 260pp. 5⅜ × 8½. 61053-5 Pa. $6.95

THE PHILOSOPHY OF MATHEMATICS: An Introductory Essay, Stephan Körner. Surveys the views of Plato, Aristotle, Leibniz & Kant concerning propositions and theories of applied and pure mathematics. Introduction. Two appendices. Index. 198pp. 5⅜ × 8½. 25048-2 Pa. $7.95

THE DEVELOPMENT OF MODERN CHEMISTRY, Aaron J. Ihde. Authoritative history of chemistry from ancient Greek theory to 20th-century innovation. Covers major chemists and their discoveries. 209 illustrations. 14 tables. Bibliographies. Indices. Appendices. 851pp. 5⅜ × 8½. 64235-6 Pa. $18.95